T0309781

PHARMACEUTICAL APPLICATIONS OF RAMAN SPECTROSCOPY

THE WILEY BICENTENNIAL–KNOWLEDGE FOR GENERATIONS

Each generation has its unique needs and aspirations. When Charles Wiley first opened his small printing shop in lower Manhattan in 1807, it was a generation of boundless potential searching for an identity. And we were there, helping to define a new American literary tradition. Over half a century later, in the midst of the Second Industrial Revolution, it was a generation focused on building the future. Once again, we were there, supplying the critical scientific, technical, and engineering knowledge that helped frame the world. Throughout the 20th Century, and into the new millennium, nations began to reach out beyond their own borders and a new international community was born. Wiley was there, expanding its operations around the world to enable a global exchange of ideas, opinions, and know-how.

For 200 years, Wiley has been an integral part of each generation's journey, enabling the flow of information and understanding necessary to meet their needs and fulfill their aspirations. Today, bold new technologies are changing the way we live and learn. Wiley will be there, providing you the must-have knowledge you need to imagine new worlds, new possibilities, and new opportunities.

Generations come and go, but you can always count on Wiley to provide you the knowledge you need, when and where you need it!

WILLIAM J. PESCE
PRESIDENT AND CHIEF EXECUTIVE OFFICER

PETER BOOTH WILEY
CHAIRMAN OF THE BOARD

PHARMACEUTICAL APPLICATIONS OF RAMAN SPECTROSCOPY

Edited By

SLOBODAN ŠAŠIĆ
Pfizer, Ltd., Sandwich, UK

WILEY-INTERSCIENCE
A JOHN WILEY & SONS, INC., PUBLICATION

Published by John Wiley & Sons, Inc., Hoboken, New Jersey
Published simultaneously in Canada

For general information on our other products and services or for technical support, please contact our Customer Care Department within the United States at (800) 762-2974, outside the United States at (317) 572-3993 or fax (317) 572-4002.

Wiley also publishes its books in a variety of electronic formats. Some content that appears in print may not be available in electronic formats. For more information about Wiley products, visit our web site at www.wiley.com.

Wiley Bicentennial Logo: Richard J. Pacifico

Library of Congress Cataloging-in-Publication Data:

Pharmaceutical applications of Raman spectroscopy / Slobodan Šašić.
p. ; cm.
Includes bibliographical references and index.
ISBN 978-0-8138-1013-3 (cloth)
1. Raman spectroscopy. 2. Drugs–Analysis. 3. Pharmacy–Technique. 4. Pharmaceutical technology.
I. Sasic, Slobodan.
[DNLM: 1. Pharmaceutical Preparations–analysis. 2. Spectrum Analysis, Raman–methods. QV 25 P5355 2008]
RS189.5.S65P43 2008
615'.1901–dc22 2007024264

Printed in the United States of America
10 9 8 7 6 5 4 3 2 1

CONTENTS

PREFACE

A review of the recent scientific literature persuasively demonstrates that Raman spectroscopy is becoming a viable industrial technique. This is best illustrated by the diverse and steadily increasing number of real-world Raman applications. Indeed, the number and quality of journal and conference publications that cover Raman spectroscopy applied to widely recognized targets highlights a strong trend towards Raman instruments becoming "standard kit." Hence, it is clear that this vibrational spectroscopy technique is no longer confined to academic laboratories and fundamental research, as has been the case for a long time, although significant advances are still being made in this area.

From the literature, a few books present a broad survey of the various aspects of industrial Raman, most notably "The Handbook of Vibrational Spectroscopy," by P. Griffiths and J. Chalmers, and "Analytical Applications of Raman Spectroscopy by M. Pelletier. However, the literature on Raman theory is far more diverse, with the excellent monograph by R. McCreery standing out as the text that most comprehensively covers all the elements of Raman spectroscopy as a technique. From the perspective of the pharmaceutical industry, there are a number of contributions that touch on the use of Raman spectroscopy in various parts of the business, but there is no single volume that collects and lists those efforts. The book in your hands is an attempt to gather and order the pharmaceutical applications of Raman in a single tome.

Regarding the pharmaceutical industry, near-infrared (NIR) spectroscopy is unquestionably the spectroscopic method of choice, as demonstrated by the number of NIR methods finding application in the pharmaceutical arena (and not only there: the food industry is an even more convincing example). The technique is well understood and employs instruments that are relatively inexpensive, easy to use and compliant with the standards of the industry. The one problem with NIR is that the spectra may appear to be too heavily overlapped and thus it is very demanding to extract useful information. The closest experimental techniques to NIR are infrared (IR) and Raman spectroscopies, the former of which will not be covered in this book. These techniques may be tested for applications on the same targets for which NIR is used. Although full comparison between the industrial positions of NIR and Raman is not really appropriate because of the much more industrially established position of NIR, it is prudent to consider Raman as an alternative. It is a technique that is progressing rapidly and is finding a role in solving real problems.

This book lists seven areas that are representative of where Raman spectroscopy can be employed in the pharmaceutical industry. Some of the applications are more developed while others are actually not far from making their initial steps into industrial laboratories. The readers will probably more easily familarize themselves with the quantitative or online applications rather than the somewhat more complex surface-enhanced Raman spectroscopy or Raman chemical imaging. The former applications are the most authoritative for the assessment of the industrial standing of Raman. The number of references and diversity of applications in these two fields best illustrate where and for what Raman spectroscopy is being used in the pharmaceutical context. The latter two methods are conceptually and mathematically more complex but are believed to carry some valuable advantages in comparison with other commonly used tools and, hence, they are making their way into pharmaceutical laboratories. The chapter on polymorphism describes the use of Raman spectroscopy in an area that is very specific and important for the formulation of the products in the pharmaceutical industry. Raman spectroscopy has actively been used in this area for some time now and is a recognized tool for use in polymorph screening. Finally, two chapters in this book address the biomedical perspectives of Raman in pharma. These chapters deal with the delivery/monitoring of active components through skin and the chemical imaging of cells. The results given there also hold clear the promise for Raman spectroscopy to become more amenable to everyday practice.

The writers of this book hope that the material presented will be of interest to those who practice Raman spectroscopy at various levels in the pharmaceutical industry, as well as to those who are beginners in the field or just pondering the possibilities of Raman. The variety of topics and level of detail explored is such that all the essential links between Raman spectroscopy and the pharmaceutical industry are covered. As a result, the authors anticipate that this publication will encourage readers from the industry to give greater consideration to Raman and that those readers from academia might gain a better understanding of the current industrial status and requirements of the technique.

Slobodan Šašić

CONTRIBUTORS

Steven E.J. Bell, Queen's University, Belfast, Northern Ireland, BT9 5AG UK S.Bell@qub.ac.uk

Peter J. Caspers, River Diagnostics BV, Dr. Molewaterplein 50, Ee 1979, 3015 GE Rotterdam, Netherlands

Kevin L. Davis, Kaiser Optical Systems, Inc., 371 Parkland Plaza, Ann Arbor, MI 48103, USA

Anne De Paepe, Materials Science Department, Pfizer Global R&D, Ramsgate Road, Sandwich, UK

Mark S. Kemper, Kaiser Optical Systems, Inc., 371 Parkland Plaza, Ann Arbor, MI 48103, USA

Fred LaPlant, 3M Corporate Analytical, Minneapolis, MN, USA flaplant@mmm.com

Ian R. Lewis, Kaiser Optical Systems, Inc., 371 Parkland Plaza, Ann Arbor, MI 48103, USA lewis@KOSI.com

Jian Ling, Bioengineering Section, Southwest Research Institute, 6220 Culebra Rd., San Antonio, TX 78238, USA jling@swri.org

Yukihiro Ozaki, Kwansei-Gakuin University, School of Science and Technology, Sanda 669, Japan ozaki@kwansei.ac.jp

Williams M. Riggs, River Diagnostics BV, Dr. Molewaterplein 50, Ee 1979, 3015 GE Rotterdam, Netherlands riggs@riverd.com

Slobodan Šašić, Pfizer, Analytical Research and Development, Ramsgate Road, Sandwich CT13 9NJ, UK slsasic@yahoo.com

W. Ewen Smith, Department of Pure and Applied Chemistry, University of Strathclyde, Glasgow G1 1XL, UK

Andre van der Pol, River Diagnostics BV, Dr. Molewaterplein 50, Ee 1979, 3015 GE Rotterdam, Netherlands vanderpol@riverd.com

1

INTRODUCTION TO RAMAN SPECTROSCOPY

YUKIHIRO OZAKI
Kwansei-Gakuin University, School of Science and Technology, Sanda 669, Japan
SLOBODAN ŠAŠIĆ
Pfizer, Analytical Research and Development, Ramsgate Road, Sandwich CT13 9NJ, UK

Raman spectroscopy may have seemed too much of a burden in the past to handle. Until 20 years ago, one had to gain some experience before becoming capable of measuring satisfactory Raman spectra. Some even said that Raman spectroscopy was "patience-testing spectroscopy," or 'a romance of patience.' But this is changing now, rapidly. While unquestionably being used in basic research for quite a long time, Raman spectroscopy is nowadays steadily gaining on importance for online monitoring of chemical reactions, analysis of food, pharmaceuticals, and chemicals, and increasingly for many other real-world applications.

Similar to infrared (IR) spectroscopy, Raman spectroscopy yields detailed information about molecular vibrations. As molecular vibrations are very sensitive to strength and types of chemical bonds, vibrational spectroscopy techniques, such as IR and Raman spectroscopy, are useful not only in identifying molecules but also in shedding light on molecular structures. In addition, IR and Raman spectra also reflect changes in the surroundings of the molecules and are thus helpful in studying intra- and intermolecular interactions.

1.1 HISTORY OF RAMAN SPECTROSCOPY

Raman spectroscopy is based on the effect of radiation being scattered with a change of frequency and its history goes 80 years back. It was in 1928 that Indian scientists

Pharmaceutical Applications of Raman Spectroscopy, Edited by Slobodan Šašić

Raman and Krishnan [Raman 1928] discovered the scattering effect that is named after Raman and that earned him a Nobel Prize in 1930. Raman spectroscopy, however, developed relatively slowly for several reasons. First, Raman experiments were not easy to carry out because of the extremely weak intensity of the Raman scattered light (roughly below 10^{-10} of that of exciting light or even weaker). Raman spectroscopy gained momentum in 1970s owing to the lasers becoming more available to researchers and thus assuming the role of primary source of excitation light, replacing the mercury-based sources of radiation. Nevertheless, the difficulty of finely adjusting optical systems was still present. Second, fluorescence from the sample severely interfered with detection of Raman scattered photons. Excitation of Raman scattering with the light of a wavelength within the visible region may concomitantly excite fluorescence from the sample or impurities contained therein. When the intensity of fluorescence is strong, the Raman signal is barely visible on top of incomparably stronger, broad fluorescence signal. The third reason is decomposition and denaturation of the samples irradiated with relatively strong laser light for a long time due to inefficient detection of scattered radiation. Due to this, despite being by nature a non-destructive method of analysis, in some cases Raman spectroscopy ended up being considered a 'fatally' destructive method. These three problems, the difficulty of measurement, fluorescence, and decomposition seriously limited applications of Raman spectroscopy despite it being theoretically quite promising. Except for a few examples, Raman spectroscopy barely found any industrial or medical application.

This situation has greatly changed over the recent two decades. It is no exaggeration to claim that a new revolution in Raman spectroscopy occurred throughout the 1990s. While it was the dissemination of laser sources that drove the revolution in 1970s, this time a combination of multiple factors revolutionized applications of Raman spectroscopy: the sophistication of laser sources, the technological progress of detectors, the development of excellent optical filters, significant improvements in software, strong progress in application of data analysis methods, and so on. With regard to applications, the emergence of near-infrared (NIR) laser sources of excitation gave an entirely different perspective on applicability of Raman spectroscopy for tackling real-world issues (including both NIR FT-Raman spectroscopy and NIR multichannel Raman spectroscopy). The introduction of NIR lasers was noticeably important because the three above-mentioned problems were significantly alleviated; in particular, excitation with NIR light largely prevented occurrence of fluorescence and eliminated light-induced decomposition. In addition, FT-Raman spectrometers do not really require fine adjustment of optical systems.

Besides Raman spectrometers with NIR excitation, those equipped with ultraviolet (UV) laser sources are also being more frequently used for solving real-world problems. In addition, of particular significance is the recent rapid development of Raman microscopes and near-field Raman devices that can be easily coupled to automated microscope stages assembling thus Raman mapping/imaging instruments.

Today's Raman instruments are compact devices with very reasonably sized key components (almost miniaturized in comparison with large devices used before the 1990s) such as lasers, spectrometers, or detectors. These instruments are rather easy to use, require little training, and allow for sophisticated experiments. The technological advances have increased reliability, repeatability, sturdiness, and confidence in the results, and have significantly improved prospects for applications in real (industrial) environments. Mitigation (not yet elimination) of the above-mentioned problems and steadily progressing technological solutions make Raman spectroscopy an emerging force among the state-of-the-art technologies to be used in various industries.

1.2 THE PRINCIPLE OF RAMAN SPECTROSCOPY

While infrared spectroscopy is based on absorption, reflection and emission of light, Raman spectroscopy is based on the scattering phenomenon. In this context, scattering occurs due to collisions between photons and molecules. Generally, a photon collides with a substance, not necessarily only with a molecule; but to simplify the treatment here, we will only consider a photon–molecule collision.

Irradiation of light with the frequency ν_0 upon a certain molecule brings a number of photons with the energy $E = h\nu_0$ to this molecule (Fig. 1.1). For instance, laser light having a wavelength of 500 nm and an optical output of 1 W emits approximately 2.5×10^{18} photons per second. These photons include photons colliding with molecules as well as those that pass without interacting with molecules. Upon irradiating carbon tetrachloride, which is a transparent liquid, we will find that about 10^{13} through 10^{15} photons collide with a molecule and change their directions among the total of 2.5×10^{18} photons.

Most photons colliding with molecules do not change their energy after the collision (elastic collision) and the ensuing radiation is called Rayleigh scattering. Rayleigh scattering consists of photons that have the same frequency as the incident light. A very small number of the photons that collide with the molecules exchange energy with them upon the collision (an example of inelastic collision). If an incident photon delivers an $h\nu$ quantum of energy to the molecule, the energy of the scattered photon

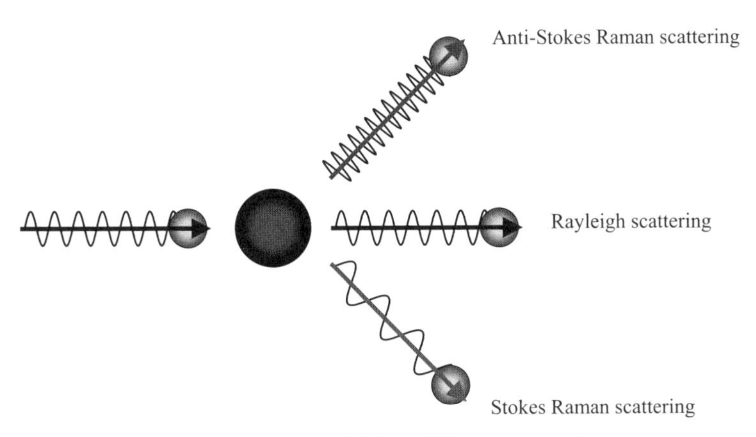

FIGURE 1.1 Rayleigh and Raman scattering.

reduces to $h(\nu_0 - \nu)$, and the frequency of the scattering photon becomes $\nu_0 - \nu$. On the contrary, when an incident photon receives the $h\nu$ energy from the molecule, the energy of the scattering photon rises to $h(\nu_0 + \nu)$, and the frequency of the scattering photon becomes $\nu_0 + \nu$. Scattering in which an incident photon exchanges energy with a molecule is known as Raman scattering. Scattered light having the frequency of $\nu_0 - \nu$ and that having the frequency of vibration $\nu_0 + \nu$ are called "Stokes Raman scattering" and "anti-Stokes Raman scattering," respectively.

By measuring Raman scattering one examines energy changes that accompany transition from one molecular energy level to another. While such a transition may occur between different electronic, vibrational or rotational energy levels, it is almost exclusively the transition between the vibrational energy levels that is associated with Raman spectroscopy. We will, therefore, treat Raman spectroscopy only as a vibrational spectroscopy technique here.

Figure 1.2 illustrates Stokes and anti-Stokes Raman scattering. Stokes Raman scattering arises from interaction between a photon and a molecule that is in the ground state, while anti-Stokes Raman scattering is due to interaction between a photon and a molecule that is in the excited state. As the molecules are normally in the ground vibrational state, Stokes Raman scattering occurs far more easily and this is why Stokes Raman scattering is usually measured.

An important factor in Raman spectroscopy is the shift ν attributable to the Raman effect and is called "Raman shift." Intensity of scattered radiation versus Raman shift forms a Raman spectrum that is unique for each individual substance. Analysis of Raman spectra, therefore, makes it possible to identify a substance and study its structure.

1.3 AN EXAMPLE OF SIMPLE RAMAN SPECTRUM: RAMAN SPECTRUM OF WATER

Let us explore a Raman spectrum of water as an example of simple Raman spectrum. Figure 1.3 shows the Raman spectrum of water measured using excitation light (Ar^+

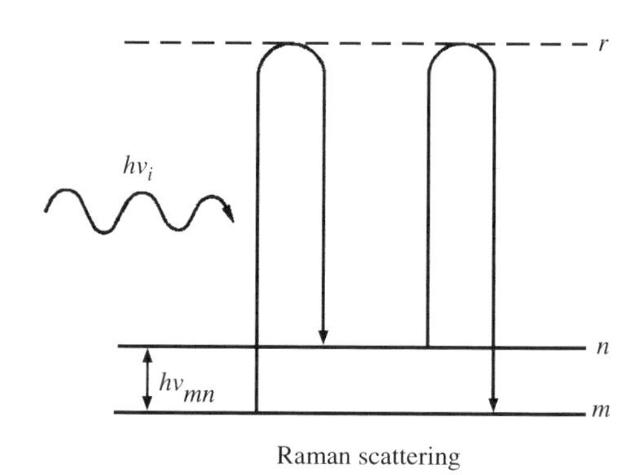

FIGURE 1.2 Stokes and anti-Stokes Raman scattering.

laser) with a frequency of 488.0 nm. The abscissa indicates three different ways of expressing the frequency of Raman spectra. Let us start with the wavelength. Since measurement of a Raman spectrum is usually a measurement of the scattered light on the longer wavelength side with respect to the excitation wavelength (because the energy of the Stokes-scattered photons is less than that of the excitation wavelength), this drawing shows the spectrum from 488.0 nm to about 600 nm. Figure 1.3 shows the absolute wavenumber right beneath the spectrum. A wavenumber is the reciprocal of a wavelength; 488.0 nm corresponds to 20492 cm^{-1}($488.0\,nm = 488.0 \times 10^{-7}\,cm; 1/(488.0 \times 10^{-7})\,cm^{-1} = 20492\,cm^{-1}$). The absolute wavenumber values decrease toward the longer wavelength side. The wavelength and absolute wavenumber readings are different for various excitation wavelengths and, therefore, inconvenient for practical use because the *X*-axis is not fixed. This ambiguity is eliminated by introducing the concept of Raman shift, the bottom scale in Fig. 1.3, that is universally being used as the scale for the Raman spectra. A Raman shift is the frequency of a band expressed as a shift from the excitation wavelength. At the position of the excitation wavelength, the Raman shift is obviously 0 cm^{-1}. The sum of a Raman shift X of a certain wavelength and the absolute wavenumber Y of this wavelength is the absolute wavenumber Z of the Raman shift ($X + Y = Z$). Therefore, $X + Y = 20,492\,cm^{-1}$ when 488.0 nm is the excitation wavelength. This can be confirmed from the readings along the second and the third horizontal axes in Fig. 1.3. The vertical axis in Fig. 1.3 indicates the Raman scattering intensity.

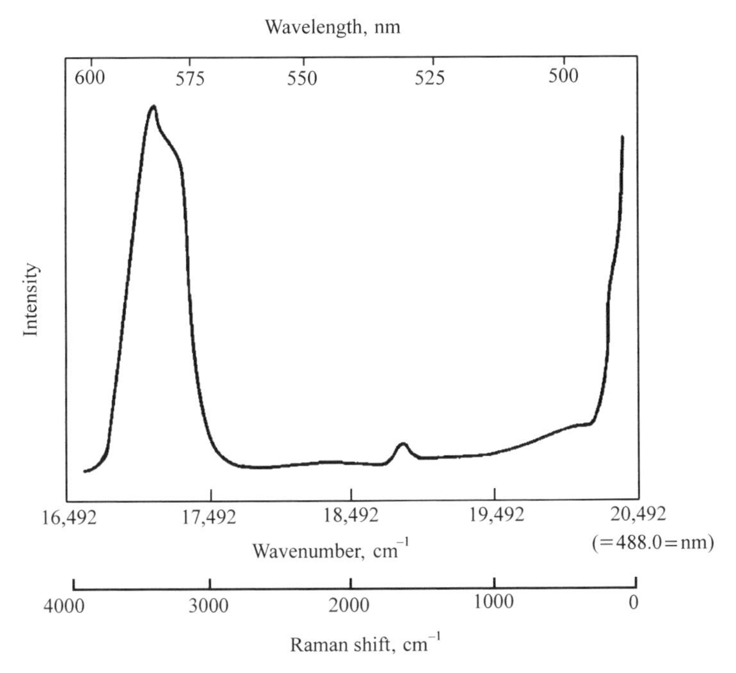

FIGURE 1.3 Raman spectrum of water.

Let us now analyze the spectrum of water in more details. The strong band around $3300\,cm^{-1}$ is due to the O–H stretching mode. The weak band around $1560\,cm^{-1}$ is assigned to the H–O–H bending mode of water. Note that the former is far more intense than the latter. This indicates that the photon–molecule energy exchange developing the peak attributable to the O–H stretching mode occurs more easily than the photon–molecule energy exchange that is characterized by the peak at $1560\,cm^{-1}$. The fact that Raman scattering of water is weak except for the region from 3500 to $3100\,cm^{-1}$ means that we can easily study molecules in aqueous solutions, which is very important for applicability of Raman spectroscopy.

1.4 CHARACTERISTICS OF RAMAN SPECTROSCOPY

Raman spectroscopy is characterized by the following features:

1. Raman spectroscopy permits acquisition of the spectra *in situ*. Monitoring of a reaction in a flask online, for example, can simply be accomplished by irradiating laser light directly upon the reactant from outside the flask (or through optical fibers and ports on the reaction vessel). There is no need to extract reactants/products from the flask and acquire their Raman spectra from vials or cuvettes.
2. Raman spectra can be measured irrespective of the state of substance, that is, regardless of whether the substance is gas, liquid, solution, solid, crystal, fiber, or film. In addition, by measuring spectra of substances in various states one can obtain information about different molecular structures of the given substance in various phases.
3. As laser is used for exciting the sample and due to high sensitivity of modern detectors, it is possible to obtain Raman spectra from very small amounts of material. This feature is of importance for local analyses and also for instruments equipped with microscopes.
4. Raman experiments can be conducted with optical fibers, which allow the spectrometer to be separated from the sample that might be, say, in a dangerous environment. This feature is very important with respect to Raman spectroscopy as a means of online or outdoor analysis.
5. A valuable application of Raman spectroscopy in fundamental research is for examination of ultrahigh speed phenomena. A combination of a pulse laser and a multichannel detector allows acquisition of time-resolved spectra even in the order of femtoseconds. Raman spectroscopy is, therefore, frequently used to study the excited states of molecules and the structures of reaction intermediates.

Raman spectroscopy is complementary to and often compared with IR spectroscopy in various aspects. Comparison with NIR spectroscopy is also appropriate with regard to industrial applications and it will be mentioned below.

These are the principal points of comparison between the Raman and IR methods:

1. Measurement of spectra from aqueous solutions is easier in Raman than in IR because of the rather poor Raman spectrum of water, which thus does not represent serious interference. Despite recent advances in FT-IR spectroscopy, Raman spectroscopy is still superior in this regard.
2. *In situ* or *in vivo* analysis as well as analyses with optical fibers are more easily carried out by Raman. This points to flexibility of Raman spectroscopy in comparison with IR.
3. The spatial resolution of Raman microspectroscopy can be high. While the spatial resolution in Raman microscopy is up to about 1 μm, and even better than that in some spatial cases, the spatial resolution of IR microscopy is up to about 10 μm.
4. Measurement of time-resolved spectra is easier with Raman. Due to restrictions originating from the detector, it is much more difficult to acquire time-resolved IR spectra in the picoseconds range.
5. There is a special sort of Raman scattering known as resonance Raman (RR) effect that has no counterpart in IR. The resonance Raman effect is an effect in which intensities of Raman bands are dramatically increased in cases where the wavelength of the excitation light overlaps with an absorption band of the probed molecule. In addition to huge signal enhancement, which significantly increases sensitivity, RR spectroscopy allows for selective examination of parts of the molecule. For instance, in an enzyme molecule with the molecular weight of tens of thousands one can selectively acquire Raman signal from an active site of that enzyme that has the molecular weight of only a few hundreds. To the contrary, IR spectroscopy only allows for acquisition of average spectra of the sample.

1.5 THE CLASSIC THEORY OF RAMAN EFFECT

The classic theory is useful for understanding the Raman effect, but it can strictly be explained only through the quantum theory (Ferraro 2003, Long 2003).

When a certain molecule is subjected to irradiation of an electromagnetic wave with a frequency of ν_0, the oscillatory electric field $\boldsymbol{E}$ of the electromagnetic wave slightly changes the distribution of electrons within this molecule. In short, dipole moment $\boldsymbol{P}$ is induced (induced dipole moment). When the electric field $\boldsymbol{E}$ is weak enough, $\boldsymbol{P}$ is in proportion to the electric field $\boldsymbol{E}$ and can be expressed as:

$$P = \alpha E \tag{1.1}$$

The symbol α denotes the electric polarizability. The polarizability may be construed as how easily a certain electron cloud becomes distorted. As the vectors $\boldsymbol{P}$ and $\boldsymbol{E}$ are generally in different directions, α is a tensor quantity. We can, therefore, rewrite Equation 1.1 as follows:

$$\begin{bmatrix} \boldsymbol{P}_x \\ \boldsymbol{P}_y \\ \boldsymbol{P}_z \end{bmatrix} = \begin{bmatrix} \alpha_{xx} & \alpha_{xy} & \alpha_{xz} \\ \alpha_{yx} & \alpha_{yy} & \alpha_{yz} \\ \alpha_{zx} & \alpha_{zy} & \alpha_{zz} \end{bmatrix} \begin{bmatrix} E_x \\ E_y \\ E_z \end{bmatrix} \tag{1.2}$$

For simplicity, let us assume both $\boldsymbol{P}$ and $\boldsymbol{E}$ are values along one coordinate axis. Substituting $E = E_0 \cos 2\pi\nu t$ in Equation 1.1, we obtain

$$P = \alpha E_0 \cos 2\pi\nu_0 t \tag{1.3}$$

We can divide the polarizability α into a term α_0 that does not change despite the molecule vibrating and a term that changes due to the molecular vibration. Defining $Q(Q = Q_0 \cos 2\pi\nu t)$ as a normal coordinate indicative of vibrational displacement of the molecular vibration, we obtain

$$\alpha = \alpha_0 + \left(\frac{\partial\alpha}{\partial Q}\right)_0 Q = \alpha_0 + \left(\frac{\partial\alpha}{\partial Q}\right)_0 Q_0 \cos 2\pi\nu t \tag{1.4}$$

Substituting Equation 1.4 in Equation 1.3 yields

$$\begin{aligned} P &= \left\{\alpha_0 + \left(\frac{\partial\alpha}{\partial Q}\right)_0 Q_0 \cos 2\pi\nu t\right\} E_0 \cos 2\,\pi\nu_0 t \\ &= \alpha_0 E_0 \cos 2\pi\nu_0 t + \frac{1}{2}\left(\frac{\partial\alpha}{\partial Q}\right)_0 Q_0 E_0 \{\cos 2\pi(\nu_0 + \nu)t + \cos 2\pi(\nu_0 - \nu)t\} \end{aligned} \tag{1.5}$$

The first term in Equation 1.5 is the product of the constant α_0 and E vibrating at the same frequency of ν_0 as that of the incident light. This term thus expresses a component of P which vibrates at ν_0. The second term contains a component that refers to vibrations at two different frequencies, $\nu_0 + \nu$ and $\nu_0 - \nu$. Therefore, the induced dipole moment P consists of the three components which vibrate at ν_0 and $\nu_0 \pm \nu$. According to the classic electromagnetics, an electric dipole having a vibrating moment emits an electromagnetic wave the frequency of which is the same as that of the electric dipole. The first term thus indicates scattering of light of the same frequency of vibration ν_0 as that of the incident light. This scattering is called the Rayleigh scattering. The second term is indicative of emission of scattered light the frequency of which has changed, $(\nu_0 \pm \nu)$, and is known as the Raman scattering.

Equation 1.5 reveals prerequisites for Raman scattering to actually occur. To give rise to Raman scattering, the factor $(\partial\alpha/\partial Q)_0 Q_0 E_0$ must not be zero. Since neither Q_0 nor E_0 is zero, the key condition is $(\partial\alpha/\partial Q)_0 \neq 0$. It follows from

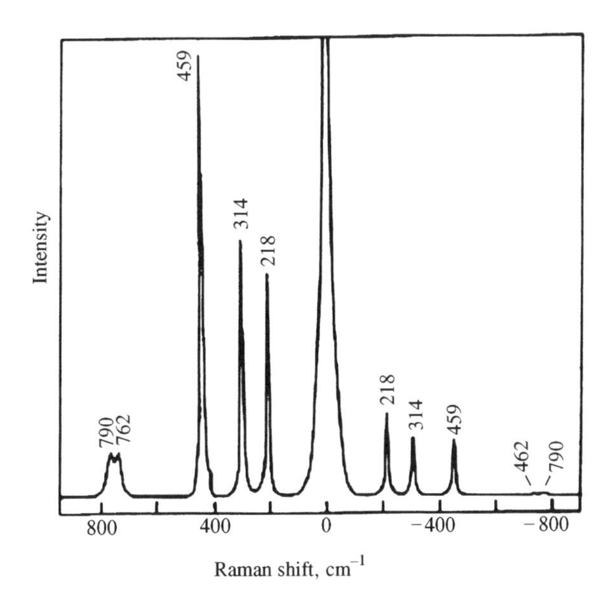

FIGURE 1.4 Stokes and anti-Stokes Raman spectra of CCl_4.

this that Raman-active are only those vibrations that change the polarizability associated with vibrational displacement of a molecule.

There are two more elements in Equation 1.5 that need to be emphasized referring to anti-Stokes Raman scattering and the ratio between Raman and Rayleigh scattering. Anti-Stokes Raman spectra are rarely collected (this has been changing recently with the advent of coherent anti-Stokes Raman spectroscopy (CARS) (Cheng and Xie, 2004)) in comparison with Stokes Raman spectra (commonly used), because the intensity of the anti–Stokes scattering depends on the population of the first excited vibrational state. According to the Boltzmann distribution, this is inversely proportional to the temperature and thus anti–Stokes scattering is of lower intensity than Stokes Raman scattering (Fig. 1.4). Both types of Raman scattering are in turn much weaker than Rayleigh scattering (normally less than 0.1%). Equation 1.5 does not explicitly reveal the ratio between the three types of scattering but, being elastic scattering, Rayleigh scattering is a much more probable event than either of the two Raman (inelastic) scatterings.

Let us now consider Raman activity of the four normal modes of CO_2 molecule. Figure 1.5 shows these four normal modes of CO_2 and Fig. 1.6 depicts changes of the polarizability α caused by the normal modes I, II, and III shown in Fig. 1.5. (It needs to be mentioned here that IR activity is determined by the change in dipole moment.) In the normal mode I that is IR-inactive, the polarizability is obviously different when the molecule is stretched in comparison to when it is shrunk, and $\partial\alpha/\partial Q$ at the equilibrium position, $(\partial\alpha/\partial Q)_0$, is not zero. It thus follows that the mode I is Raman-active. With respect to the normal mode vibrations II, IIIa, and IIIb (all of which are IR-active), the values α at the both ends of the vibrations

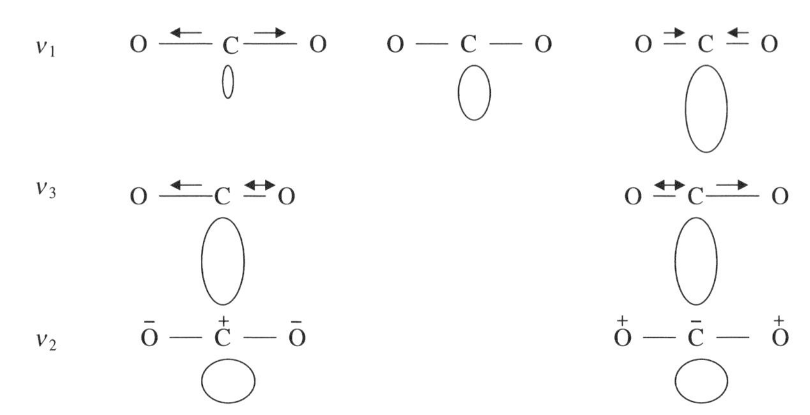

FIGURE 1.5 Changes of the polarizability ellipsoid for the vibrations of CO_2 molecule. The v_1 vibration is Raman active because of the change in the polarizability ellipsoid for the two extreme displacements. The other two vibrations (v_2 is degenerate) are not Raman active because of the same shape of the ellipsoid for two extreme displacements. On the other hand, the dipole moment does not change for v_1 and this vibration is thus inactive in IR, while the other two are active as the dipole moment obviously varies. The example shown illustrates the mutual exclusion principle that holds for molecules that have center of symmetry.

are equal to each other in any one of these modes, and at the equilibrium position, $(\partial\alpha/\partial Q)_0$ is zero. Therefore, these modes are Raman-inactive. This example illustrates the rule of mutual exclusion of IR and Raman that is valid for molecules with the center of symmetry.

The overlap between IR and Raman spectra, can be substantial despite the difference in the mechanism of their origin. The allowed bands in the two types of spectra are determined by the selection rules. In general, symmetric stretching or breathing vibrations (large movements of electron clouds) are active in Raman but not in IR, while the bending or asymmetric vibrations (significant changes in dipole moments that are not accompanied by large shifts of the electrons) are strong in IR but weak in Raman. Both types of spectra are characterized by sharp and distinct bands.

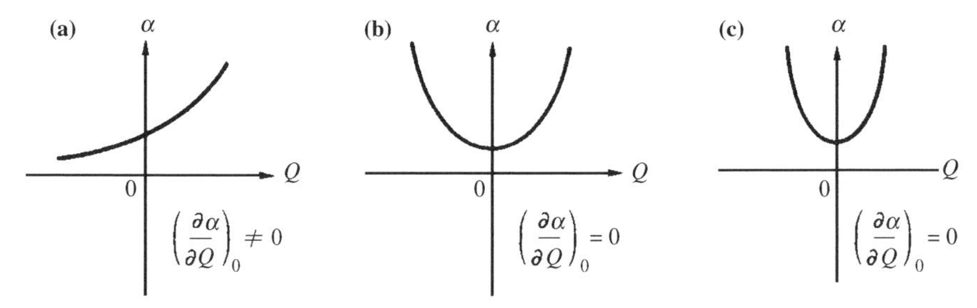

FIGURE 1.6 Changes in the polarizability α caused by the normal modes (a)I, (b)II, (c)IIIa and IIIb of CO_2.

IR spectra become progressively more complicated as the size of the probed molecule increases due to variety of vibrations that are normally easily detected. On the contrary, due to a much lesser intensity of Raman spectra, many of the bands from complicated biological samples may be below the limit of detection.

1.5.1 Polarization Properties of Raman Scattering

While the polarization property of incident light changes during the process of scattering, in general, the symmetry of a normal vibration determines to what extent it changes. In short, as the polarization property of a Raman band is closely relevant to the symmetry of a normal mode, knowledge about the polarization property of a Raman band helps us in assigning that band.

Figure 1.7 shows an example of polarized Raman spectra (CCl_4). In Fig. 1.7, the symbols $I_{\perp}$ and $I_{\parallel}$ are, respectively, indicative of a Raman scattering intensity component polarized vertically to and parallel to the electric field vector $\boldsymbol{E}$ of incident light. It is self-explanatory from Fig. 1.7(a) that the symmetry of a normal vibration significantly changes the degree of polarization of a Raman band.

The polarization property of a Raman band is generally expressed by the degree of depolarization ρ that is defined as a ratio between $I_{\perp}$ and $I_{\parallel}$.

$$\rho = I_{\perp}/I_{\parallel} \tag{1.6}$$

The reason why the property of polarized incident light (in general, incident light is strongly polarized) changes during the process of scattering is because α is a tensor quantity. $I_{\perp}$ and $I_{\parallel}$, (i.e. ρ) can be expressed as a combination of α components.

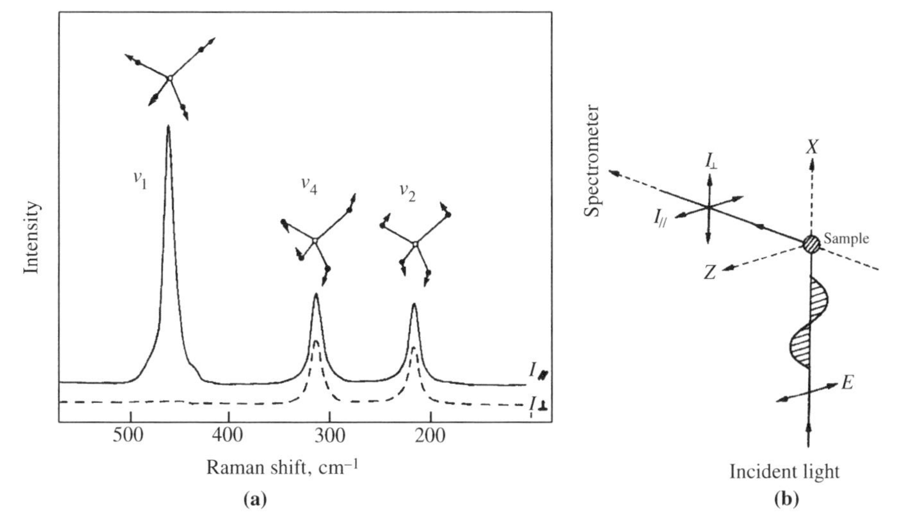

FIGURE 1.7 (**a**) Polarized Raman spectra of carbon tetrachloride and the vibration modes ν_1, ν_4, and ν_2. (**b**) Relationship between polarization of incident light ($\boldsymbol{E}$) and those of Raman scattered light components $I_{\perp}$ and $I_{\parallel}$.

Although detailed explanation will be skipped, according to Placzek (Long, 2003), ρ is expressed by

$$\rho = \frac{3G^s + 5G^a}{10G^0 + 4G^s} \tag{1.7}$$

with that

$$G^0 = \tfrac{1}{3}(\alpha_{xx} + \alpha_{yy} + \alpha_{zz})^2 \tag{1.8}$$

$$\begin{aligned} G^s =& \tfrac{1}{3}\{(\alpha_{xx} - \alpha_{yy})^2 + (\alpha_{yy} - \alpha_{zz})^2 + (\alpha_{zz} - \alpha_{xx})^2\} \\ &+ \tfrac{1}{2}\{(\alpha_{xy} + \alpha_{yx})^2 + (\alpha_{yz} + \alpha_{zy})^2 + (\alpha_{xz} + \alpha_{zx})^2\} \end{aligned} \tag{1.9}$$

$$G^a = \tfrac{1}{2}\{(\alpha_{xy} - \alpha_{yx})^2 + (\alpha_{yz} - \alpha_{zy})^2 + (\alpha_{zx} - \alpha_{xz})^2\} \tag{1.10}$$

G^0, G,^{s} and G^a are called an "isotropic," a "symmetric anisotropic," and "antisymmetric anisotropic" components, respectively (Note G^0, G^s, $G^a \geqq 0$). In the case of ordinary (nonresonant) Raman scattering, a polarizability tensor is always symmetric. Hence, $G^a = 0$ holds true. Therefore, we can rewrite the Equation 1.7 as follows:

$$\rho = \frac{3G^s}{10G^0 + 4G^s} \tag{1.11}$$

When vibration is a non-totally symmetric vibration, $\alpha_{\gamma\gamma} = 0$, and γ and $\gamma'(\gamma \neq \gamma')$ satisfy the relationship $\alpha_{\gamma\gamma} \neq 0$. Hence, $G^0 = 0$ and $G^s > 0$. In this condition, we obtain $\rho = 3/4$ from Equation 1.11. Meanwhile, $0 \leqq \rho < 3/4$ since $G^0 > 0$ and $G^s \geqq 0$ in the case of a totally symmetric mode. A Raman band that provides the degree of depolarization $\rho = 3/4$, is called a "depolarized band," whereas a Raman band that provides the degree of depolarization $0 \leqq \rho < 3/4$ is called a "polarized band." $G^a \neq 0$ often holds true with respect to RR scattering Eq. (1.8). As this occurs, a band expressing anomalous depolarization expressed as $\rho > 3/4$ appears. When $\alpha_{xy} = -\alpha_{yx}$ and other components are zero, we obtain $G^0 = G^s = 0, G^a \neq 0$, and hence, $\rho \rightarrow \infty$. A band that provides anomalous depolarization is called an "anomalously polarized band."

The polarization property of Raman bands can greatly aid assignment of the bands in a spectrum, because the polarization property of a Raman band is closely related to the symmetry of a vibration, that is, structure of the part of the molecule that gave rise to the band.

1.6 THE QUANTUM THEORY OF RAMAN SCATTERING

Although the classic theory provides a thorough view of Raman scattering, there are some effects that the classic theory alone cannot explain. For instance, without

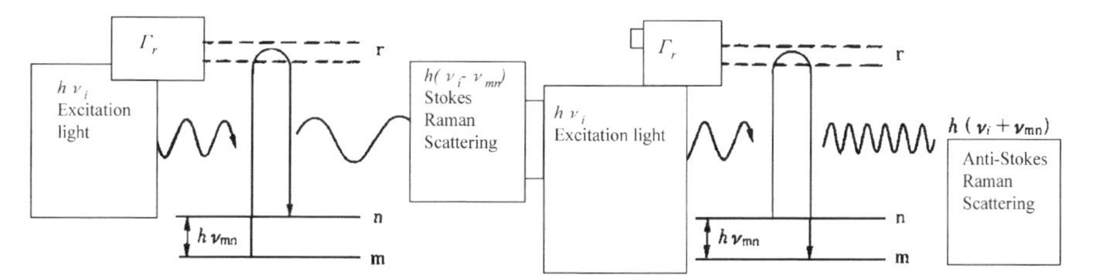

FIGURE 1.8 Mechanism of Raman scattering.

solutions from the quantum theory, the different intensities between Stokes and anti-Stokes Raman scattering cannot be explained. In explaining Raman scattering by the quantum theory, we first consider the energy levels of molecules (Fig. 1.8). As light interacts with a molecule, which is in the state m (or n), the molecule is "excited" to a higher energy state r, as the first step of the process. It needs to be clarified here that the "excitation" does not mean that this molecule is excited to a real excited state; the light perturbates the molecule, which then assumes a state with r-like property. In other words, the state r is not an actually existing state but is only a nominal (virtual) state. The second step is emission of a photon, followed by transition of the molecule to the state n (or to a state m when the initial state is n). We should not view these two processes as two independent processes but as one simultaneous process.

According to quantum mechanical calculation, the total intensity I_{mn} of Raman scattered light corresponding to the transition from the state m to the state n can be expressed as follows:

$$I_{mn} = \frac{128\pi^5}{9c^4}(\nu_i \pm \nu_{mn})^4 I_i \sum_{\rho\sigma} |(\alpha_{\rho\sigma})_{mn}|^2 \tag{1.12}$$

In this formula, the symbol I_i denotes the intensity of incident light (having the frequency of vibration ν_i) while the symbol ν_{imn} denotes a Raman shift. The symbol $\alpha_{\rho\sigma}$ denotes the $\rho\sigma$ component of scattering tensor. The formula above expresses that the Raman scattering intensity is proportional to the intensity I_i of incident light, the biquadratic of the frequency of ν_i of the scattered light $\pm\nu_{mn}$, and the square of the polarizability $\alpha_{\rho\sigma}$. The fact that the Raman scattering intensity is proportional to the biquadratic of the frequency of the scattered light is called the "ν^4 Rule."

The scattering tensor $(\alpha_{\rho\sigma})_{mn}$ in Equation 1.12 is defined as follows:

$$(\alpha_{\rho\sigma})_{mn} = \frac{1}{h}\sum_r \left[\frac{\langle m|\mu_\sigma|r\rangle\langle r|\mu_\rho|n\rangle}{\nu_{rm} - \nu_i + i\Gamma_r} + \frac{\langle m|\mu_\rho|r\rangle\langle r|\mu_\sigma|n\rangle}{\nu_{rn} + \nu_i + i\Gamma_r}\right] \tag{1.13}$$

In this formula, the symbol r encompasses all quantum mechanical eigenstates of molecules, while the symbol ν_{rm} denotes the frequency of the m-to-n transition. Denoted at $< m|\mu_\alpha|r >, < r|\mu_\rho|n >, \ldots$ are components of the transitional electric dipole moment, μ_ρ is an electric dipole moment operator along the direction ρ. The symbol Γ_r is a damping constant in the state r. Equation 1.13 is an important formula that explains Raman scattering and is known as the "Kramers–Heisenberg–Dirac dispersion expression." The content of this formula will be described below in relation to resonance Raman scattering.

1.7 CROSS SECTION

Raman efficiency of a scatterer is usually characterized by cross section, σ, which heavily depends on $d\alpha/d\mathrm{Q}$. σ is proportional to the probability of an incident photon to be scattered as a Raman shifted photon. Conceptually, it can straightaway be related to the absorption coefficient from Beer's law that is often met in analytical chemistry/analytical spectroscopy. The unit of σ is cm^2/molecule that is simple to convert into the more familiar term of $\mathrm{mol}^{-1}\ \mathrm{cm}^{-1}$. Raman intensities also depend on various experimental parameters such as collection geometry, polarization, and wavelength of the incident light, but the cross sections tend to be the major indicator of the intensity of Raman scattering given that the mentioned parameters vary with experimental setup and are difficult to keep constant.

Raman intensity linearly depends on σ according to

$$I_{\mathrm{R}} = I_0 \sigma_{\mathrm{j}} D\mathrm{d}z \tag{1.14}$$

where I_{R} and I_0 are Raman intensity and laser intensity, respectively, expressed in photons/second, σ_{j} refers to the cross section at the wavenumber j, with D being the density of scatters and dz the path length of the laser in the sample. Since σ_{j} equals $\sigma_{\mathrm{j}}{}^0\ (\bar{\upsilon}_0 - \bar{\upsilon}_{\mathrm{j}})^{-4}$ where $\sigma_{\mathrm{j}}{}^0$ is the frequency-independent cross section and $(\bar{\upsilon}_0 - \bar{\upsilon}_{\mathrm{j}})$ is the absolute frequency of the scattered light (in reciprocal centimeters), the final form of I_{R} is

$$I_{\mathrm{R}} = I_0 \sigma_{\mathrm{j}}^0 (\bar{\upsilon}_0 - \bar{\upsilon}_{\mathrm{j}})^{-4} D\mathrm{d}z \tag{1.15}$$

from which one concludes that intensity of a Raman band linearly depends on the cross section, density, path length, and the fourth power of the scattered light. However, this equation is based on expressing power in watts but since spectrometers count photons, it is more exact to introduce P_{R} in units of photons per second that is related to I_{R} as

$$P_{\mathrm{R}} = I_{\mathrm{R}}/hc(\bar{\upsilon}_0 - \bar{\upsilon}_{\mathrm{j}}) \tag{1.16}$$

so that the final expression for the photon counting systems is

$$P_{\mathrm{R}} = P_0 \sigma_{\mathrm{j}}^0 \hat{\upsilon}_0 (\bar{\upsilon}_0 - \bar{\upsilon}_{\mathrm{j}})^{-3} D\mathrm{d}z \tag{1.17}$$

that is only slightly different from Equation 1.15 [McCreery, 2000]. Since collection geometry significantly affects the observed Raman intensity, σ should ideally be counted along all the directions around the sample but this is normally not practiced. Rather, the differential Raman cross section, β, is used that is defined as σ for the employed solid angle of collection, Ω

$$\beta = d\sigma_j/d\Omega \tag{1.18}$$

1.7.1 Magnitude of Raman Cross Section

Tables 1.1 and 1.2 list several cross sections from selected substances. The results were obtained in various laboratories (McCreery, 2000).

The data for the absolute cross sections convincingly illustrate that symmetric, multiple bonds with high electron density and mobile electrons are likely to yield strong Raman bands, while the single bonds between the atoms with a significant difference in electron affinity, and therefore high partial charge and localization of electrons, are not likely to be strong Raman scatterers. An example of the former is, say, the C≡N bond while the C–H bond or $CHCl_3$ belongs to the latter case. Conjugated bonds (delocalized π stretches) tend to be quite strong scatterers. This is best illustrated with the band of liquid benzene at 992 cm^{-1} that is of significantly stronger intensity than the rest of the entries in Table 1.1 (both gas and liquid). Comparison with cyclohexane that does not contain benzene ring is particularly noteworthy. The ring stretching vibration in the region around 1000 cm^{-1} is important with respect to pharmaceutical materials, because Active Pharmaceutical Ingredients (APIs) usually contain a band in that region that, provided it is not overlapped, can be an excellent and easy-to-use spectroscopic indicator of the API.

Raman scattering cross section strongly increases with delocalization of π electrons. Table 1.2 shows that the Raman intensity of the ring-stretching band of benzene at 992 cm^{-1} is many times weaker than the corresponding band of anthracene. Of particular note is the scattering efficiency of conjugated system of π electrons of β-carotene. For the employed laser light, the two listed bands exhibit resonant Raman effect that makes their Raman responses almost four orders of magnitude

TABLE 1.1 Absolute Raman Cross Sections (from McCreery, 2000).

Sample	Raman shift (cm^{-1})	β (cm^2 / sr molecule × 10^{30})
Benzene (liquid)	992	27
	3060	45.3
Benzene (gas)	992	7
Cyclohexane (liquid)	802	5.2
O_2 (gas)	1555	0.58
CCl_4 (gas)	459	4.7
$CHCl_3$	3032	4.4
	667	6.6

The 514.5 nm excitation line employed for all the spectra from which the data below were obtained.

TABLE 1.2 Relative Raman Cross Sections (from McCreery, 2000).

Sample	Raman shift (cm^{-1})	β (cm^2 / sr molecule $\times 10^{30}$)
SO_4^{2-} (aqeous)	981	9.9
H_2O (liquid)	1595	0.11
ClO_4^- (aqeous)	932	12.7
Cyclohexane (in benzene)	801	11.9
Glucose (aqeous)	1126	5.6
β-Carotene (in benzene)	1520	1.1×10^7
Benzene	992	28.6
Naphthalene (in benzene)	1382	82
Anthracene (in benzene)	1402	540
CCl_4	459	16.9

The 514.5 nm excitation line employed for all the spectra from which the data below were obtained. All the values were obtained by direct comparison with a benzene band measured under the same conditions.

stronger than the normal Raman bands. Also, vibrations of a molecule as a whole create strong Raman bands. A good example of this is the accordion mode of the saturated hydrocarbon chain (when the hydrocarbon chain as a whole stretches and shrinks).

Various illustrations of Raman activity of electron-rich bonds can be found in the literature. For example, Raman spectra are used to monitor the structure of inorganic complexes and coordination of the molecules of the solvent to the central cation. The shifts of electrons in the complexes between the large, bulky Hg^{2+} cation with the easily polarizable thiocyanate SCN ligand (bonded through S) in different solvents is followed via shifts of several stretching bands in the complex, Hg–S, C≡N and C=S, the first one being the most while the last one being the least intensive (Šašić et al., 1998). The decrease of charge of Hg^{2+} due to coordination of a ligand molecule reduces the energy of the Hg–S bond, which can be followed via shift of the Hg–S Raman band to lower wavenumbers.

On the contrary, water is a very suitable solvent for Raman spectroscopy as its stretching and bending bands are weak across a very large portion of the Raman spectrum (100–3000 cm^{-1}).

One can also notice that Raman activity of stretching vibrations is better than that of bending vibrations, because the intensity of Raman scattering is proportional to movements of electron clouds.

The above considerations refer mostly to the normal Raman effect. In case the energy of the excitation line of the laser is close to an electronic transition of the molecule, significant amplification of the signal is noticed and the effect is known as resonance Raman scattering. Another case of amplification of Raman signal is when a scattering molecule is adsorbed onto a surface of a metal particle (Ag or Au), or roughened metal surface. Raman spectra of such molecules are termed "surface-enhanced Raman spectra" (SERS).

The Raman cross sections are quite small in comparison with other light-emitting processes that can occur in the molecule. For example, the cross section of fluorescence is several orders of magnitude higher than that of Raman scattering. In cases where fluorescence occurs, Raman spectra are barely visible on top of strong, broad, and featureless fluorescence signal. The problem of elimination of fluorescence has been tackled in various ways: by employing the excitation line of energy insufficient to cause electron transition (say 782 or 1064 nm), or by software operations that mathematically subtract relatively simple fluorescence signal that generally can be fitted with polynomials of varying degrees. Fluorescence represents a significant problem in Raman spectroscopy of biological materials because of various organic matrices unavoidably fluorescing.

1.8 RELEVANCE TO PHARMACEUTICALS

The above discussions on the Raman cross sections and accompanying illustrations reveal why Raman spectroscopy can be a very useful technique for pharmaceutical research particularly during development and manufacturing. Significant difference in Raman cross section of the structures with delocalized electrons and the single bonds among common atoms such as C–H, C–O or O–H can actively be exploited in mixtures between APIs that commonly have benzene rings and various excipients the spectra of which often do not overlap with the strongest stretching vibration of the ring in the region around 1000 cm^{-1}. Of course, this is a very general observation and some overlap certainly occurs, but it is reasonable to assume that more often than not the APIs can be monitored via their Raman bands without significant interference from the excipients.

Chemical formulae and Raman spectra of several frequently used excipients are shown in Figure 1.9 together with some APIs to illustrate chemical complexity of the latter in comparison with the former. Although Raman spectra of the excipients are quite rich, most of them do not have strong Raman response at 1000 cm^{-1} thus freeing a spectral window around that wavenumber through which APIs can be followed. Besides, Figure 1.9 shows that there are large spectral regions that are not covered with the bands of excipient and thus can be used for monitoring API bands. Although all listed substances display quite complex Raman spectra, the overlap among them is not complete (and is unlikely to ever be because of the sharpness of the Raman bands) and thus one may hope to come across unobstructed Raman bands of API, subject to satisfactory intensities. As it will be shown later on through examples from practice, this allows for active participation of Raman spectroscopy in monitoring APIs at various stages of manufacturing or development. Furthermore, it needs to be mentioned here that using sophisticated data analysis tools allows for reliable analysis of overlapped peaks, so that lack of interference is not such a rigorous condition for achieving satisfactory results although it is very desirable.

With regard to this, it is worth mentioning the Raman study by Hausman et al. [2005] in which quite a low loading of API (azimilide dihydrochloride 1%) was

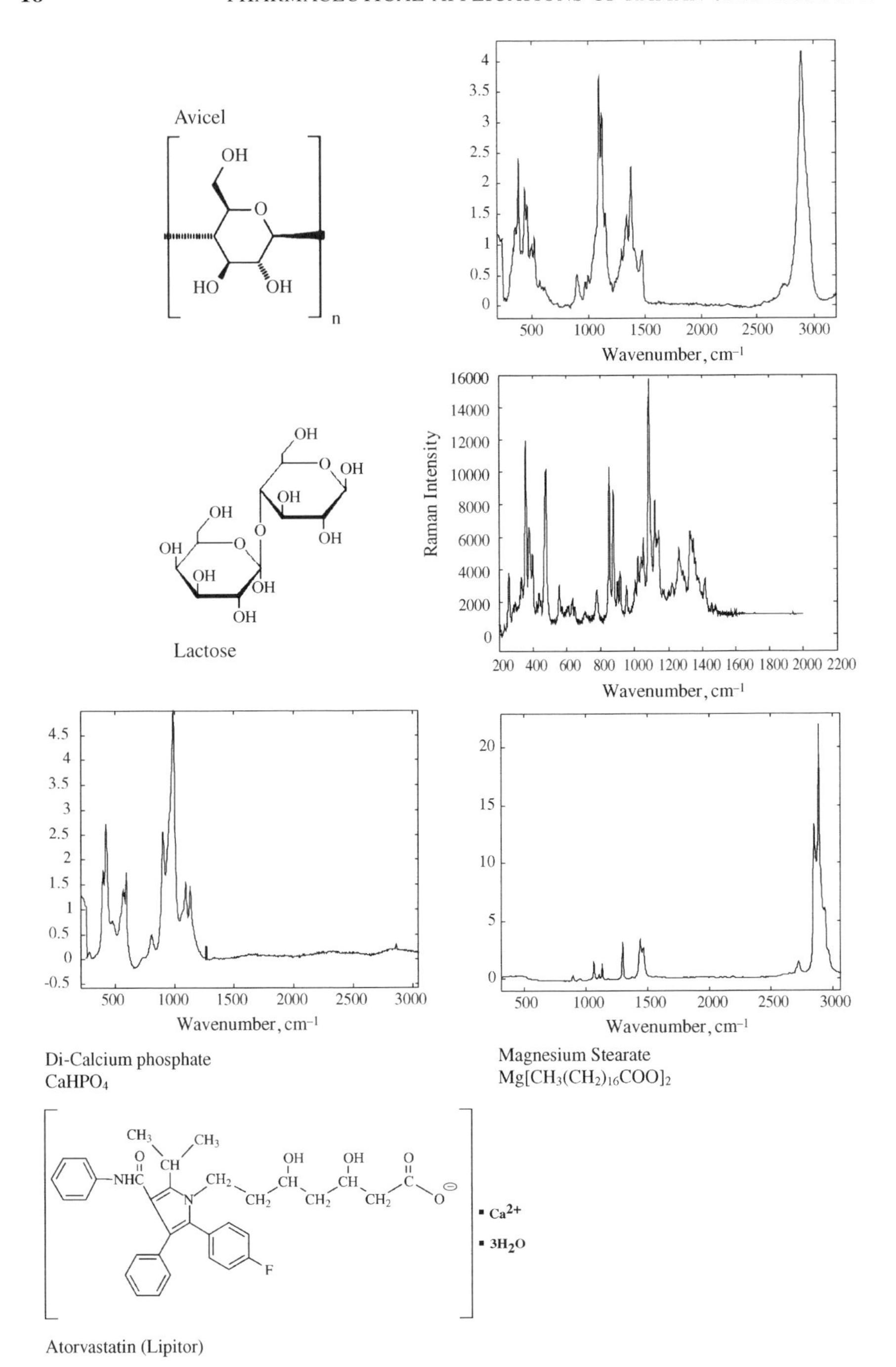

FIGURE 1.9 Chemical formulae and Raman spectra of some commonly used pharmaceutical excipients, and a few recognizable APIs to illustrate their chemical complexity in comparison with the excipients.

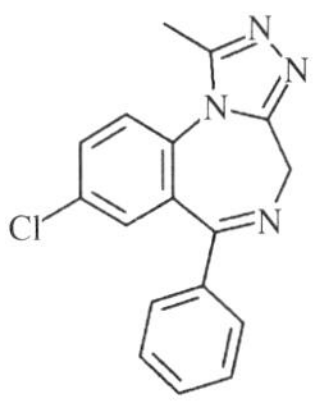

Triazolobenzodiazepine (Alprazolam)

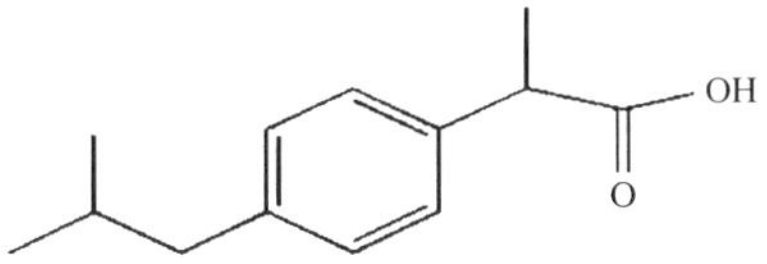

Ibuprofen

FIGURE 1.9 *(Continued)*

monitored online for blend homogeneity in a matrix dominated by lactose (95.5%). The nonoverlapped, combined azimilide ring breathing bands at 1600 and 1620 cm^{-1} were clearly detected and used for monitoring the homogeneity of the blend. Moreover, an assumption is made that even 0.1% of azimilide blend can be monitored by Raman spectroscopy.

Similarly, during the determination of spatial distribution of API in Alprazolam tablets via Raman microscopy it is found that despite API accounting for only 0.4% w/w of the tablet, quite strong Raman signal is obtained from the API from some spots on the surface of the tablets (Fig. 1.10) (Šašić, 2007a). For comparison, magnesium stearate is also present in approximately the same amount but its Raman response is much weaker and could not be detected, while the Raman response of 78% w/w of lactose was not much stronger, which is in a sharp contradiction with its much higher concentration.

Figure 1.11 shows an example of the formulation in which the loading of the API is substantial (> 20%). The spectrum of the API obtained by a Raman microscope from the surface of a tablet is much stronger than the spectra of the excipients with its ring-stretching band at 1000 cm^{-1} being almost entirely nonoverlapped. Interestingly, dicalcium phosphate (DCP) is present in the nearly same concentration but its Raman signal is only a weak shoulder to the API band at 1000 cm^{-1}. Although microcrystalline cellulose (Avicel) is about twice more abundant than the API, its Raman bands are obviously much weaker in comparison with the strong band of the API.

Of particular note may be the comparison between the Raman and NIR spectroscopies. The latter has become a frequently used technique in some industries, notably in pharmaceutical and food industry mainly because of relatively inexpensive instrumentation, ease of installation and handling, and compliance with health and safety regulations. The entire near-infrared region (NIR) of the electromagnetic spectrum encompasses light with wavelengths ranging from 0.7 to 2.5 μm (14,286 – 4000 cm^{-1}). NIR spectra predominantly correspond to harmonics of

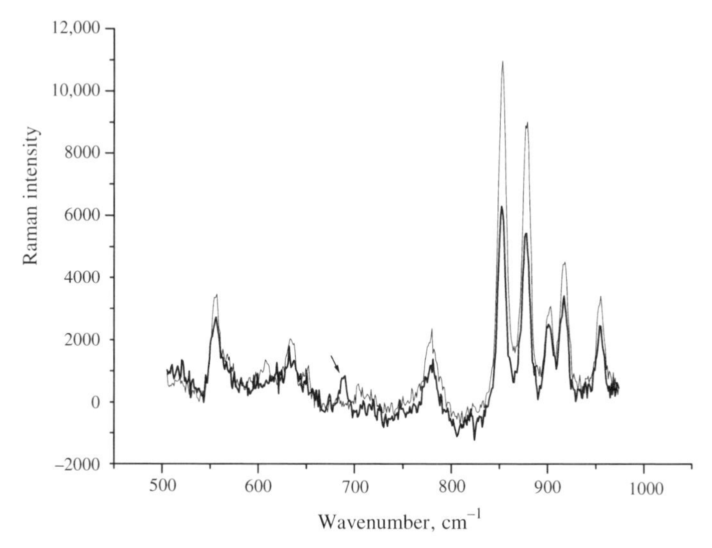

FIGURE 1.10 The spectra from two spots on the surface of a tablet that contain lactose (predominant excipient, thin line) and API (thick line), respectively. Despite accounting for only 0.4% w/w, the Raman signal of API is visible by naked eye at 678 cm^{-1}. If chemometrics is applied to a set of such spectra [Šašić, 2007a,b], there are good prospects to retrieve the Raman signal of API more clearly. The Raman signal of lactose that is about 80 times more abundant than API is shown for comparison. The ubiquitous presence of the lactose spectrum in all the spectra from the surface of the tablet is unavoidable.

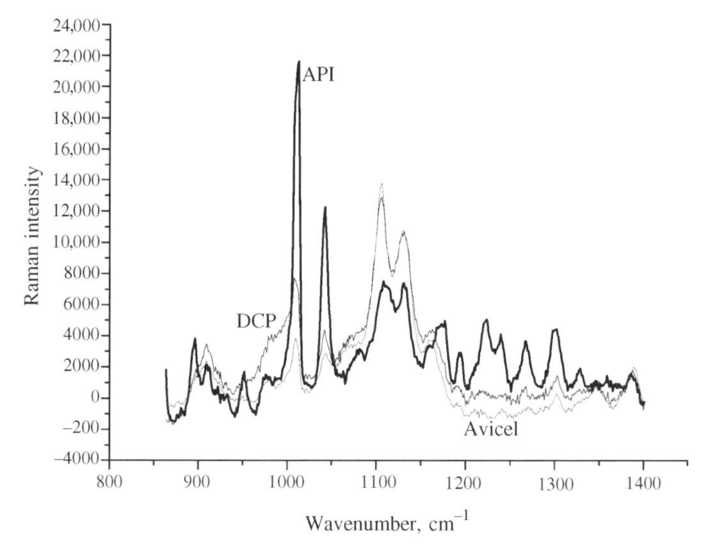

FIGURE 1.11 Three Raman spectra from the surface of a high-loading API tablet featuring the difference in Raman response from API, Avicel and DCP. The characteristic peaks of API and DCP are marked in the figure, while the bands of Avicel are always present and the marked spectrum of Avicel only illustrates absence of API and DCP bands.

overtones and combinations of fundamental vibrational transitions more frequently associated with (mid-)infrared spectroscopy. Overtone and combination absorptions are principally seen for CH, OH, and NH molecular groups. The ambiguity of these groups within biological systems and the physical nature of these transitions result in very complex, overlapping spectra that are in most cases analyzed by chemometrics.

The NIR spectra of solid materials used in development and manufacturing facilities are not that complex and this is an additional reason for using NIR spectroscopy for process monitoring purposes, quality control, and so on.

The NIR spectra of some frequently used excipients are shown in Fig. 1.12. The spectra from the surface of a tablet (or from a blend) that contains those excipients will contain visible spectral indications of the API only for formulations with high concentrations of API. The overlap among the pure component spectra is substantial and hence the spectral contributions from various components are usually difficult to recognize in the NIR spectra of the tablet. Although S/N ratio for NIR spectra is much better than that for the Raman spectra, the overlapping is so severe that spectral identification of minor components may be unachievable despite chemometrics means used. On the contrary, the major problem with the Raman spectra of the sample is in the weakness of S/N ratio. The Raman spectra of the pure components overlap much less and if the quality of the experimental spectra can be improved, for example, by denoising via principal component analysis, the spectral traces of the minor components can potentially be retrieved. With regard to Fig. 1.12, the signal of sodium starch glycolate (Explotab) in the matrix of spectra from the tablet is retrievable in Raman (after principal component analysis) but not

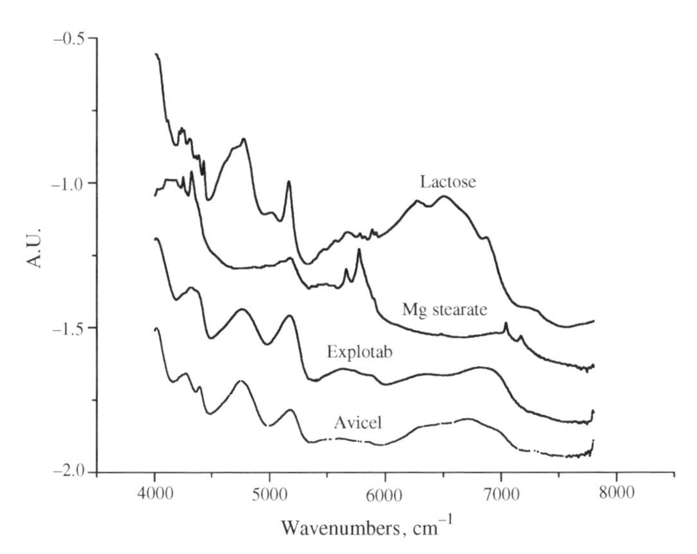

FIGURE 1.12 NIR spectra of some common pharmaceutical excipients. DCP has no NIR signal in the wavenumber range shown.

in NIR due to a substantial overlap between the NIR spectra of Explotab and Avicel (Šašić, 2007b).

1.9 RESONANCE RAMAN EFFECT

Let us first look at a specific example to see what is the resonance Raman effect. Figure 1.13 shows a Raman spectrum of met hemoglobin with coordination of F-(met HbF) measured using excitation wavelength of 488.0 nm. The wavelength of 488.0 nm is located within the absorption band of a heme group and therefore exciting the sample with this wavelength yields a Resonance Raman spectrum that can be attributed to the heme group (Nagai et al., 1980). In short, the bands in Fig. 1.13 all originate from the heme group. Figure 1.14 shows a Raman spectrum of met HbF measured using excitation light at 200 nm (Copeland et al., 1985). This spectrum is considerably different from the spectrum shown in Fig. 1.13 and includes no Raman band from the heme group. The bands appearing in this spectrum are due to either amide groups or tyrosine (Tyr) and phenylalanine (Phe) residues. We can draw this conclusion because the excitation wavelength in Fig. 1.14 directly overlaps with the strong absorption bands of an amide group, residues of Tyr, Phe, and so on. As we can see from Figs. 1.13 and 1.14, Raman spectra may change substantially depending on the excitation wavelength employed. If we excite a Raman spectrum of met HbF in a region free from any absorption band, for example, at 1000 nm, the resulting Raman spectrum will be nonresonant and there will be no bands with substantially increased intensities. However, if the excitation wavelength is partially absorbed by the sample, the vibration modes of the part of the molecule that absorbs may be dramatically

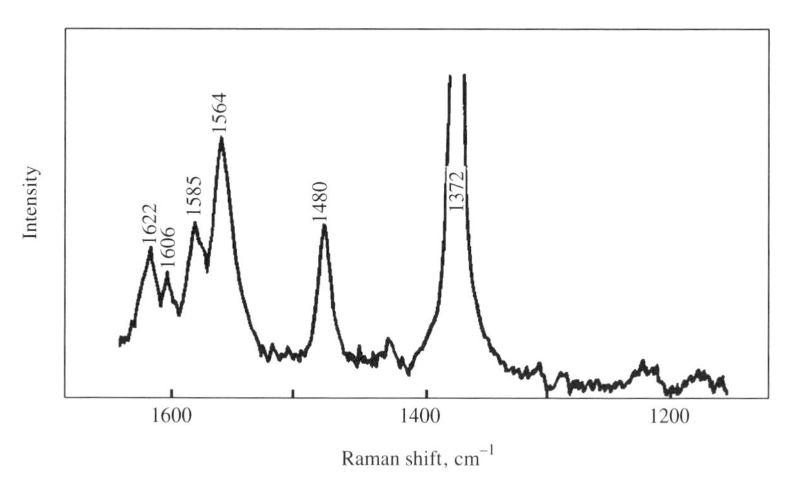

FIGURE 1.13 Resonance Raman spectrum of met HbF (excited at 488.0 nm) (reprinted with the permission of Nagai et al.; copyright (1980) Elsevier Science Publisher).

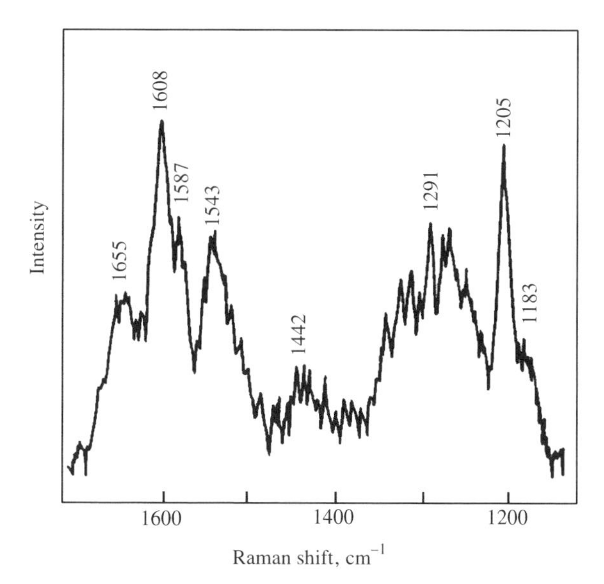

FIGURE 1.14 Resonance Raman spectrum of met HbF (excited at 200 nm)(reprinted with the permission of Copeland et al.; copyright (1985) American Chemical Society).

changed producing thus resonant Raman spectrum that is apparently different from the normal Raman spectrum.

Let us now give some more theoretical thought to the resonance Raman effect. Where there is resonance Raman scattering, the energy of incident light is approximately equal to the electronic transition energy, which permits us to reillustrate Figs. 1.8, 1.13 and 1.14. That is, in cases where resonance Raman scattering occurs, the state r is an actual electronic excited state. The phenomenon that the closer the energy $h\nu_i$ of the incident light becomes to the electronic transition energy $h\nu_{\mathrm{rm}}$, exceptionally stronger the intensity of Raman scattering becomes is obvious from Equations 1.12 and 1.13. As the incident light becomes closer to the electronic transition energy $(\nu_{\mathrm{i}} \simeq \nu_{\mathrm{irm}})$, the denominator in Equation 1.13 becomes extremely small and the scattering tensor $(\alpha_{\rho\sigma})_{\mathrm{mn}}$, therefore, becomes abnormally large.

A closer inspection of Equation 1.13 reveals relation between the widths of the absorption band and magnitude of the intensity increase. The denominator in Equation 1.13 includes the damping constant Γ_r that is indicative of the energy width in the excited state that corresponds to the bandwidth of the absorption band. From Equation 1.13 it follows that the sharper the absorption band, the smaller Γ_r, or the greater the scattering tensor, the stronger the intensity of Raman scattering. It is possible to measure the excitation profile of the RR effect by following the intensity of the collected spectra as a function of employed excitation wavelength. It is always desirable to determine the excitation profile because it provides precious findings about the symmetry of vibration that can help to better the RR effect.

The dramatic increase of the intensity of Raman scattering is the crucial element of the resonance Raman effect but is not the only feature of this effect. Other unique features include anomalous depolarization, increased intensities of overtones and combination modes, and so on.

With regard to applications, resonance Raman spectroscopy is advantageous in the following aspects:

1. Great increase in sensitivity allows for measurements of Raman spectra from very diluted samples. Acceptable Raman spectra can be obtained even from the sample with poor optical homogeneity.
2. Proper selection of the excitation wavelength makes it possible to selectively study a particular part contained in a giant molecule whose molecular weight is very large (a heme group in a heme protein, for instance). Similarly, selected molecules in complex mixtures can be pinpointed (such as carotenoid in a reaction center in photosynthetic bacteria.
3. Resonance Raman spectroscopy provides information about the excited electronic states. Direct measurements of resonance Raman spectra of the excited electronic states are feasible.

Resonance Raman spectroscopy has usually been linked only with using visible light to excite the sample, but other sources of excitation have also been employed (such as UV and NIR) recently leading to the appearance of UV and NIR RR spectroscopy, the former being useful for bioanalytical investigations (e.g. for analysis of DNA and proteins) due to high sensitivity, lack of fluorescence, and suitability for use in aqueous solutions (Smith and Dent, 2004).

1.10 INSTRUMENTATION FOR RAMAN SPECTROSCOPY

The basic elements of Raman instruments will only briefly be described here. More details about them are given at various places in this book. Excellent treatment of this topic can be found in the monograph by McCreery (2000).

Four major elements of a Raman instrument are: the laser, the spectrometer, the detector, and the optical setup.

1.10.1 Lasers

The most important question that needs to be addressed with regard to what laser to use for a particular experiment is whether or not the samples analyzed fluoresce. In case the probability of fluorescence is high, the NIR lasers are the excitation source of choice. There are several types of the laser lasing at sufficiently high wavenumbers so that the effect of fluorescence is diminished. The most convenient choice is to use lines in the region of 780 nm from a diode laser. These lasers are based on optical emission from semiconductor junctions. As these are usually characterized by compact design, small cost, and allowing for adequate Raman response from a variety of samples, the popularity of these lasers appears to be increasing. Another option is to employ the Nd:YAG laser that provides 1064 nm line. These lasers are used in FT instruments. In principle, these are solid-state, pumped lasers in which the lasing medium is Nd^{3+} hosted inside a rod of yttrium aluminium garnet (YAG).

For samples that weakly fluoresce, the lasers with the lines in the range of 470–650 nm are preferred. The lasers that provide these lines are often Ar^{+}- based. Other types include He–Ne or Kr^{+} lasers. All these types are capable of providing relatively high power.

The key features of these two groups of lasers are as follows: The lasers with visible excitation normally provide much better signal-to-noise ratio of Raman spectra, are coupled with cheaper detectors because of strong flux of Raman photons that simplifies detection; the only drawback being excitation of fluorescence that may overwhelm the Raman signal. Also, samples can potentially be damaged if unacceptably high power density is used. The lasers with NIR excitation provide poorer Raman signal on the detector, which necessitates the use of more sophisticated and optimized detectors. Instruments equipped with such lasers are less sensitive than the instruments that use visible excitation, but normally the background tends to be significantly lower and thus detection of the Raman signal is more straightforward.

As mentioned above, the NIR lasers are frequently used for experiments with biological samples, because a number of components present in such samples usually fluoresce and there is smaller risk of damage to the sample. On the contrary, owing to much higher sensitivity and simpler hardware requirements, the laser with visible excitation may be used for analyzing nonbiological materials, which may be of interest in pharmaceutical development and manufacturing in particular, subject to health and safety clearance. The health and safety concerns are probably the major drive for steadily increasing popularity of NIR laser sources despite them not being harmless yet invisible.

1.10.2 Spectrometer

The primary function of the spectrometer is to allow for separation of the scattered light according to wavenumbers leading thereby to the appearance of the Raman spectrum. There are two key types of spectrometers: dispersive and nondispersive. The dispersive spectrometers are based on diffracting the scattered radiation by a grating. These spectrometers are used with the systems equipped with visible and up to 785 nm excitation. Their major characteristics are sensitivity (high S/N ratio), no need for high laser power to be used, and no moving parts. The (slight) shortcomings are trade-off between the spectral resolution and wavenumber coverage and varying resolution across the spectrum. These spectrometers are very popular and frequently used, and are manufactured in several modes of which the single grating one is the simplest yet quite an efficient configuration. As the name suggests there is only one grating in this configuration that is equipped with efficient stray light rejection filters in order to be functional. The problem of stray light is a major obstacle to efficient working of a spectrometer and it originates in the probability of the reflected laser light to enter the optical system. As the intensity of that source is much stronger than that of Raman scatter, the latter becomes undetectable. To prevent this, various filters that do not transmit the excitation light can be positioned inside the system. Frequently used filters are dielectric notch and edge filters, holographic notch filters, and absorption filters (McCreery, 2000).

The performance of dispersion spectrometers is judged by their resolution and spectral coverage, *f* number, transmission, and stray light rejection (which is of particular importance in analyzing the low wavenumber region $< 200\,cm^{-1}$).

Nondispersive Raman spectrometers are almost exclusively associated with FT modulation of the signal and do not include physical separation of the wavenumbers.

These spectrometers are used in FT-Raman instruments with the excitation at 1064 nm. Their major advantage is excellent wavenumber precision across wide range of wavenumbers and therefore the instruments with such spectrometers are frequently used for identification of substances or for registering small shifts. The major shortcoming is relatively low sensitivity compared to the dispersive configuration.

1.10.3 Detectors

Charge coupled devices (CCDs) are overwhelmingly popular in modern Raman instruments and thus only their characteristics will be presented here.

CCD cameras are produced from silicon (or other photosensitive semiconductors) and register Raman signal via electron–hole pairs produced by photons of sufficient energy, usually in the 200–1100 nm range. CCD cameras consist of two-dimensional arrays of pixels (e.g., 256×1024) that each can be considered as an independent detector. The horizontal pixels are calibrated so as to correspond to the wavenumber axis, while the vertical pixels actually measure the strength of the Raman signal. In other words, the image on the CCD camera is an electronic picture of Raman signal that is then converted into a spectrum.

One of the most important features of a CCD camera is its quantum efficiency curve that displays the probability of generating a photoelectron versus the energy of radiation. Such curves usually peak at around 600 nm and zero after 1000 nm (or under 250 nm) limiting thus the wavelength range of the lasers that can be used.

CCDs can further be characterized by a number of elements:

(i) *Gain*: It describes transformation from electrons to counts.

(ii) *Dark current*: This refers to the spontaneous formation of electron–hole pairs which are not produced by external radiation but rather by a thermal effect. This effect is thus very dependable on temperature and roughly doubles for each 5°C (McCreery, 2000).

(iii) *Readout noise*: This is an unavoidable, electronic imprecision in counting electrons that becomes important for very weak signals when its magnitude may become comparable with that of the signal itself.

(iv) *Binning*: It represents adding or averaging signal along the vertical pixels with the purpose of increasing the intensity of the signal. This way the spatial information along that axis is lost (which may be of importance in spectra obtained by Raman microscopes) but significant gains in intensities may be achieved.

(v) *Types of CCDs*: There are front- and back-illuminated CCDs, as well as deep depletion CCDs. The first configuration refers to electrical circuits positioned on the front surface of the camera thus blocking partially the area available for photoreaction. The back-illuminated CCDs do not suffer from this issue but are usually more difficult to produce, fragile, and more expensive. Deep-depletion devices have their quantum efficiency curve adjusted (shifted toward longer wavelengths) to increase sensitivity in the NIR region.

(vi) *Cosmic rays*: This is actually not a function of a CCD, because the appearance of cosmic rays is another unavoidable effect that occurs because of cosmic radiation hitting/being registered on the CCD camera. These events are easily recognizable as sharp, very narrow peaks that can be eliminated with software operations.

1.10.4 Optics

Depending on the way the light is delivered and collected from the sample, there are 180° and 90° configurations. Alternatively, optical fibers are used to excite the sample and collect the Raman scatter.

In 180° configuration, the laser beam and (back-) scattered light are on the same axis. There are various ways to combine mirrors and lenses to form this setup. This is a very popular option because there is no need for extra alignment of the laser beam and collection optics. The 90° configuration obviously implies that there is an angle between the excitation beam and collected Raman scatter. This configuration is no more of practical significance.

The fiber optic technology strongly advanced during the 1990s when it matured as a viable option for delivering and collecting light in various real-world applications. Raman signal/laser light in the range 500–1000 nm is very amenable for applications of optical fibers because of the total reflection between the core and the cladding of the fiber, which makes the losses of light inside the fibers minimal, even in very long fibers. The usually used multimode fibers are normally sized between 50 and 600 μm, are quite rugged, and can be easily coupled to the laser or set to collect back scattered light. The additional elements that are used with the fibers are band-pass and band-reject filters that can be placed at various points. The former specifies the wavelength of the laser light that enters the fiber, while the latter eliminates the stray light on the spectrometer end.

The fiber optic probes interface excitation and collection fiber and the sample. Those compact devices can be located at a large distance from the laser and spectrometer and are extremely suitable for use in industrial environments. There are filtered and unfiltered probes, based on presence of the background signal (coming from the fiber) that is not particularly strong but still in some situations may mask the Raman signal. The unfiltered probes are small, simple, and relatively inexpensive, and are produced in a variety of *n*-around-1 configurations in which '1' stands for the excitation fiber and "*n*" for the surrounding collection fibers. The filtered probes have optical band-pass and band-reject filters. Of particular importance are probe heads that are small devices that accommodate fibers, filters, and focusing optics in a compact housing.

REFERENCES

Cheng JX, Xie XS, 2004. Coherent anti-Stokes Raman microscopy: instrumentation, theory and applications. *J. Phys. Chem. B*, **108**, 827.

Copeland RA, Dasgupta S, Spiro TG, 1985. *J. Am. Chem. Soc.*, **107**, 3370.

Ferraro JR, Nakamoto K, Brown CW, 2003. *Introductory Raman Spectroscopy*, 2nd edition. Academic Press, New York.

Hausman DS, Cambron RT, Sakr A, 2005. Application of Raman spectroscopy for on-line monitoring of low dose blend uniformity. *Int. J. Pharm.*, **298**, 80.

Long DA, 2003. *The Raman Effect: A Unified Treatment of the Theory of Raman Scattering by Molecules*. Wiley, Chichester, UK.

McCreery RL, 2000. *Raman Spectroscopy for Chemical Analysis*. Wiley, New York.

Nagai K, Enoki Y, Kitagawa T, 1980. *Biochim. Biophys. Acta*, **624**, 304.

Raman CV, Krishnan KS, 1928. A new type of secondary radiation. *Nature*, **121**, 501.

Šašić S, 2007. A Raman mapping of low-content API pharmaceutical formulations. I. Mapping of Alprazolam in Alprazolam/Xanax tablet. *Pharm Res.*, **24**, 58.

Šašić S, 2007b An in-depth analysis of Raman and near-infrared chemical images of common pharmaceutical tablets. *Appl. Spectrosc.*, **61**, 239.

Šašić S, Jeremić M, Antić-Jovanović A, 1998. Raman study of solvent–solute interactions in solutions of mercury(II) - thiocyanate complexes. *J. Raman Spectrosc.*, **29**, 321.

Smith E, Dent G, 2004. *Modern Raman Spectroscopy: A Practical Approach*. Wiley, Chichester, UK.

2

QUANTITATIVE ANALYSIS OF SOLID DOSAGE FORMULATIONS BY RAMAN SPECTROSCOPY

Steven E.J. Bell

Queen's University, Belfast, Northern Ireland, BT9 5AG, UK

2.1 INTRODUCTION

For much of its considerable history (it is almost 80 years since the first observation of the effect in 1928), Raman spectroscopy has been an experimentally challenging technique; so it has primarily been used for probing the structures of materials in order to gain a deeper understanding of their properties. Essentially, Raman spectroscopy was a technique of last resort that was used only when other more straightforward spectroscopic analysis methods had failed or were inappropriate. However, this chapter will discuss why this is no longer the case and how improvements in instrument performance now mean that Raman measurements can be made more rapidly than established methods for quantitative pharmaceutical analysis, while still providing the accuracy and precision required for analysis of solid dosage forms. This discussion will necessarily touch on aspects of instrumentation and experimental procedures, because the field is still sufficiently new that standard practices have not yet been fully established; so some familiarity with the general experimental methodologies is useful in interpreting the results from published studies.

In principle, it would have been possible to apply Raman methods to both qualitative and quantitative analysis of pharmaceutical solid dosage forms right from its earliest days. However, in the era before the introduction of lasers as the excitation sources, spectral accumulation times were typically measured in hours so it was

Pharmaceutical Applications of Raman Spectroscopy, Edited by Slobodan Šašić

simply not practical to use Raman scattering rather than conventional analysis techniques. Under such conditions, trying to establish even a simple calibration curve of peak height versus composition would be a major undertaking since several spectra would need to be recorded and these could take days to accumulate. Even with the introduction of the laser excitation, when scanning spectrometers and photomultiplier (PMT) detectors were used the time required to record a full spectrum even for favorable samples was at least several minutes.

In order for Raman spectroscopy to become competitive with other spectroscopic techniques, never mind the established methods for quantitative analysis of solid dosage form pharmaceuticals such as HPLC, several different advances in the technology needed to be made. One of the most significant of these was the introduction of spectrometer/detector systems which allowed simultaneous acquisition of the Raman signal at multiple cm^{-1}. These were either Fourier transform instruments or dispersive spectrometers where the PMT was replaced by a multichannel detector, initially a Vidicon or diode array but now almost universally a CCD. The introduction of instruments with multichannel detectors run under computer control not only dramatically reduced the time required to acquire spectra, but also gave the ability for users to manipulate digital spectral data, including simple processes such as spectral subtraction and baseline removal. FT instruments also had an advantage over previous spectrometers that they operated at a much longer excitation wavelength than was available for dispersive instruments. This was important since the long wavelength excitation dramatically reduced problems with fluorescence backgrounds that were often encountered when visible wavelength excitation was used with dispersive instruments. The most recent development has been the widespread adoption of far red diode lasers (operating around 800 nm) into dispersive spectrometers; again, this long wavelength helps to reduce fluorescence problems and has made dispersive instruments competitive with FT spectrometers. This is particularly obvious for the current generation of dispersive instruments that typically have higher sensitivity than earlier models and can acquire spectra of useful quality within seconds rather than a few minutes, which is typical for FT instruments.

The topics that need to be covered in depth within a chapter dealing with quantitative analysis of pharmaceuticals have changed significantly from even a decade ago. In particular, the vast majority of the published work in the past decade was carried out using commercial instrumentation, a trend likely to continue since all the major analytical instrumentation manufacturers now offer Raman spectrometers suitable for use in pharmaceutical applications. Although Raman microscopes with small spot sizes (1–50 μm) and excellent spatial resolution are widely used in pharmaceutical research, the quantitative studies described here mostly use macroscopic systems (spot sizes 50 μm –3 mm) because this reduces sampling problems (vide infra). These systems are rapid and convenient to use, routinely allowing measurements in a few minutes per sample at most and now evolving to high throughput noncontact analysis at rates of >1 sample/s.

The range of pharmaceutical materials which have been subjected to qualitative Raman analysis is now extremely large and the number of quantitative studies has also increased dramatically in recent years, which is one of the reasons that this chapter is not structured as a comprehensive, compound-by-compound review of the various solid dosage form pharmaceutical samples, mostly tablets, that have been quantitatively analyzed since the first reports in the early 1990s (Deeley et al., 1991). There are already several excellent reviews of the general area of pharmaceutical Raman studies (Bugay, 2001; Frank, 1999; Pinzaru et al., 2004; Smith and Dent, 2005; Stephenson et al., 2001; Vankeirsbilck et al., 2002; Wartewig and Neubert, 2005) and a list the compounds which have been studied up to 2002 is available (Vankeirsbilck, 2002). However, a short summary table is included here to cover examples of studies that have been published 2002–2006. Table 2.1 is

TABLE 2.1 Overview of Quantitative Raman Studies on Pharmaceutical Materials Published 2002–2006.

Active compound	Physical form	Spectrometer	Analysis	References
Acetylsalicylic acid, acetaminophen	Tablets	FT	M	Szostak and Mazurak (2002)
Acyclovir	Blister pack	FT	U	Skoulika and Georgioa (2003)
Ambroxol	Tablets	FT	M	Szostak and Mazurak (2004)
Ambroxol	Tablets	Dispersive (785 nm)	M	Hwang et al. (2005)
Calcitonin	Powder	Dispersive (670 nm)	U	Vehring, (2005)
Captopril, prednisolone	Tablets	FT	M	Mazurek and Szostak (2006)
Carbamazepine	Powder	FT	M & U	Strachan et al. (2004)
Chlorophenicol palmitate	Solid	FT	U	Gamberini et al. (2006)
Chlorophenicol palmitate	Tablet	Dispersive (633 nm)	–	Lin et al. (2006)
Dipyrone	Tablets	FT	U	Izolani et al. (2003)
Diltiazem hydrochloride	Tablets	FT	U	Vergote et al. (2002)
Indomethacin	Tablets	FT	M	Okumura and Otsuka (2005)
Mannitol	Powder	FT	M	Auer et al. (2003)
Mannitol	Powder	FT	U	Roberts et al. (2002)
Nitrofurantonin, theophylline, caffeine, carbamazepine	Powder	FT	M	Rantanen et al. (2005)
Paracetamol	Powder	FT	U	Al-Zoubi et al. (2002)
Ranitidine	Tablets	FT	M	Pratiwi et al. (2002)
5-*p*-Fluorobenzoyl-2-benzimidazolecarbamic acid, methyl ester	Powder	FT	M	De Spiegeleer et al. (2005)

Univariate and multivariate data analysis are denoted U and M, respectively.

primarily intended to give a general impression of the range of samples for which quantitative Raman analysis methods have been developed and the methods currently being employed. No strenuous effort has been made to make this a comprehensive list; partly because such tables very quickly become obsolete but mostly because researchers who require compound-specific information will undoubtedly use electronic literature searching methods whether a table contains their compound of interest or not.

Finally, one very encouraging feature of the quantitative Raman studies, which have been undertaken, is that there is an increasing tendency to not only to quantify and then tabulate the accuracy and precision of the experimental procedures developed but also to compare these to the values found in pharmacopoeias for the established methods of carrying out the same analysis (Gamberini et al., 2006, Skoulika and Georgiou, 2003, Szostak and Mazurek, 2002). In many cases, the established methods are based on chromatography (particularly HPLC) rather than spectroscopy, but it is important to note that the Raman methods are being tested against the actual methods which they would hope to replace, rather than against other spectroscopic methods that are also being advanced as potential replacements for the standard analytical techniques.

2.2 QUANTITATIVE ANALYSIS

The fundamental basis for quantitative Raman analysis is that, all other things being equal, the intensity of Raman signal from a compound will increase in direct proportion to the quantity present in the probed volume of the sample. In principle, it is possible to determine (through experiment) the Raman scattering cross section for any compound of interest and to use it to determine the concentration of the compound in a sample from the absolute signal height (McCreery, 2002). Although this sounds no more difficult than determining the concentration of fluorophores in a sample using known quantum yields, the technical challenges are much more extreme. In particular, the intensity of the observed signal can alter because of small changes in numerous instrumental parameters (laser power, sampling position relative to focus of laser, etc.), so it is often found that the absolute signal from the same sample measured on the same commercial instrument will vary by at least a few percent if it is removed and replaced, even if this is done in the course of a single day. As a result of this reproducibility problem, the absolute Raman intensities are seldom used for quantitative analysis.

It is easy to circumvent these problems by measuring the bands of interest relative to other bands in the sample, since both will rise and fall in concert as the experimental conditions change. In pharmaceutical formulations, there are typically numerous bands from the other constituents in the tablets, particularly excipients, which the signal of the API can be measured against. For example, Fig. 2.1. shows the systematic growth of bands due to salicylic acid in a calibration set of model tablets prepared with salicylic acid contents 0–15% (w/w) in mannitol. Measuring API against excipient is often particularly appropriate, because the analytical

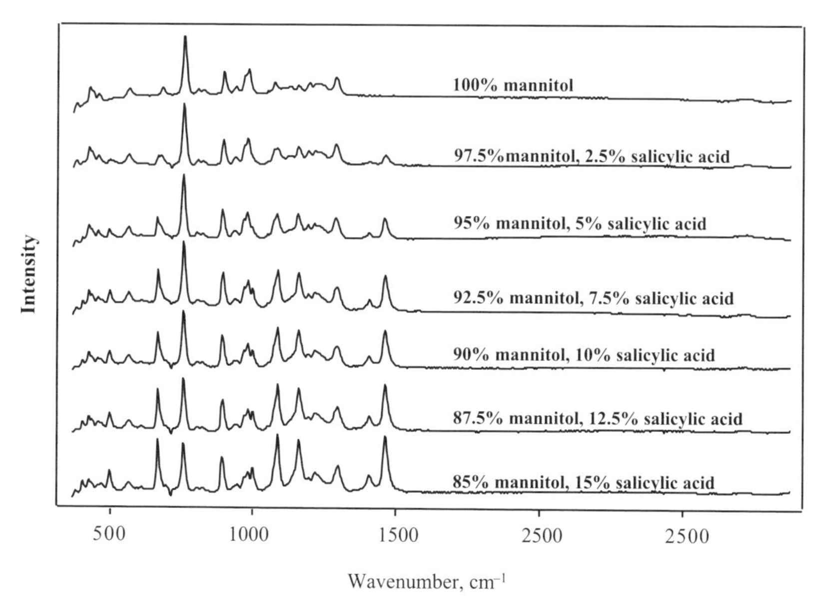

FIGURE 2.1 Illustration of the systematic growth of bands due to an API in a calibration set of model tablets prepared with salicylic acid (0–15%, w/w) in mannitol.

problem is to determine the relative proportions of the various constituents in any case. The alternative approach of introducing a standard reference band by adding a material chosen to have intense, sharp, noninterfering bands to the sample is time-consuming and normally unnecessary; indeed, it is often found to degrade analytical performance (Mazurek and Szostak, 2006; Szostak and Mazurek, 2002, 2004).

The two broad approaches to quantitative data analysis are univariate and multivariate (Pelletier, 2003). The traditional univariate approach is to measure either the heights or peak areas of characteristic bands of the compound of interest and of the reference material. Calibration plots of relative band intensity against composition are then constructed and used to predict the composition of test samples. This method is still widely used and has the advantages of being straightforward and robust. However, more recently multivariate data analysis methods are being applied much more widely. These methods attempt to capture the variance within the entire spectra (or a range within them) in terms of a much reduced set of latent variables and excellent descriptions are available in the literature (Cooper, 1999; Shraver, 2001). The key difference from univariate methods is that this general type of analysis includes data from numerous points across the whole recorded spectral range (shown through loading plots, see Fig. 2.2) and is thus less sensitive to noise at any given point. The main driving force for the recent blossoming of these techniques has been the fact that they are essential for extracting quantitative data from NIR spectra, which has led to the development of easy-to-use and powerful

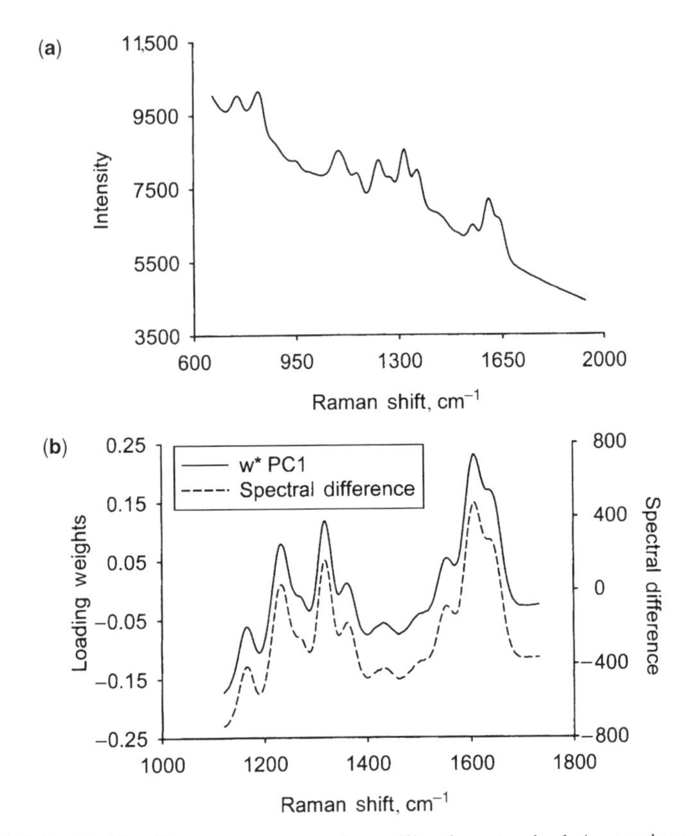

FIGURE 2.2 (**a**) The Raman spectrum of a calibration standard Acetaminophen tablet taken from a series which were prepared at different API levels. (**b**) A comparison of the difference trace obtained by subtracting the high and low API spectra with the loading weight of the first principal component obtained through a multivariate (PLS) calibration. This PLS component has obviously captured the expected variance in the data. Reprinted from Wikstrom et al. (2006).

software that can equally well be used for analysis of Raman data. This, along with the universal availability of desktop computers sufficiently powerful to carry out the large numbers of arithmetical operations required, means there is no real barrier to adopting these much more sophisticated data analysis techniques and to take advantage of the increased accuracy and precision that multivariate data analysis can provide (see Strachan et al., 2004 for example).

Even a cursory examination of the literature on the quantitative analysis of solid dosage forms reveals that many of the reports describe a generic type of study in which the main objective is to establish the relative proportion of API (or a particular polymorphic form of the API) and excipient in simple tablets (or power mixtures) within conventional laboratory settings (Table 2.1). Quantitative studies have been carried out on systems ranging from high dose formulations to cases where the

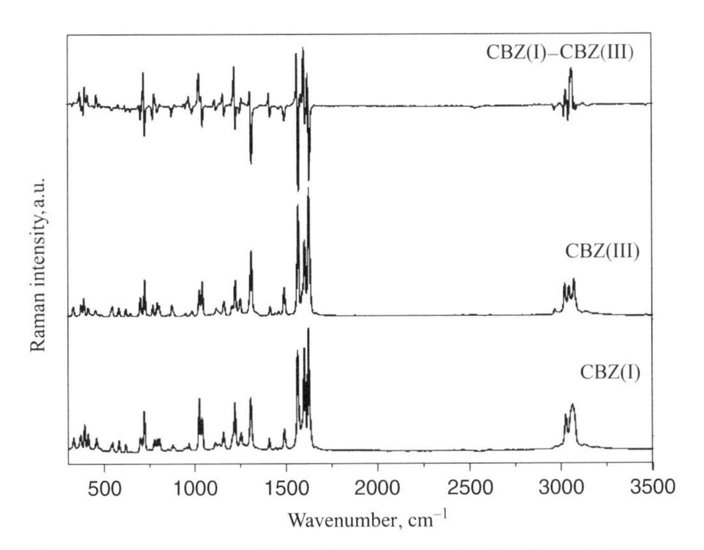

FIGURE 2.3 FT-Raman spectra of two CBZ polymorphs that have similar spectra shown along with the spectrum created by subtracting them to highlight the differences present. Reprinted from Strachan et al. (2004).

API was present as <1% of the total tablet mass (Taylor and Langkilde, 2000). In addition, these studies also covered conformance testing in which the analytical problem was to confirm that undesired components (polymorphs) were present only below specified levels (Gamberini et al., 2006; Pratiwi et al., 2002).

For our purposes, it is more useful to consider the effect of changes to the experimental procedures on the accuracy and precision of the analysis than to try to use raw statistics on published studies to draw general conclusions about the absolute accuracy and precision which can be achieved using Raman methods. One reason why it is so difficult to draw general conclusions about quoted prediction errors from the literature is that the factors that have determined them are often not discussed. Moreover, it cannot be assumed that all formulations are alike and variations in the amount of API, spectral congestion and interference from other constituents, particle size, and so on, will have an impact on the accuracy of the analysis. In this situation, information that allows the relative importance of these various factors to be judged for any given method is useful because it allows the potential precision and performance of a new method to be predicted with some confidence.

Since the detection and quantification of different polymorphic forms of the API have become topics of intense scrutiny and indeed a whole chapter of this book is devoted to polymorphism, it might be expected that a separate section of polymorphs would also be included here. However, despite the fact that with polymorphs Raman methods are typically being investigated as an alternative to X-ray powder diffraction, rather than HPLC, for example, this does not make any significant difference to the way in which the Raman experiments are carried out. So, for the purposes of this discussion, different polymorphs are simply regarded as different constituents within the solid dosage form and treated in the same way as any other component. The only proviso being that with polymorphs the differences between the forms will always be much smaller than between chemically unrelated compounds and for some compounds the differences between polymorphs will be very small indeed, increasing the analytical challenge (Mehrens et al., 2005; Strachan et al., 2004). For example, Fig. 2.3 shows FT-Raman spectra of two carbamazepine (CBZ) polymorphs and the spectrum created by subtracting them to highlight the differences present.

In this chapter, experimental factors which are important in generic experiments, such as resolution and sampling, are discussed first. The additional factors that need to be discussed for nonstandard samples (powders, patches etc.) are discussed afterwards since this allows us to avoid unnecessary duplication of arguments which apply to both tablets and loose powders, for example.

2.3 INSTRUMENTAL PARAMETERS

The literature is well supplied with detailed treatments of the experimental parameters that need to be considered in the design (and operation) of Raman spectrometers, particularly McCreery's outstanding monograph (McCreery, 2000). Here it is more appropriate to approach this area from the point of view of an instrument user and to assume that many of the decisions have already been taken in the design of the commercial instrumentation available. This leaves a much smaller subset of topics that need to be treated to provide the necessary background.

2.3.1 Sources of Noise

Noise in Raman spectra can be divided into two broad categories: one that arises from the instrument and one that is inherent within the spectra. In dispersive instruments, the noise on the signal will be a combination of the electronic noise generated by reading the detector (typically $<10\ e^-$ in scientific grade CCDs and assumed to be constant) and shot noise on the spectra, which is the statistical uncertainty associated with recording any type of random event. In this case, detected photons are the events that give photon shot noise (equal to the square root of the number of photons detected). In addition, thermally generated electrons in the detector mimic the signal from photons and the noise on this "dark signal" is also equal to the square root of the signal. It is instructive to consider that with detected signal of 10^4 cts acquired in 2 s the dark signal is negligible, the read noise is $\sim 10\ e^-$ and the photon shot noise is 100 cts. Even with this short accumulation, the noise on the signal is dominated by the photon shot noise, which is only 1% of the intensity of the signal. Increasing accumulation time gives a linear increase in the signal while the noise grows as the square root of the signal, meaning that the S/N grows as the square root of the accumulation time. For FT spectrometers the arguments are more complex but a similar result is obtained.

In earlier studies of Raman spectroscopy, S/N ratios were a major preoccupation amongst experimentalists, because obtaining spectra with useful S/N was so difficult. In modern studies, there is often little or no discussion of S/N ratios in the data and there are very few studies in which the effects of changes in the acquisition times on the prediction errors of otherwise identical experimental procedures have been carried out. Presumably, this is because the sensitivity of modern spectrometers is such that very high S/N spectra of pharmaceutical tablets can be acquired in a few minutes at most (for FT spectrometers) and within seconds for dispersive instruments. Under these conditions, researchers can afford to choose acquisition times that yield spectra with S/N levels so high that noise is assumed to have little effect on the precision of the measurements (illustrated in Fig. 2.4). This is a rational approach in laboratory–based experiments where only 10s of spectra need to be acquired to give reference data and test samples and where an additional few minutes acquisition is irrelevant compared to the time required for sample preparation and data analysis. Moreover, it is very easy to run spectra at a series at different acquisition times and to check that accuracy and precision are not being sacrificed for the sake of a trivial amount of extra acquisition time. This can be useful; for example, Niemczyk et al. (1998) found in a study of the bucindolol content in gel capsules that the SEP of 0–100 mg dosage forms was 3.63 mg with 30 s accumulation times, but it could be reduced to 2.58 mg simply by increasing integration times to 90 s.

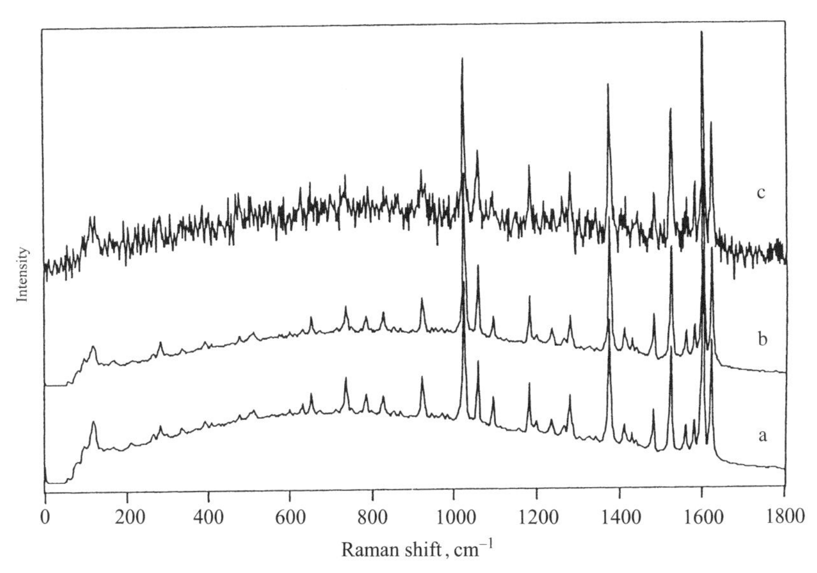

FIGURE 2.4 Raman spectra of acetonitrile acquired using 785 nm excitation and a CCD detector with accumulation times of (**a**) 5 s, (**b**) 1 s, and (**c**) 0.01 s. Noise levels are sufficiently low in the 1 and 5 s spectra that they appear identical. Reprinted from Pelletier (1999).

For high throughput online/in-line studies, the balance is entirely different and acquisition times need to be as short as possible, consistent with obtaining acceptable analytical models. In some cases, this can be as low as <1 s per sample. Since such studies need to operate closer to the limits, the effect of changing acquisition times is an important experimental variable, which needs to be tested and optimized (Vergote et al., 2004; Wikstrom et al., 2006).

2.3.2 Range and Resolution

Resolution is an important factor in Raman measurements, particularly for the analysis of multicomponent materials such as solid dosage forms where complex spectra result from combining constituents that have their own rich set of Raman bands. This results in spectra with large numbers of overlapping, partly overlapping, and distinct vibrational bands; and the extent to which these bands can be separated depends directly on the instrument resolution. Unfortunately, the resolution of dispersive Raman spectrometers can be (and is often) defined several different ways, so making meaningful comparisons between published data with these instruments can be very difficult. Moreover, specifying a generally applicable optimum instrument resolution is difficult, because all samples are not the same; this means that a more pragmatic approach may be more useful. For example, in studies of Si wafers the shifts in band position associated with stress are only of the order of 0.1 cm^{-1}, so high resolution instruments are required to study these samples (Manotas et al., 1999). However, even in solid dosage pharmaceuticals the natural linewidth of the Raman bands are typically >4 cm^{-1}, so using high resolution instrumentation of the type used for Si would merely give an excellent reproduction of this natural linewidth and lower resolution would be perfectly adequate. Nonetheless, there is a point at which sacrificing resolution will adversely affect accuracy and precision. So, it is useful to consider some of the more important features, which determine where this point lies and how instruments achieve appropriate resolution.

In FT-IR instruments, there is little to discuss; the limiting resolution is ultimately determined by the interferometer and the linewidth of the laser (since the Raman band cannot be narrower than the source which gives rise to it), but resolution (in terms of spectrometer resolution) is typically software selectable down to that limit. For studies of solid dosage forms using FT spectrometers, resolutions between 1 and 8 cm^{-1} are typical. For studies of polymorphs where the band positions may differ by only a few cm^{-1}, it is rational to use 1 cm^{-1} resolution (Langkilde et al., 1997); but in most published studies, the reasons for choosing a particular resolution are often not explicitly discussed and may be set as much by local practice as a conscious decision. In part, this is probably because although there is always a balance to be struck between sensitivity and resolution (any attempts to increase resolution will be at the expense of longer exposure times to achieve the same S/N) the balance is not finely poised. This is particularly true for low sample throughput laboratory-based experiments, since any loss of signal due to increasing resolution can be compensated by extending the accumulation time. Similarly, since the natural linewidth of the bands in the samples is usually

$> 4\,cm^{-1}$ and often $> 8\,cm^{-1}$, the effect of switching from 2 to $4\,cm^{-1}$ resolution is not very dramatic. Moreover, the key quality parameter for the spectra is whether it is possible to extract quantitative analytical information from the spectra, and this is not directly related to the width of bands; indeed, it is common to digitally smooth spectra before measuring the bands in the spectra, which effectively broadens the Raman bands in any case. An example of how insensitive to spectrometer resolution that calibration models can be was shown by Szostak and Mazurak (2002), who found only a small difference in the precision of measurements of the acetylsalicylic acid content of tablets taken at 16 and $2\,cm^{-1}$ resolution (RSEP = 1.52% at $16\,cm^{-1}$, 0.99% at $2\,cm^{-1}$).

With dispersive instruments the "resolution" (for simplicity taken here to be the FWHM of a Raman band with an infinitely sharp natural linewidth) is a convolution of the laser linewidth and the spectrometer/detector combination. In the scanning instruments, which have now all but disappeared from commercial applications, the measured width of a line will alter in a rational fashion with the width of the input slit and narrowing the slit will give a predictable increase in resolution at the cost of lower signal levels. With multichannel detectors (almost universally CCDs) the spectrum is dispersed across the detector and the theoretical minimum width for a Raman band (normally not achievable) is one detector pixel. This cannot be improved by reducing the input slit width so it is important to have the width of each pixel to be as small a cm^{-1} as possible. Some manufacturers quote "resolution" as the cm^{-1} width of a single pixel in their spectrometer. This is a poor measurement, firstly because this value may change across the cm^{-1} range and secondly because even with infinitely sharp input lines the signal will normally not lie within a single pixel at a normal input slit setting. Moreover, the width of any Raman line is also determined by that of the excitation laser so that spectra recorded at high spectrometer resolution can still have poor resolution if the excitation laser is not sufficiently narrow.

The most important point with CCD-based systems is that the detector has a fixed overall width and number of pixels, so there will be a compromise between the total spectral range covered and the cm^{-1} of each pixel; for example, in a CCD with 1162 pixels mounted on a spectrometer whose grating has been set to cover a range from 150 to $3500\,cm^{-1}$ each pixel will be on average $3\,cm^{-1}$ wide.

Numerous technical solutions have been used to circumvent the range versus resolution problem. The most obvious method is "scan and stitch" where the grating is rotated to cover different regions while spectra are recorded from each region in turn (McCreery, 2000). The resulting spectra are then mathematically merged after acquisition. This has several disadvantages, the most obvious being the potential for creation of "stitch marks" at the boundaries between adjacent acquisition regions. The second, often overlooked, disadvantage is that when the Raman scattering from one spectral region is being accumulated, the scattering from the others is not, so longer accumulation times are required. A more sophisticated variant of this approach is to record a series of spectra over a limited region but to shift the central cm^{-1} in small steps between each acquisition. In this protocol, there will be a considerable overlap between consecutive spectra and the signal from the same

cm^{-1} will be recorded several times as the range is altered. The resultant spectrum is generated by summing the contributions from each of the relevant spectra for each cm^{-1} position. This scanning/multichannel method has the advantage that it effectively eliminates stitch artifacts at the expense of increased complexity in the instrument and, of course, photons that arrive at the wrong time are lost in the same way as "scan and stitch."

Finally, there are techniques that make use of the fact that CCDs with rectangular active areas are widely available, which raises the possibility of recording more than one spectrum on the detector by offsetting data channels vertically. Commercial systems that use specially prepared diffraction gratings with two sets of lines to direct two different spectral ranges onto the detector simultaneously have been available for several years. A more extreme version of this concept is the use of echelle spectrometers, which contain two diffraction gratings, oriented so that they disperse light in orthogonal directions. This arrangement leads to the creation of a series of spectra, each covering a different range, stacked one above the other in the detector plane (McCreery, 2000). In use, the data are stitched together in much the same way as "step and scan." This can give a $4\times$ increase in the effective number of detector pixels allowing simultaneous collection over scatter over the full Raman spectral range (100–3600 cm^{-1}) at a resolution $<1\ cm^{-1}$ per pixel.

2.3.3 Wavenumber Calibration

Since quantitative Raman measurements rely, at the most basic level, on measurement of photons at appropriate wavenumber it is obvious that accurate wavenumber calibration is essential for long-term stability of calibration models, since it is only by correctly identifying the position of a peak that it is possible to return to it and measure its intensity at a later date. When manual measurements of peak heights are being used to develop univariate models, calibration errors that displace the peak maximum to the adjacent pixel are easy to detect and correct for, but such manual measurements are likely to become increasingly rare as multivariate calibrations become dominant. With multivariate calibrations the model will be based on factors derived from a training set; if the model is then applied to data with a slightly offset cm^{-1} scale, inappropriate weighting values are applied to the data. One approach that may help to increase the robustness of the models (apart from the obvious one of trying to maintain rigorous control over calibration) is to include the effects of calibration drift in the training data by running repeat sets on different days. This is a good example of the advantage of using training data that reflects as accurately as possible the range of variables that will be encountered in the test data.

The most obvious method for calculation of cm^{-1} shift would be the determination of the absolute position of both the excitation laser and the scattering signal. With Fourier transform instruments the interferometer is calibrated through an internal HeNe reference, so absolute shift can be measured directly. With dispersive spectrometers it is possible to calibrate the system in nm using low pressure emission lamps, whose band positions are known with very high precision, and to calculate the shift by subtraction. A very complete description of this process and the effect of undersampling has been published (Mann and Vickers, 2001). However, with both home-built and commercial dispersive instruments calibration is now typically carried out by recording the spectra of materials with known Raman shifts (McCreery, 2002). For spectrometers whose dispersion characteristics are well understood (such as in commercial instrumentation), determination of a single peak position, for example, from a Si reference sample, may be sufficient to establish the spectrometer calibration. With home-built CCD-based systems the same approach can be used (Holy, 2004), but it is conventional to use a series of peaks to establish the calibration curve (cm^{-1} vs. pixel: Vankeirsbilck et al., 2002) in home-built spectrometers, and some commercial instruments also use this approach. The cm^{-1} value for a series of calibration standards (Anon, 1996) have now been published and using these an accuracy of ± 1 cm^{-1} should be readily achievable. However, it is worth noting that although the introduction of these standards has certainly helped to increase the consistency of reported band positions, differences of even 3 or 4 cm^{-1} between band positions reported by different laboratories are normally not regarded as significant and seldom draw comment.

The long-term stability of dispersive systems is compromised by thermal drift in which temperature changes cause physical expansion/contraction of the components and changes in the internal alignment, which can lead to drift of 0.05 cm^{-1}/°C (Bowie et al., 2000). This is a significant source of the day-to-day calibration drift in conventional dispersive Raman spectrometers, which means that they need not only to be checked but also adjusted on a daily basis if not located in a temperature-controlled environment (see Fig. 2.5). In addition, diode lasers are prone to lasing

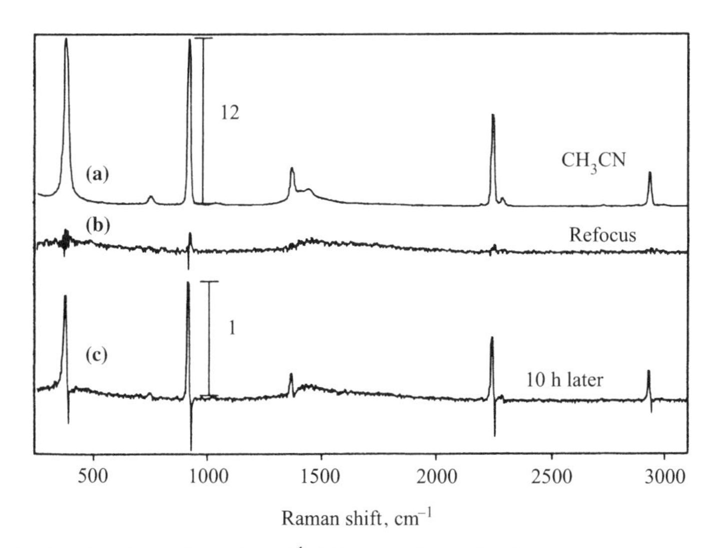

FIGURE 2.5 An illustration of cm^{-1} drift in a dispersive Raman spectrometer. (**a**) The original spectrum of acetonitrile, (**b**) the difference signal recorded by subtracting two spectra obtained before and after a single refocusing step. (**c**) The difference spectrum generated by subtracting two spectra of acetonitrile taken 10 h apart. Reprinted from Bowie et al. (2000).

on different cavity modes with slightly different frequencies, which again causes an apparent shift in the Raman band positions and will compromise calibration models whether univariate or multivariate.

Vehring (2005) has demonstrated a useful method for reducing these problems by measuring the laser position at the start of each experimental run to give a coarse correction and then making final adjustments postacquisition. By interpolating between neighboring data points it was possible to shift the spectra relative to each other in much smaller increments than integer numbers of pixels; for example, spectra could be shifted by 0.133 pixels. Similarly, McCreery has shown that by using center of gravity methods to estimate band positions to an accuracy better than 1 pixel, it was possible to calibrate spectra routinely with a precision better than 1 pixel. For example, when the calibration of a spectrometer was checked using 4-acetamidophenol 23 times in 60 days the standard deviation in the position of the 767 cm^{-1} peak was just 0.14 cm^{-1} (McCreery, 2000).

2.4 EXPERIMENTAL CONSIDERATIONS

2.4.1 Fluorescence

The experimental conditions needed to record the Raman scattering from a sample (high laser intensity, efficient collection of radiation from the sample, and high sensitivity detector) are very similar to those one would choose in designing an experiment to record a weak fluorescence spectrum. It is therefore hardly surprising that Raman signals often contain a detectable fluorescence background that typically appears as a broad underlying signal and can arise either from one of the known constituents in the sample or, more commonly, from a highly fluorescent adventitious impurity. The extent to which this is a problem is principally determined by the relative intensities of the background and the Raman signal but, of course, the inherently low Raman scattering probabilities of most samples mean that even what would be regarded as weak fluorescence will give a significant signal under the high collection efficiency conditions of a Raman spectrometer. If the ratio of background to signal is small, such as shown in the lower traces of Fig. 2.6 then it is normally possible to separate the Raman signals from the background using any one of a large range of spectral processing methods. Since this is a common process it is treated here first before moving on to the more challenging problem of analyzing samples with very high background intensities.

Within the entire field the most common method of removing backgrounds is to manually fit a broad function, which follows the shape of the underlying fluorescence profile, typically by selecting points that fall on sections of the trace that are identified as having no Raman bands and then fitting a polynomial to these "baseline" points. Subtraction of the background signal then gives the "pure" Raman spectrum. The main problem with this approach is that it requires manual input and judgment by the user, which makes it time-consuming and introduces an element of subjectivity into the procedure. For this reason numerous methods for

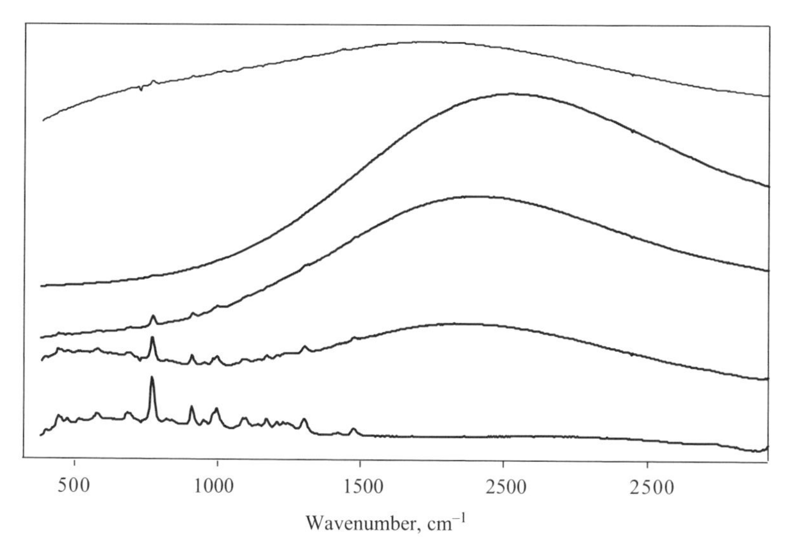

FIGURE 2.6 An illustration of the loss of the Raman signals of API (Aspirin) and excipient in the presence of increasing levels of fluorescence.

automatic background subtraction have been developed. The major approaches including the noise median method, use of artificial neural networks, and Fourier transform methods, have been compared in a very large study using synthetic data in which the strengths and weaknesses of each method were discussed (Schulze, 2005). For example, Fig. 2.7 shows the results of automatically identifying the positions of peaks in the squared first derivative of the spectrum and then fitting a polynomial function to the positions identified as being free of peaks (the threshold-based classification method). It is clear from the figure that the method worked acceptably well under one of the sets of test conditions, but with lower S/N artificial test data the estimated polynomial diverged significantly from the known artificial background function.

The background subtraction methods discussed above are often adequate for cases where the background is not so large as to obscure the Raman bands of interest. However, when the ratio of background to Raman signals becomes larger (such as in the upper traces in Fig. 2.6) subtracting away a background function is not the solution to the problem for two reasons. The first is that it becomes extremely difficult to determine the form of the background function with an acceptable accuracy, because small errors in background fitting may be significant compared with the size of the Raman bands. The second reason is the random shot noise, which is generated by the fluorescence signal will remain, even if the baseline offset can be subtracted away, so that the remaining Raman signal sits among the noise generated by the background. For example, Fig. 2.8 shows that even if the background is completely cancelled by subtracting two sequentially acquired spectra the increased noise in the spectral region with the largest fluorescence signal is readily apparent.

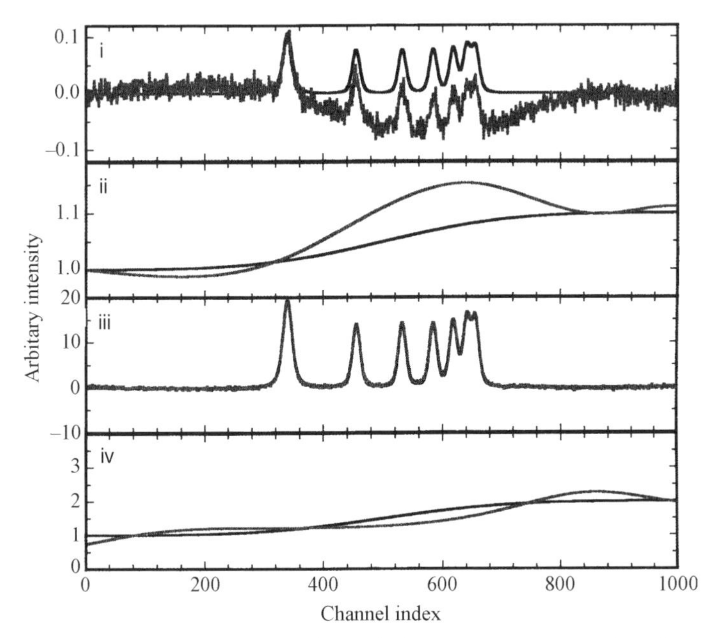

FIGURE 2.7 The results of threshold-based classification as a method for estimating baselines under Raman spectra illustrated for two different sets of input spectra. Two of the panes compare the estimated and actual baselines, the other two compare the true baseline-free spectra with the result of subtracting off the estimated baseline from a spectrum with artificially added fluorescence. Upper traces have S/N = 10, signal/background = 0.1, baseline slope = 0.1, lower traces have S/N = 100, signal/background = 10, baseline slope = 1.0. Reprinted from Schulze et al. (2005).

The best approach to problem fluorescence is to attempt to eliminate it at source and in recent years this has increasingly meant using long excitation wavelengths (normally ca. 800 nm for dispersive instruments and 1064 nm in FT spectrometers). The rationale, often borne out in practice, is that the vast majority of fluorescent compounds do not absorb at these long wavelengths or, if they do absorb, they do not emit with a high quantum yield in the spectral region the Raman signal lies in. It has been estimated that at 633 nm perhaps 10% of samples fluoresce, but this drops to 5% at 785–815 nm and 1–2% at 1064 nm (Wartewig and Neubert, 2005) and although these figures will vary significantly depending on sample type, they illustrate the advantage of moving to as long an excitation wavelength as possible. There are technical limits to the excitation wavelengths that can be used with Si-based detectors and there is a cut-off at around 800 nm excitation, which arises from the need to keep high wavenumber signals within the sensitive wavelength range of the detectors (200–1100 nm). For example, a band at 3600 cm^{-1} obtained with 800 nm excitation lies at 1094 nm. In the future, the experimental multichannel detectors with better long wavelength performance that are currently being investigated for medical applications with 1064 nm excitation may well also become available for pharmaceutical applications (Min et al., 2005).

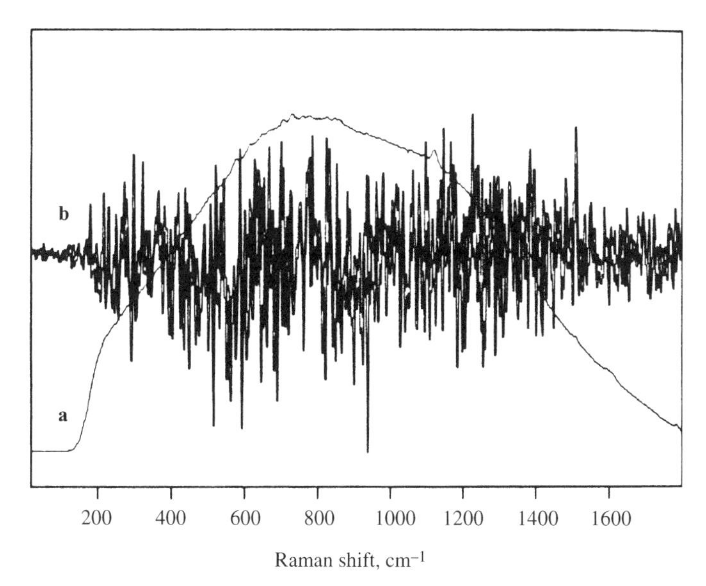

FIGURE 2.8 (a,b) An illustration of the effect of shot noise due to fluorescence on a Raman signal. Trace A is the Raman spectrum of Kapton and has some Raman bands, which are sufficiently strong to be visible on top of the large fluorescence background. Subtraction of two sequentially acquired spectra has the effect of cancelling the background signal, which is present in both but shows the random shot noise associated with it. This noise (shown in (B)) is much larger at the centre of the trace where the fluorescence intensity was high. Reprinted from Bowie et al. (2000).

Unfortunately, long wavelength excitation does not necessarily lead to complete elimination of problem fluorescence. Although there has not been any systematic study of the proportion of colored tablets and capsules which fluoresce even with long wavelengths excitation, it is known that some common colorants such as iron oxide and organic dyes such as Alphazurine FG are fluorescent in the near-infrared (Romero-Torres et al., 2006). Moreover, it is possible that many other samples do fluoresce even with far-red excitation, but if the fluorescence problems lead to the Raman studies being abandoned, such cases will not usually be reported. This means it is useful to have other methods for fluorescence rejection available. One possibility is to move to shorter rather than longer wavelengths and to use UV laser excitation. This approach has been demonstrated to be effective for model forensic samples and pharmaceuticals but sample degradation problems do arise when UV lasers are focused onto organic samples (Sands et al., 1998, Thorley et al., 2006).

A more traditional approach to removing fluorescence signals is to use the differences in temporal response of Raman scattering and fluorescence emission. Since there is no measurable delay between the incoming excitation photon and creation of the Raman scattered photon, the Raman signal always follows the intensity of the excitation source while fluorescence will rise with the pulse but decay after. If the detector is "on" during the time when the excitation pulse is incident on the sample, the Raman scattering signal will be detected but any fluorescence that falls outside this window is rejected. The proportion of fluorescence that is rejected depends on the relative duration of the signal collection window compared to the decay time of the fluorescence, if the ratio is small then dramatic levels of fluorescence rejection are possible. The best systems now have 1 ps pulses and detector gate on/off times <4 ps (Matousek et al., 2001), and with homogeneous solutions and films this system gives excellent rejection of even relatively short fluorescence; for example, for Coumarin 480 that has a 3.3 ns lifetime the suppression factor was found to be ~2400. However, for powders it was found that multiple scattering effects tended to blur the time resolution since both the laser beam and scattered light undergo refraction and reflection at air/particle interfaces before finally emerging from the sample. On a ns timescale such increased pathlength is insignificant, but on ps timescales increased pathlengths of even a few mm are significant, because light takes 1 ps to travel 0.3 mm in air and even longer in materials with high refractive indices. With powdered *trans*-stilbene, as a model analyte it was found that even with 1 ps laser pulses scattering 30% of the peak value could be detected even 100 ps after excitation (Everall et al., 2001). Of course, this temporal broadening of the Raman signal severely degrades the system performance, as it is necessary either to collect the Raman scattering over a longer duration (which increases the amount of fluorescence collected) or to keep a narrow gatewidth that retains rejection but means only a small proportion of the Raman scattering is detected. Nonetheless, this system has been tested with cocaine and it was found that it worked well for higher purity samples (86% cocaine hydrochloride), but it was less effective with a more heavily cut (75% pure) cocaine sample (Littleford et al., 2004).

One alternative to removing fluorescence at source by changing the experimental conditions is to recognize that since the noise associated with the fluorescence is the problem and the S/N improves as the square root of the number of detected photons, accumulation of sufficiently high signals should thus ultimately lead to acceptable S/N ratios. With modern CCD-based Raman spectrometers very large signals can be accumulated in reasonable times. Unfortunately, even with long accumulations, apparent noise is often found on what would be expected to be a relatively smooth featureless background. This apparent noise arises from irregularities in the detector response and it can be very significant at high signal levels. For example, if the responses of adjacent pixels vary by as little as ±1% of the incident signal, in a case where the signal is 1,000,000 cts, the shot noise on the signal is $\sqrt{1,000,000} = 1000$ cts, but 1% detector irregularity is 10,000 cts. It appears that with CCD detectors the problem of shot noise is simply replaced by that of irregular detector response. However, this irregular response can be removed by generating not one spectrum but two spectra, which are slightly shifted with respect to each other, either by changing the excitation wavelength or simply by rotating the diffraction grating to translate the spectrum on the detector. Subtraction of these pairs of spectra removes the irregular response features since they are fixed and appear in the same position in both spectra. (Bell et al., 1998). In addition, most of the fluorescence is also eliminated. Figure 2.9 illustrates this procedure. In these

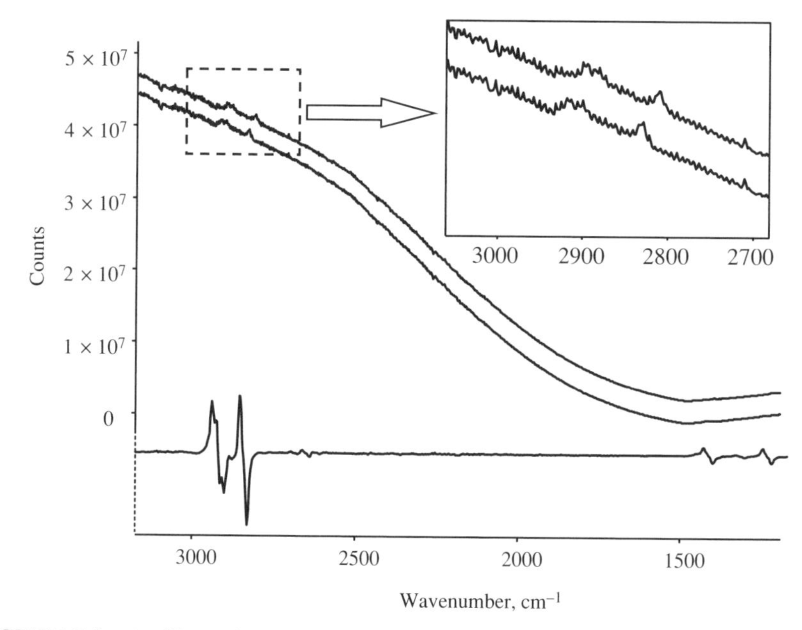

FIGURE 2.9 An illustration of the principle of subtracted shifted Raman spectroscopy (SSRS). A pair of highly fluorescent Raman spectra (top), is recorded at slightly shifted grating positions. Authentic Raman bands move with the grating rotation but the irregular detector response features remain in the same position. Subtraction of the pair of spectra removes these irregularities along with most of the fluorescence. Removing the remainder generates the derivative-like spectrum at the bottom of the trace.

subtracted spectra, the Raman bands have a derivative-like shape, but these signals can be reconstructed to generate a conventional representation of what the signal would look like if the shifting and subtraction had not taken place. Alternatively, the subtracted derivative-like data have been used as the input for standard PLS calibrations with liquid mixtures and at background fluorescence levels up to 100 × larger than the most intense Raman signals. Under these conditions it was found that the SEP for 0–100% methanol was just 3.3% (O'Grady et al., 2001). With model 0–15% (w/w) salicylic acid tablets that had been doped with fluorescent dyes to provide intense backgrounds, PLS calibration gave prediction errors of 0.6% even for the sets of tablets which included highly fluorescent (100:1, fluorescence: Raman) samples (O'Grady et al., 2007).

2.4.2 Sampling

Even a cursory scan through the literature relating to quantitative pharmaceutical analysis shows that in a very large proportion of the studies sampling errors are cited as the main limitation on the precision of the measurements. With homogeneous liquid samples there is little difficulty obtaining data that accurately reflect the composition of the sample. However, for solid dosage forms this is by no means a simple problem.

In fact, there are two distinct types of measurement that can be carried out: either determination of the overall composition in terms of the relative proportions of each of the constituents (this chapter) or Raman mapping in which spatial distribution of the constituents is determined (see Chapter 6). Although, in principle, mapping or imaging can be regarded as an extended series of composition measurements simply carried out at different points on the tablet surface, in practice it is very seldom used to obtain high precision composition data and tends more to classify particular regions as being predominantly one or other of the constituents present. Of course, summing the 100s to 1000s of individual spectra recorded in a mapping experiment would give a cumulative spectrum with extremely high S/N, because the effective accumulation time would typically be several seconds per point multiplied by the total number of spectra recorded, and this could be minutes to hours.

For the purposes of Raman measurements, the parameter that determines whether a sample can be regarded as homogeneous or heterogeneous is the size of the individual domains of each of the constituents within the sample compared with the diameter of the probing laser spot (or, more accurately, the probed volume). If the laser spot is many times larger than the individual domains, the spectrum that is recorded will be an average of many domains and will give an indication of the overall composition. If the domains are, for example, 20 μm microcrystalline particles of API and excipient, a Fourier transform instrument with a 1 mm spot diameter (and >1 mm penetration depth) will sample and record the averaged spectra of the hundreds of API and excipient particles in the beam. However, a microscope-based system with a spot size of 1 μm might detect only pure API or excipient in the same sample and therefore give an entirely erroneous picture of the composition. For example, Fig. 2.10 shows a series of spectra obtained

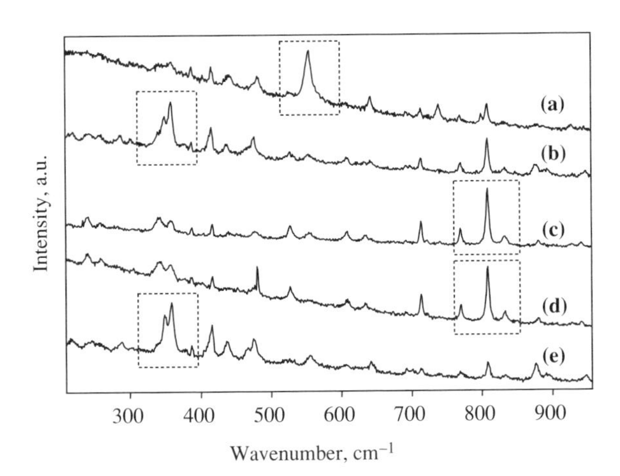

FIGURE 2.10 Raman spectra of a three component MDMA, lactose, caffeine tablet obtained with a Raman microscope (3 μm spot). The sequence shown is part of a single row of data taken from an 8 × 8 grid on 200 μm spacing. (**a**) predominantly caffeine, (**b**) and (**e**) predominantly lactose (**b**) also shows some MDMA, (**c**) and (**d**) MDMA. Reprinted from Bell et al. (2004).

with a Raman microscope for a three component MDMA, lactose and caffeine tablet, where the spectra are dominated by signals from just one of the constituents.

Even the relatively large diameter spot sizes used in FT spectrometers (100–2000 μm) are typically not sufficient to eliminate sampling errors in tableted pharmaceuticals, either because the grain sizes of the individual constituents are so large that only a few are sampled in the beam or because there are gross inhomogeneities in the samples, in which case a large proportion of the sample must be probed. The latter situation will typically occur in tablets with the API localized in high concentration regions up to mm diameter (see Johansson et al., 2005) and, of course, much of the recent Raman imaging work has centerd on characterizing such inhomogeneity. This problem is not confined to tablets. In a study of the relative proportions of crystalline and amorphous indomethacin in powder mixtures, it was found that even though the particle size was <150 μm and the FT spectrometer was operated at the 1 mm maximum beam diameter (i.e., much larger than the particle size so that several particles could be sampled in a single experiment), the predominant source of error appeared to be mixing samples (Taylor and Zografi, 1998). This problem was particularly noticeable when there were approximately equal amounts of both forms in the sample, so with 55% crystalline samples the mean value 54% was acceptably accurate but the standard deviation was a very large 4% [i.e., 7.3% relative standard deviation (RSD)] while one sample from the 30 tested gave a value of 44.8%.

Most importantly, the absolute precision of measurements of tablets can also be disappointing if they are sampled at a single point, even with an FT instrument. For example, Auer et al. (2003) found in a study of two polymorphic forms of mannitol that form I could be determined at 30.8% with a standard deviation of 3.8% (i.e., a low precision, similar to that found for indomethacin powder). However, Dyrby et al. (2002) found somewhat better precision in prediction of an unspecified pharmaceutical tablet where the cross validated prediction error was 0.56% (w/w). In a study of Acyclovir tablets (400 and 200 mg samples) held in their blister packs, where the sample was required to remain static, precision of the measured API content was on average found to be *ca* 1% but the RSD ranged from 0.4 to 4.0% (Skoulika and Georgiou, 2003). A similar precision was found for determination of the amount of hydrochloride in tablets prepared with a mixture of free base and hydrochloride forms of the same drug (Williams et al., 2004). The dramatic effect particle size can play when static sampling is used has been illustrated for mannitol polymorphs, where samples were sieved to give two sets with particle sizes <125 μm and 125–500 μm (Roberts et al., 2002). It was found that analysis of standard 50% (w/w) samples gave 54.2 ± 16.1% and 49.0 ± 1.5% for the larger and smaller particle sizes, respectively. It was clear that sampling errors were unacceptably high when a 100 μm laser spot was used to analyze samples with > 125 μm the particles.

In all these examples with static samples and normal particle dimensions (i.e., excluding 125–500 μm), the standard deviations in the w/w proportion of the constituent of interest in the tablet were ca. 1–3% and this is a reasonable figure for guidance. However, since the errors associated with different types of samples do vary significantly, it is possible to find examples in the literature where the precision of the measurements is much better than this 1–3% figure, even with stationary samples. For example, Vergote et al. (2002) found the drug content of diltiazem hydrochloride as 63.28 ± 0.26 mg in experiments were spectra of five different points on the tablets were averaged. Since these were 200 mg tablets, the error was just 0.13% of the total tablet mass. Presumably, this is the simple result of the samples being extremely homogeneous, the effect of taking five sampling points on each tablet instead of one would have helped to reduce the sampling error but only by $\sqrt{5}$, which is significantly less than the improvement observed.

An obvious way to get around subsampling problems is to move the sample under the beam so that an averaged representative spectrum is obtained. Traditionally, this is achieved by rotating the sample in some way so that the laser probes a circular track in the sample. This gives no problems with dispersive instruments and can have a very significant effect on the overall success of the calibration model. For example, Hwang et al. (2005) found that in quantifying the ambroxol content (8–16%) of ambroxol/lactose tablets, the standard error of a cross validation was $> 3 \times$ lower (1.04–0.3%) when the tablets were rotated compared to when they were stationary.

For FT instruments, sample rotation helps reduce subsampling and has an added advantage as it reduces sample heating, which can be a problem when powerful lasers are directed onto stationary samples within FT Raman spectrometers. It has been found that a 1.5 W laser can heat pharmaceutically relevant materials such as microcrystalline cellulose by up to 60°C if the sample is stationary (Johansson et al., 2002). One small problem associated with rotating samples in FT instruments is that the rotation can give rise to periodic variations in detected signal superimposed on top of the interferogram and these can appear as additional bands after Fourier transformation. This effect only occurs if high rotation speeds ($> \sim$ 1650 rpm) are used so it can easily be avoided by using slower rotation, typically 10s of RPM (De Paepe et al., 1997).

Very low absolute and relative prediction errors have been achieved using rotating samples with FT spectrometers; recent reports included determination of captopril and prednisolone with ca. 3% relative standard errors of prediction for samples where the APIs were only ca. 4–16% of the total tablet mass (Mazurek and Szostak, 2006). Similarly, in ambroxol tablets where again the API is typically $< 15\%$, the ambroxol content could be determined with ca. 3.0% RSEP, that is, in a 200 mg tablet the ambroxol content could be determined as 29.6 ± 0.4 mg (Szostak and Mazurek, 2004).

Given the results shown above, it is clear why sample rotation is so often chosen as the best way to improve the overall precision of calibrations by reducing subsampling errors and why it should always be considered when sampling options are being investigated. However, rotating bulky samples such as tablets in blister packs is difficult and in studies using Raman microscopes it is hard to keep spinning samples within what is often a very narrow depth of focus throughout the entire rotation. Where the sample cannot be moved the obvious solution to poor sampling is to increase the laser spot size. In systems where the laser focus and collection are carried out by independent optical systems (i.e., systems with a 90° collection excitation geometry), it is possible simply to increase the spot size by defocusing the laser but still to maintain optimum focus on the collection optical train. However, this is a poor strategy, because the purpose of the collection optics is to form an image of the irradiated area at the entrance to the spectrometer. If the spectrometer entrance is a slit, then moving from a tightly focused spot on the sample to a larger one will simply mean that the Raman scattering will then focus as a large spot on the spectrometer entrance slit and consequently most of the light will not enter the spectrograph. With 180° backscattering systems, the type most widely used for analysis of solid samples, moving the sample out of the laser focus is doubly wasteful since it defocuses the collection optics as well as the irradiating laser spot.

For slit-based systems a better solution is to increase the area of the sample which is probed by focusing the laser to a line on the sample or by keeping the laser spot focused but having it oscillate rapidly back and forth, which again effectively writes a line of laser intensity on the sample surface (this option is available on many Raman microscopes). In both these cases, the line of laser intensity (and corresponding Raman scattering signal) can be arranged to pass through the entrance slit so the probed volume can be increased without loss of signal. Stationary samples can also be probed in 180° illumination systems by inserting a mirror after the final focusing lens. Altering the orientation of the mirror alters the point on the sample where the laser is focused but the collection optics remain in alignment. This means that the mirror can move the probing laser over the sample without loss of signal. We have used a mirror mounted on a rotating spindle at a small angle to the rotation axis to write elliptical paths on the surface of samples (Bell et al., 2000a).

An obvious alternative to line focus is the use of grid sampling, something which is becoming increasingly popular with the advent of motorized, automated mapping stages. In essence, if the laser spot is smaller than the individual domains in the sample then the grid sampling simply becomes a method for determining which of a relatively small set of individual domains is being probed. Indeed, it is possible to simply count the numbers of grid points that are predominantly one of the tablet constituents and to estimate the relative proportions directly from the relative numbers of pixels. For example, in a study of chloroamphenicol paliatate polymorphs, which used spatially directed agglomeration method to identify clusters with a particular composition, Lin et al. (2006) found that they could relate the volume fraction of each component in the tablet to the number of pixels allocated to the clusters of either of the two polymorphs in their samples. Using this method, tablets with 25, 50, and 75 wt.% crystal form I were estimated at 27.1%, 52% and 75.6%, respectively.

One advantage of grid sampling is that it leads naturally to a numerical model of the sampling error that can be used to underpin all the foregoing observations regarding sampling strategies (Bell et al., 2004). It is obvious that with solid tablets composed of a blend of microparticulate constituents distributed throughout the tablet, the sampling error decreases as the size of the particles decreases, more grid points are sampled or the diameter of the laser spot is increased. For example, Fig. 2.11 compares data obtained with a macro system (100 μm spot) and micro system (3 μm spot) and the reduction in sampling error with the macrosystem is very apparent. Although the relationship between the various factors is not obvious, it can be derived since the statistical basis for determining the proportion of objects

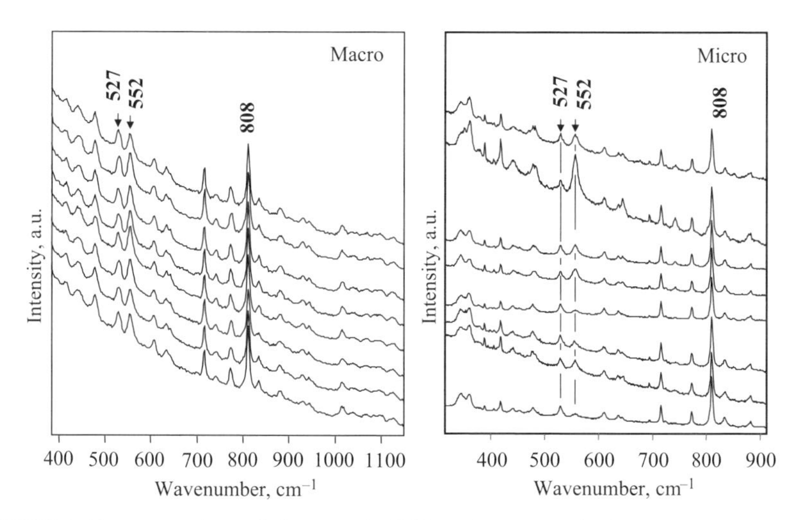

FIGURE 2.11 comparison of the reproducibility of macro and microRaman spectra of a seized ecstasy tablet. All spectra are sums of eight spectra taken from an 8×8 grid. (**a**) MacroRaman system (0.5 mm spacing, 2 s per point), (**b**) MicroRaman ($50 \times$ objective, 200 μm spacing, 20 s per point). Note poor reproducibility in the relative intensities of the 552 cm^{-1} (caffeine) and 527 cm^{-1} (MDMA) bands in the microspectra due to sampling errors. Reprinted from Bell et al. (2004).

in a population with a particular property by testing a random sample is well understood and can be applied to this problem.

The simplest case is where the tablets are made of a compressed mixture of discrete particles of drug (A) and a single excipient (B) that are probed by a laser spot that is much smaller than the particles (typically with a microscope). In this case, the property that is measured is essentially whether the sampled spot is drug or excipient (Lin et al., 2006) and essentially binary data are obtained. If 50% of the surface is composed of API and the other 50% the excipient, this would be reflected in a long run average where 50% of the measured spectra would be found to be type A and 50% type B. However, there will be a sampling error associated with this process that depends on the relative abundances of the various species present and, critically, on the number of points where spectra are recorded. It is self-evident that if two spectra are measured for the 50/50 (A/B) case the only possible outcomes are AA, BB or AB, the probability being 1/4, 1/4, and 1/2, respectively. This means that there is a 25% chance that the sample will be characterized as entirely A, 25% entirely B, and only a 50% chance that the correct result will be obtained. This, of course, limits the precision of the measurements but the situation improves if more points are sampled.

In the general case, using standard nomenclature, the problem is to determine, π, the proportion of the members of a population of objects that have a particular property (e.g., being drug rather than excipient), from a sample, size n (i.e., number of grid points probed), of that population. The proportion of objects in the sample with the property is denoted p and its value is determined by the experiments in which a series of grid points are probed. If the process is carried out numerous times the long-run average value p, μ_p, will be equal to the population value, π. However, the values of p determined in shorter experiments will not always be the same; it is known that they will be distributed about π and the standard deviation in the results, σ_p, will be given by

$$\sigma_p = \sqrt{\frac{\pi(1 - \pi)}{n}} \tag{2.1}$$

Equation 2.1 gives a statistical basis for predicting the uncertainty in the drug/excipient ratio measured in experiments, which record the number of "drug" points/total number of points sampled. Importantly, under these ideal conditions σ_p can be determined rigorously from Equation 2.1 and the distribution will be approximately normal. For example, in the case of tablets with 25% drug being probed by an infinitely small spot, the long-run average of p will always be 0.25 but if $n = 8$ (i.e., eight grid points used), σ_p will be 0.15 so the average values determined from experiments with just eight points will display considerable scatter, increasing the number of grid points to 64 reduces σ_p by a factor of $\sqrt{8}$ to 0.054.

When a larger spot is used several domains may be probed, but this is equivalent to recording data at more points and so, all other things being equal, the precision will improve as the square root of the number of individual domains included at each point. The improvement will be even larger if the system also probes a finite depth into the sample and this additional averaging can be very significant particularly in powders and tablets where multiple reflections can dramatically increase the interaction length with the sample (see discussion of spatially offset Raman spectroscopy (SORS) in Section 2.5.1).

The next step from static/grid sampling/rotation is to probe large areas of the samples, rather than points or lines, because this maximizes the number of individual domains in the probed volume. With FT instruments the spot size can typically be set to 1 mm but, in dispersive instruments, the spot sizes are typically < 100 μm. Sampling can be improved in dispersive spectrometers by using a "PhAT" fiber optic probe in which illumination and collection are achieved through a large bundle of fibers, giving an approximately 3 mm sampling spot. When a probe of this type was used to monitor a wet granulation process it gave a significantly lower RSD in measured rate than a conventional 300 mm immersion probe, although it was only slightly better than a noncontact probe with a 600 mm spot (RSDs 31%, 12%, and 10%, respectively; Wikstrom et al., 2005).

More generally, large area illumination can be achieved either by rotating the sample under a line-focused beam or by translating a rotating sample laterally during the signal accumulation so that the laser traces out a cycloid on the surface. In a test of the latter method on a wet granulated pharmaceutical, it was found that the prediction error of this illumination method was lower than point irradiation but that rotating the sample gave a similar improvement (Johansson et al., 2005). With model Acetaminophen tablets McCreery (2000) showed that relative standard deviations of a standard signal fell significantly 13.0 – 6.0 – 1.5%. on moving from point (50 μm spot) focus to line focus and then combining line focus with rotation.

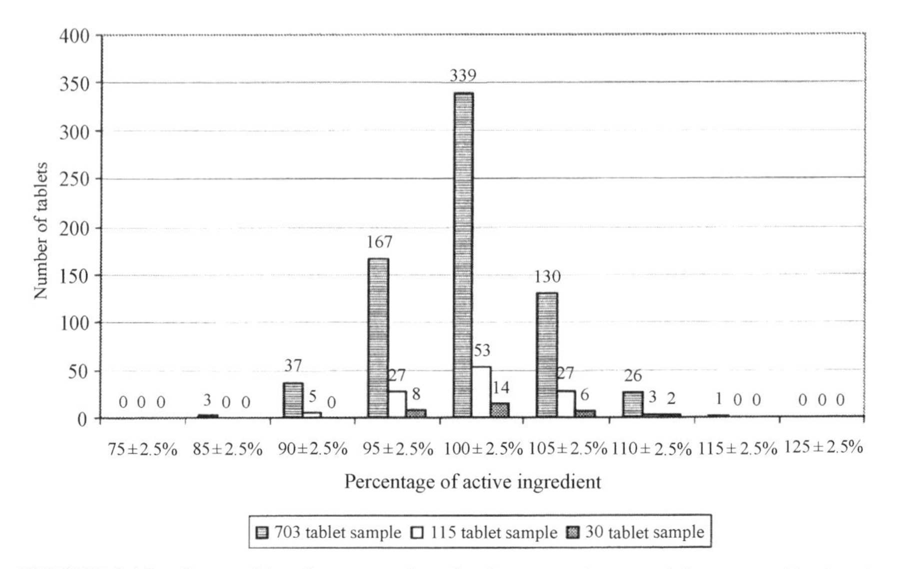

FIGURE 2.12 Composition frequency data for three sample sets of the same tablet batch. The largest sample set contained 703 tablets and establishes the composition spread at a high confidence but is still 24 × faster than testing just 30 tablets using wet chemical methods (Bonawi-Tan and Williams, 2004).

The ultimate method for sample averaging is to record the data as they pass under a probe on a conveyor and this has now been achieved. Wikstrom et al. (2006) have used a specially constructed low resolution spectrometer to look at stationary tablets mounted under a rotating laser beam and compared these data to an experiment in which the tablets passed under the beam on a conveyor. A large number of stationary tablets (703) were measured to establish a calibration and the sampling statistics (see Fig. 2.12) and the same number were recorded on the conveyor (Bonawi-Tan and Williams, 2004). Despite the high throughput, 150 tablets/min, the RSD was found to be 4%.

2.5 NONSTANDARD SAMPLES

The vast majority of drugs consumed in solid dosage forms are carefully tableted pharmaceuticals with rigorously controlled compositions and much of the research in the use of quantitative Raman methods for solid dosage forms has rightly concentrated on this predominant form. However, there are two areas where the methods developed for conventional pharmaceutical tablets have been adapted with considerable success. These are in the study of other dosage forms for conventional pharmaceuticals, which includes powders and controlled release materials such as patches, and in the study of illicit drug materials where the identities of the active and excipients are not necessarily known.

2.5.1 Powders

For simple powder mixtures the approach can be essentially the same as adopted for tablets, with the exception that along with the normal noncontact probing it is also possible to immerse a fiber optic probe in the sample. In either case, representative sampling remains a major issue. Rantanen et al. (2005) have carried out a comparative study of model systems, which were binary mixtures of nitrofurantoin, theophylline, caffeine, and CBZ, using both a conventional immersion probe and a noncontact monitoring (in this case a PHaT probe). Interestingly, attaching a mixing blade to the immersion probe (60 μm spot) and rotating the sample vial during collection gave only a small improvement in RMSEP (4.3% vs. 4.5%) over a manual procedure of moving the probe by hand during accumulations; but, of course, the mixing blade method has a distinct advantage as it requires no human intervention. Moving to a stationary but large 3 mm spot reduced the RMSEP to 3.3% but even this large size was obviously not sufficient to guarantee representative sampling, because combining rotation along with large spot size gave a further reduction to 2.3%.

One successful application to powder samples was in the characterization of model respirable pharmaceutical powders, which were spray-dried, small particle diameter (5 μm) mixtures of salmon calcitonin and mannitol in their three polymorphic forms (Vehring, 2005). The samples were contained in a conical aluminum cavity with a base diameter of 2 mm, which reflected scattered laser radiation back into the powder to increase the signal and led to a more homogeneous radiation field in the sample. The sensitivity of the dispersive system was such that the band from a single tyrosine moiety in the calcitonin (32 amino acids) was detected with high S/N in a sample that was 15% calcitonin by mass.

Similarly, De Spiegeleer investigated mixtures of three benzimidazole drugs in lactose where the total drug content was just 5%. Despite the use of static sampling at five fixed points good precision was obtained, for example, in the 50/50 samples of two polymorphs, the percentage drug content was found to be $51 \pm 13\%$ and $53 \pm 14\%$ (De Spiegeleer et al., 2005).

Powders have also been studied inside intact gel capsules (Niemczyk et al., 1998). Indeed, the powders were held in clear gel capsules that were in turn contained in blister packs. With the gel capsules the various sampling tricks for powders were not available and in this case for bucindolol capsules with nominal 12.5 mg API the standard deviation for repeat measurements on the same capsule was ca. 1 mg while average values for different, nominally identical, capsules gave differences as large as 2.6 mg despite high S/N ratios in the data (see Fig. 2.13). Even more challenging samples have been studied. For example, Severdia and Siek (2002) investigated samples where the API was contained in an opaque gel capsule with an apparent absorbance more than an order of magnitude higher than that of transparent gelatin at the excitation wavelengths commonly available (785 and 1064 nm). Despite the fact that the opaque capsule reduced the laser intensity on the way in and the escape of the scattering signal into the detection system, the precision was comparable to that found above for transparent tablets. With 10

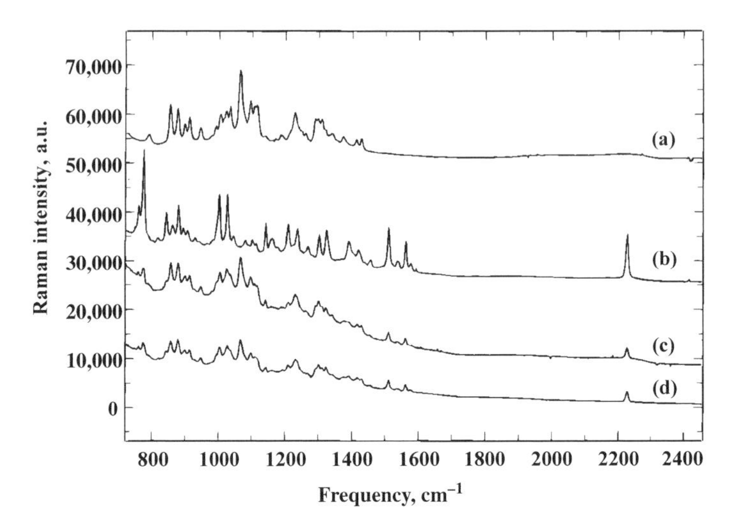

FIGURE 2.13 Raman spectra of bucindolol in gel capusules (**a**) excipient, (**b**) bucindolol, (**c**) 50 mg bucindolol formulation as a powder mixture, and (**d**) 50 mg bucindolol formulation recorded in a gel capsule. Spectra have been offset for clarity. A 785 nm dispersive instrument was used to collect the spectra. Reprinted from Niemczyk et al. (1998).

capsules containing 10 mg/g API the average predicted concentration was 10.49 mg/g with a standard deviation of 1.05 mg/g.

A very interesting new concept has recently emerged for measuring the Raman signals from deep within diffusely scattering samples by using SORS. The technique arose as a refinement of the ultrafast Kerr gating methods discussed above. The first step was the realization that due to the very short pulse and gate durations in this experiment, diffuse scattering through turbid media occurred on longer timescales than the system's time response. This meant that it is possible to excite the sample with a laser pulse and record the Raman scattering at a series of times after excitation. In such an experiment, the scattering at increasing times would correspond to scattering that had traveled a longer diffusive path and therefore would carry information on the regions deeper into the sample. This concept was successfully demonstrated by Matousek et al. (2005b) (see Fig. 2.14), but the enormous complexity and expense of the equipment required would preclude its wide adoption. However, this work led to a chapter and simpler variant, the SORS method that involves collecting light from regions that are laterally displaced from the point of excitation. Here the diffuse scattering in turbid media is again exploited but in this case the concept is that signals collected from points more distant from the excitation will carry more information about the deeper layers in the sample, as shown in Fig. 2.15 (Matousek et al., 2005a). Of course, the intensity of the Raman signal falls rapidly with distance, but this problem was overcome by using a fiber Raman probe in which the central illumination fiber was surrounded by a concentric circle of collection fibers that were arranged in a linear array at the distal end to match the input slit of the spectrometer (Matousek et al., 2006). In a model study, a layer of trans-stilbene powder was buried under a 1-mm layer of 20 μm polymethyl methacrylate

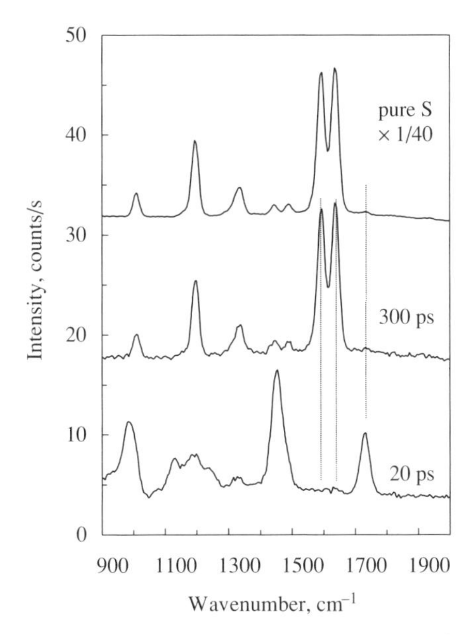

FIGURE 2.14 A series of Kerr-gated Raman spectra collected from a two-layer system consisting of a layer of trans-stilbene (S) powder buried under 1 mm of 20 μm PMMA spheres. The time delays between excitation and collection are as marked. At short delay the spectrum is dominated by the upper (PMMA) layer, at longer time scattering from trans-stilbene dominates. Reprinted from Matousek et al. (2005b).

(PMMA) beads. It was found that the signal due to the buried stilbene layer, which would be expected to be dominated by the Raman bands from the PMMA overlayer, could be increased ~ 19 × with respect to the PMMA by using SORS. Although work in this area has been confined to powder samples, it would be expected to work equally well for tableted materials that are also diffusely scattering. This means that the technique has obvious potential not only as a method for depth profiling of solid dosage forms but also as a means of recording spectra that have originated from large volumes within the solid.

2.5.2 Other Solid/Semisolid Dosage Forms

For other drug delivery systems and devices, Raman methods are typically used as a characterization tool for research purposes rather than being envisaged as a routine analytical technique. There have been several studies aimed at determining the form of the API or excipients in dosage forms ranging from polymer-based transdermal drug delivery patches (Dennis et al., 2004) to polymer microparticles (Savolainen et al., 2003). Such studies are clearly important if there is a strong interaction between the various constituents in a given dosage form, particularly since such interactions can affect the pharmacokinetics. Interactions between PVP and various other constituents including indothacin (Taylor and Zografi, 1997), ketoprofen (de Carvalho et al., 2006) and Gantrez (poly(methyl vinyl ether)-maleic anhydride) (Hao et al., 2004) have been

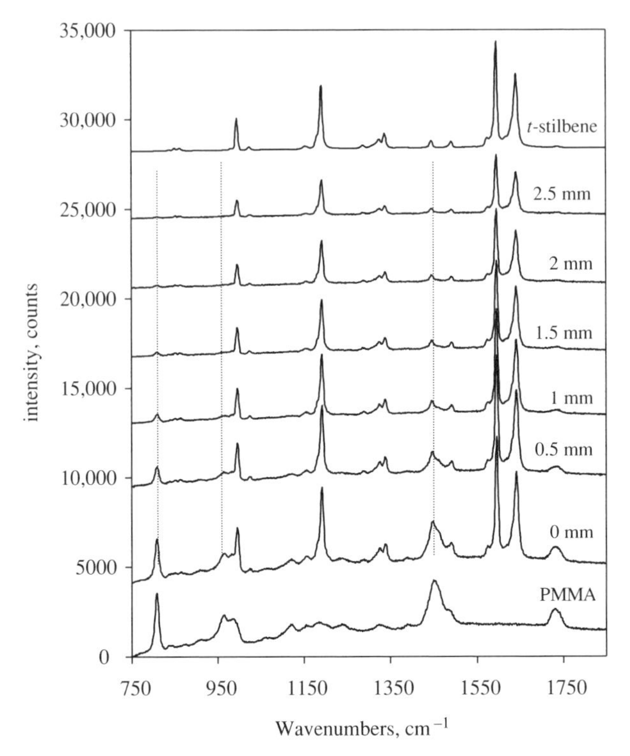

FIGURE 2.15 A series of spatially offset Raman spectra collected form a two-layer system consisting of a layer of trans-stilbene (S) powder buried under 1 mm of 20 μm PMMA spheres. The spatial offsets between excitation and collection are as marked. Line markers are to guide the eye to the PMMA bands, which decrease in intensity with increasing spatial offset. Reprinted from Matousek et al. (2005a).

characterized, as have been the complex interactions between a ternary mixture of three polymers and chlorhexidine in a model bioadhesive antimicrobial gel designed for treatment of infections of the oral cavity (Jones et al., 2000).

Even in the absence of specific interactions, information on the location of the various constituents in the device is important, so that Armstrong et al. (1996) used a FT Raman microscope to construct a profile of the ostradiol distribution in a transdermal drug delivery system. Similarly, the concentration profile of a low concentration API, TMC120, a potent antiviral, has been determined across the cross sections of both reservoir and core/shell silicone intra-vaginal rings intended for anti-HIV use (see Fig. 2.16; Bell et al., 2007).

2.5.3 Forensic Samples

Forensic applications of Raman spectroscopy have tended to concentrate on the identification of illicit drugs (Bell et al., 2000b,c; Ryder, 2002) or in detecting counterfeit pharmaceuticals (Deisingh, 2004). However, some quantitative studies have

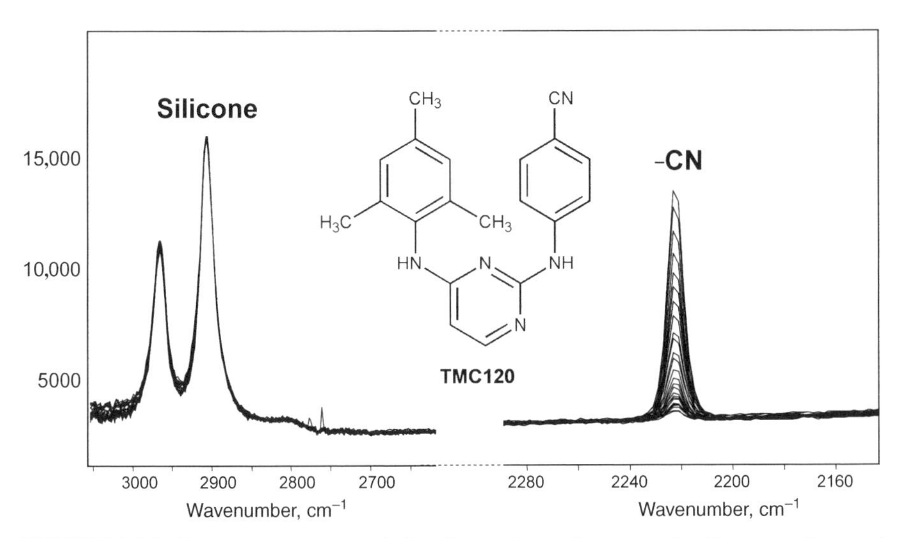

FIGURE 2.16 Raman spectra recorded at 50 μm intervals across the diameter of a core/shell silicone intravaginal ring loaded with TMC120, a potent antiviral API. Spectra have been scaled to the silicone bands at *ca.* 2900 cm^{-1} while the concentration profile of the TMC120 is shown by the intensity of the –CN band at *ca.* 2220 cm^{-1} which falls with increasing distance from the core. Reprinted from Bell et al. (2007).

been carried out on illicit drugs mixed with diluents (Leger and Ryder, 2006; Ryder et al., 2000) and the relative proportions of different drugs on the surfaces of banknotes have been determined by systematically sampling the surface in a manner similar to that described in Section 2.4.2 (Noonan et al., 2005). Quantitative measurements of drug excipient ratios and hydration levels have also been extensively used in chemical profiling of seized ecstasy tablets for drug distribution intelligence purposes (Bell et al., 2000b, 2003).

2.6 CONCLUSIONS

There are now few technical obstacles to much more widespread adoption of Raman methods for quantitative analysis of solid dosage forms. The published examples show that the technique can give rapid results with no sample preparation and with accuracy and precision competitive with the established methods in the pharmacopoeias. For the simple generic analytical task of determining the relative proportions of an API (or mixture of APIs) and excipient, either FT or dispersive spectrometers with long wavelength excitation work well. Accumulation times are typically shorter with the dispersive systems but the FT spectrometers have been the predominant technology for these applications and the majority of recent studies continue to use them since accumulations take a few minutes at most in any case. The major determining factor in the accuracy and precision of measurements of solid dosage forms is the choice of a sampling method that gives spectra, which reflect the bulk, rather than local, composition of the samples.

The next major challenge, apart from gaining wider acceptance for the laboratory-based quantitative Raman methods described here, is to exploit the speed of the technique for in-line analysis of much larger numbers of samples than has previously been possible. This has the potential to dramatically alter the way in which QC is carried out and the first experiments demonstrating that this is technically possible are just starting to appear.

REFERENCES

Al-Zoubi N, Koundourellis JE, Malamataris S, 2002. *J. Pharm. Biomed. Anal.*, **29**, 459–467.

Anon, 1996. *Guide for Raman Shift Standards for Spectrometer Calibration.* ASTM, West Conshohocken, PA

Armstrong CL, Edwards HGM, Farwell DW, Williams AC, 1996. *Vib. Spectrosc.*, **11**, 105–110.

Auer ME, Griesser UJ, Sawatzki J, 2003. Qualitative and quantitative study of polymorphic forms in drug formulations by near infrared FT-Raman spectroscopy. *J. Mol. Struct*, **661–662**, 307–317.

Bell SEJ, Bourguignon ESO, Dennis A, 1998. Analysis of luminescent samples using subtracted shifted Raman spectroscopy. *Analyst*, **123**, 1729–1734.

Bell SEJ, Bourguignon ESO, Dennis AC, Fields JA, McGarvey JJ, Seddon KR, 2000a. Identification of dyes on ancient Chinese paper samples using the subtracted shifted Raman spectroscopy method. *Anal. Chem.*, **72**, 234–239.

Bell SEJ, Burns DT, Dennis AC, Matchett LJ, Speers JS, 2000b. Composition profiling of seized ecstasy tablets by Raman spectroscopy. *Analyst*, **125**, 1811–1815.

Bell SEJ, Burns DT, Dennis AC, Speers JS, 2000c. Rapid analysis of ecstasy and related phenethylamines in seized tablets by Raman spectroscopy. *Analyst*, **125**, 541–544.

Bell SEJ, Barrett LJ, Burns DT, Dennis AC, Speers SJ, 2003. Tracking the distribution of "Ecstasy" tablets by Raman composition profiling: A large scale feasibility study. *Analyst* **128**, 1331–1335.

Bell SEJ, Beattie JR, McGarvey JJ, Peters KL, Sirimuthu NMS, Speers SJ, 2004. Development of sampling methods for Raman analysis of solid dosage forms of therapeutic and illicit drugs. *J. Raman Spectrosc.*, **35**, 409–417.

Bell SEJ, Dennis AC, Fido LA, Toner CF, Malcolm K, Wolfson AD, 2007. Characterisation of silicone elastomer vaginal rings containing HIV microbicide Tmc120 by Raman spectroscopy. *J. Pharm. Pharmacol.*, **59**, 203–207.

Bonawi-Tan W, Williams JAS, 2004. Online quality control with Raman spectroscopy in pharmaceutical tablet manufacturing. *J. Manuf. Syst*, **23**, 299–308.

Bowie BT, Chase DB, Griffiths PR, 2000. Factors affecting the performance of bench-top Raman spectrometers. *Appl. Spectrosc.* **54**, 164A–172A.

Bugay DE, 2001. Characterization of the solid-state: spectroscopic techniques. *Adv. Drug Deliv. Rev.*, **48**, 43–65.

Cooper JB, 1999. Process control applications for Raman spectroscopy in the petroleum industry. In *Analytical Applications of Raman Spectroscopy*, Pelletier MJ, (ed.) Blackwell Science Oxford: pp. 193–223.

de Carvalho L, Marques M, Tonmkinson J, 2006. Drug–excipient interactions in ketoprofen: A vibrational spectroscopic study. *Biopolymers*, **82**, 420–424.

De Paepe ATG, Dyke JM, Hendra PJ, Langkilde FW, 1997. Rotating samples in FT-Raman spectrometers. *Spectrochim. Acta Part A. Mol. Biomol. Spectrosc.*, **53**, 2261–2266.

De Spiegeleer B, Seghers D, Wieme R, Schaubroeck J, Verpoort F, Slegers G, Van Vooren L, 2005. Determination of the relative amounts of three crystal forms of a benzimidazole drug in complex finished formulations by FT-Raman spectroscopy. *J. Pharm. Biomed. Anal.*, **39**, 275–280.

Deeley CM, Spragg RA, Threlfall TA, 1991. A comparison of Fourier transform infrared and near-infrared Fourier transform Raman spectroscopy for quantitative measurements: An application in Polymorphism. *Spectrochim Acta Part A. Mol Biomol. Spectrosc.*, **47**A, 1217–1223.

Deisingh AK, 2004. Pharmaceutical counterfeiting. *Analyst*, **130**, 271–279.

Dennis AC, McGarvey JJ, Woolfson AD, McCafferty DF, Moss GP, 2004. A Raman spectroscopic investigation of bioadhesive Tetracaine local anaesthetic formulations. *Int. J. Pharma.*, **279**, 43–50.

Dyrby M, Engelsen SB, Norgaard L, Bruhn M, Lundsberg-Nielsen L, 2002. Chemometric quantitation of the active substance (containing C N) in a pharmaceutical tablet using near-Infrared (NIR) Transmittance and NIR FT-Raman spectra. *Appl. Spectrosc.*, **56**, 579–585.

Everall N, Hahn T, Matousek P, Parker AW, Towrie M, 2001. Picosecond time-resolved raman spectroscopy of solids: Capabilities and limitations for fluorescence rejection and the influence of diffuse reflectance. *Appl. spectrosc.*, **55**, 1701–1708.

Frank CJ, 1999. Review of pharmaceutical applications of Raman spectroscopy. In *Analytical Applications of Raman Spectroscopy*, MJ Pelletier (ed.), Blackwell Science Oxford: pp. 224–275.

Gamberini MC, Baraldi C, Tinti A, Rustichelli C, Ferioli V, Gamberini G, 2006. Solid state characterization of chloramphenicol palmitate. Raman spectroscopy applied to pharmaceutical polymorphs. *J. Mol. Struct.*, **785**, 216–224.

Hao JS, Chan LW, Shen ZX, Heng PWS, 2004. Complexation between Pvp and Gantrez polymer and its effect on release and bioadhesive properties of the composite Pvp/Gantrez Films. *Pharm. Dev. Technol.*, **9**, 379–386.

Holy JA, 2004. Determination of spectrometer detector parameters from calibration spectra and the use of the parameters in spectrometer calibrations. *Appl. Spectrosc.*, **58**, 1219–1227.

Hwang MS, Cho S, Chung H, Woo YA, 2005. Nondestructive determination of the ambroxol content in tablets by Raman spectroscopy. *J. Pharm. Biomedical Analysis*, **38**, 210–215.

Izolani AO, de Moraes MT, Tellez CA, 2003. Fourier transform Raman spectroscopy of drugs: Quantitative analysis of 1-phenyl-2,3-dimethyl-5-Pyrazolone-4-methylaminomethane sodium sulfonate (dipyrone). *J. Raman Spectrosc.*, **34**, 837–843.

Johansson J, Pettersson S, Taylor LS, 2002. Infrared imaging of laser-induced heating during Raman spectroscopy of pharmaceutical solids. *J. Pharm. Biomed. Anal.* **30**, 1223–1231.

Johansson J, Pettersson S, Folestad S, 2005. Characterization of different laser irradiation methods for quantitative Raman tablet assessment. *J. Pharm. Biomed. Anal.*, **39**, 510–516.

Jones DS, Brown AF, Woolfson AD, Dennis AC, Matchett LJ, Bell SEJ, 2000. Examination of the physical state of chlorhexidine within viscoelastic, bioadhesive semisolids using Raman spectroscopy. *J. Pharm. Sci.*, **89**, 563–571.

Langkilde FW, Sjoblom J, TekenbergsHjelte L, Mrak J, 1997. Quantitative FT-Raman analysis of two crystal forms of a pharmaceutical compound. *J. Pharm. Biomed. Anal.*, **15**, 687–696.

Leger MN, Ryder AG, 2006. Comparison of derivative preprocessing and automated polynomial baseline correction method for classification and quantification of narcotics in solid mixtures. *Appl. Spectrosc.*, **60**, 182–193.

Lin W-Q, Jiang J-H, Yang H-F, Ozaki Y, Shen G-L, Yu R-Q, 2006. Characterisation of Chloroamphenicol Palmitate drug polymorphs by Raman mapping with multivariate image segmentation using a spatial directed agglomeration clustering method. *Anal. Chem.*, **78**, 6003–6011.

Littleford RE, Matousek P, Towrie M, Parker AW, Dent G, Lacey RJ, Smith WE, 2004. Raman spectroscopy of street samples of cocaine obtained using Kerr gated fluorescence rejection. *Analyst*, **129**, 505–506.

Mann CK, Vickers TJ, 2001. The quest for accuracy in Raman spectra. In *Handbook of Raman Spectroscopy, from the Research Laboratory to the Process Line*, Lewis IR and Edwards HGM (eds.) Marcel Dekker New York: pp. 251–274.

Manotas S, Agullo-Rueda F, Moreno JD, Martin-Palma RJ, Guerrero-Lemus R, Martinez-Duart JM, 1999. Depth-resolved microspectroscopy of porous silicon multilayers. *Appl. Phy. Lett.*, **75**, 977–979.

Matousek P, Towrie M, Ma C, Kwok WM, Phillips D, Toner WT, Parker AW, 2001. Fluorescence suppression in resonance Raman spectroscopy using a high-performance picosecond Kerr gate. *J. Raman Spectrosc.*, **32**, 983–988.

Matousek P, Clark IP, Draper ERC, Morris MD, Goodship AE, Everall N, Towrie M, Finney WF, Parker AW, 2005a. Subsurface probing in diffusely scattering media using spatially offset Raman spectroscopy. *Appl. Spectrosc.*, **59**, 393–400.

Matousek P, Everall N, Towrie M, Parker AW, 2005b. Depth profiling in diffusely scattering media using Raman spectroscopy and picosecond kerr gating. *Appl. Spectrosc.*, **59**, 200–205.

Matousek P, Draper ERC, Goodship AE, Clark IP, Ronayne KL, Parker AW, 2006. Non-invasive Raman spectroscopy of human tissue in vivo. *Appl. Spectrosc.*, **60**, 758–763.

Mazurek S, Szostak R, 2006. Quantitative determination of captopril and prednisolone in tablets by FT-Raman spectroscopy. *J. Pharm. Biomed. Anal*, **40**, 1225–1230.

McCreery RL, 2000. *Raman Spectroscopy for Chemical Analysis.* Wiley-Interscience, New York.

McCreery RL, 2002. Photometric standards for Raman spectroscopy. In *Handbook of Vibrational Spectroscopy*, Chalmers JM and Griffith PR (eds.) Wiley Chichester.

Mehrens SM, Kale UJ, Qu XG, 2005. Statistical analysis of differences in the Raman spectra of polymorphs. *J. Pharm. Sci.*, **94**, 1354–1367.

Min YK, Yamamoto T, Kohda E, Ito T, Hamaguchi H, 2005. 1064 nm Near-infrared multichannel Raman spectroscopy of fresh human lung tissues. *J. Raman Spectrosc.*, **36**, 73–76.

Niemczyk TM, Delgado-Lopez MM, Allen FS, 1998. Quantitative determination of bucindolol concentration in intact gel capsules using Raman spectroscopy. *Anal. Chem.*, **70**, 2762–2765.

Noonan KY, Beshire M, Darnell J, Frederick KA, 2005. Qualitative and quantitative analysis of illicit drug mixtures on paper currency using raman microspectroscopy. *Appl. Spectrosc.*, **59**, 1493–1497.

O'Grady A, Dennis AC, Denvir D, McGarvey JJ, Bell SEJ, 2001. Quantitative Raman spectroscopy of highly fluorescent Samples using pseudosecond derivatives and multivariate analysis. *Anal. Chem*, **73**, 2058–2065.

O'Grady A, Dennis AC, Denvir D, McGarvey JJ, Bell SEJ, 2007. Unpublished results.

Okumura T, Otsuka M, 2005. Evaluation of the microcrystallinity of a drug substance, indomethacin, in a pharmaceutical model tablet by chemometric FT-Raman spectroscopy. *Pharm. Res.*, **22**, 1350–1357.

Pelletier MJ, 2003. Quantitative analysis using Raman spectroscopy. *Appl. Spectrosc.*, **57**, 20A–39A.

Pelletier MJ, 1999. Analytical applications of Raman spectroscopy. Blackwell Science, Oxford.

Pinzaru SC, Pavel I, Leopold N, Kiefer W, 2004. Identification and characterization of pharmaceuticals using Raman and surface-enhanced Raman scattering. *J. Raman Spectrosc.*, **35**, 338–346.

Pratiwi D, Fawcett JP, Gordon KC, Rades T, 2002. Quantitative analysis of polymorphic mixtures of ranitidne hydrochloride by Raman spectroscopy and principal component analysis. *Eur. J. Pharm. Sci.*, **54**, 337–341.

Rantanen J, Wikstrom H, Rhea FE, Taylor LS, 2005. Improved understanding of factors contributing to quantification of anhydrate/hydrate powder mixtures. *Appl. Spectrosc.*, **59**, 942–951.

Roberts SNC, Williams AC, Grimsey IM, Booth SW, 2002. Quantitative analysis of mannitol polymorphs. FT-Raman spectroscopy. *J. Pharm. Biomed. Anal.*, **28**, 1135–1147.

Romero-Torres S, Perez-Ramos JD, Morris KR, Grant ER, 2006. Raman spectroscopy for tablet coating thickness quantification and coating characterization in the presence of strong fluorescent interference. *J. Pharma. Biomed. Anal.*, **41**, 811–819.

Ryder AG, 2002. Classification of narcotics in solid mixtures using principal component analysis and Raman spectroscopy. *J. Forensic Sci.*, **47**, 275–284.

Ryder AG, O'Connor GM, Glynn TJ, 2000. Quantitative analysis of cocaine in solid mixtures using Raman spectroscopy and chemometric methods. *J. Raman Spectrosc.*, **31**, 221–227.

Sands HS, Hayward IP, Kirkbride TE, Bennett R, Lacey RJ, Batchelder DN, 1998. UV-excited resonance Raman spectroscopy of narcotics and explosives. *J. Forensic Sci.*, **43**, 509–513.

Savolainen M, Herder J, Khoo C, Lovqvist K, Dahlqvist C, Glad H, Juppo AM, 2003. Evaluation of polar lipid-hydrophilic polymer microparticles. *Int. J. Pharma.*, **262**, 47–62.

Schulze G, Jirasek A, Yu MML, Lim A, Turner RFB, Blades MW, 2005. Investigation of selected baseline removal techniques as candidates for automated implementation. *Appl. Spectrosc.*, **59**, 545–574.

Severdia AG, Siek K, 2002. Fourier transform Raman spectroscopy for identification and differentiation between dosage strengths of an active drug in white opaque hard gelatin capsules. *Appl. Spectrosc.*, **56**, 545–548.

Shraver JM, 2001. Chemometrics for Raman spectroscopy. In *Handbook of Raman Spectroscopy, from the Research Laboratory to the Process Line*, Lewis IR, Edwards HGM, (eds.) Marcel Dekker New York: pp. 275–306.

Skoulika SG, Georgiou CA, 2003. Rapid, noninvasive quantitative determination of acyclovir in pharmaceutical solid dosage forms through their poly(vinyl chloride) blister package by solid-state fourier transform Raman spectroscopy. *Appl. Spectrosc.*, **57**, 407–412.

Smith WE, Dent G, 2005. *Modern Raman Spectroscopy a Practical Approach.* Wiley, Chichester.

Stephenson GA, Forbes RA, Reutzel-Edens SM, 2001. Characterization of the solid state: Quantitative issues. *Adv. Drug Deliv. Rev*, **48**, 67–90.

Strachan CJ, Pratiwi D, Gordon KC, Rades T, 2004. Quantitative analysis of polymorphic mixtures of carbamazepine by Raman spectroscopy and principal components analysis. *J. Raman Spectrosc.*, **35**, 347–352.

Szostak R, Mazurek S, 2002. Quantitative determination of acetylsalicylic acid and acetaminophen in tablets by FT-Raman spectroscopy. *Analyst*, **127**, 144–148.

Szostak R, Mazurek S, 2004. FT-Raman quantitative determination of ambroxol in tablets. *J. Mol. Struct.*, **704**, 229–233.

Taylor LS, Langkilde FW, 2000. Evaluation of solid-state forms present in tablets by Raman spectroscopy. *J. Pharm. Sci.*, **89**, 1342–1353.

Taylor LS, Zografi G, 1997. Spectroscopic characterisation of interactions between Pvp and indomethacin in amorphous molecular dispersions. *Pharm. Res.*, **14**, 1691–1698.

Taylor LS, Zografi G, 1998. The quantitative analysis of crystallinity using FT-Raman Spectroscopy. *Pharm. Res.*, **15**, 755–761.

Thorley FC, Baldwin KJ, Lee DC, Batchelder DN, 2006. Dependence of the Raman spectra of drug substances upon laser excitation wavelength. *J. Raman Spectrosc.*, **37**, 335–341.

Vankeirsbilck T, Vercauteren A, Baeyens W, Van der Weken G, Verpoort F, Vergote G, Remon JP, 2002. Applications of Raman spectroscopy in pharmaceutical analysis. *Trends in Anal. Chem.*, **21**, 869–877.

Vehring R, 2005. Red-excitation dispersive Raman spectroscopy is a suitable technique for solid-state analysis of respirable pharmaceutical powders. *Appl. Spectrosc.*, **59**, 286–292.

Vergote GJ, Vervaet C, Remon JP, Haemers T, Verpoort F, 2002. Near-infrared FT-Raman spectroscopy as a rapid analytical tool for the determination of diltiazem hydrochloride in tablets. *Eur. J. Pharm. Sci.*, **16**, 63–67.

Vergote GJ, De Beer TRM, Vervaet C, Remon JP, Baeyens WRG, Diericx N, Verpoort F, 2004. In-line monitoring of a pharmaceutical blending process using FT-Raman spectroscopy. *Eur. J. Pharm. Sci.*, **21**, 479–485.

Wartewig S, Neubert RHH, 2005. Pharmaceutical applications of Mid-IR and Raman spectroscopy. *Adv. Drug Deli. Rev.*, **57**, 1144–1170.

Wikstrom H, Lewis IR, Taylor LS, 2005. Comparison of sampling techniques for in-line monitoring using Raman spectroscopy. *Appl. Spectrosc.*, **59**, 934–941.

Wikstrom H, Romero-Torres S, Wongweragiat S, Williams JAS, Grant ER, Taylor LS, 2006. On-line content uniformity determination of tablets using low-resolution Raman spectroscopy. *Appl. Spectrosc.*, **60**, 672–681.

Williams AC, Cooper VB, Thomas L, Griffith LJ, Petts CR, Booth SW, 2004. Evaluation of drug physical form during granulation, tabletting and storage. *Int. J. Pharm.*, **275**, 29–39.

3

SURFACE ENHANCED RESONANCE RAMAN SCATTERING

W. Ewen Smith
Department of Pure and Applied Chemistry, University of Strathclyde, Glasgow G1 1XL, UK

Raman scattering has key advantages for use in pharmaceutical and biomolecular analysis. For example, the weak signal of water enables aqueous solutions to be used easily and the Raman spectrum of many analytes gives a sharp and clearly defined pattern of peaks that can be used to provide molecular identification *in situ* in a biological matrix or buffer. However, the scope of the technique is limited by two key disadvantages. Firstly, of all the photons scattered from a molecule, only one in every 10^6–10^8 photons is Raman scattered. This makes it an inherently weak process and increasing the laser power to increase the scattering can result in sample damage. With the use of modern microscope systems, the Raman spectra of small quantities of solid samples can be obtained, but for distributed samples, such as analytes in solution, only concentrated solutions give good Raman scattering. A second disadvantage of Raman scattering is that fluorescence very often interferes with the detection of the signal. Raman scattering uses a monochromatic source of radiation to produce scattering that is shifted in frequency. It is normally recorded on the low energy side as Stokes Raman scattering. This process is similar to the one that can occur in fluorescence. The weak nature of the effect means that high laser powers tend to be used compared to those that would be used in fluorescence. As a result, weakly fluorescent molecules or impurities that fluoresce can provide sufficient signal to dominate the spectrum and prevent the effective detection of Raman scattering.

Pharmaceutical Applications of Raman Spectroscopy, Edited by Slobodan Šašić

Surface-enhanced Raman scattering (SERS) can help in overcoming both these disadvantages, since it provides a very significant increase in scattering efficiency per molecule and the process of adsorption of the analyte on a metal surface also promotes fluorescence quenching. The method has yet to be adopted for widespread use in pharmaceutical analysis, but improvements in equipment and methodology suggests that there is significant potential to obtain unique and informative measurements using this technique.

SERS uses standard Raman spectroscopy equipment. To obtain surface enhancement, the analyte is adsorbed onto a suitably roughened surface of a suitable metal and the spectrum of the analyte on the surface is recorded. The effect was first discovered during experiments to use Raman scattering to detect a layer of pyridine on a silver electrode surface (Fleischman et al., 1974). It was found that, before the electrode was roughened by cycling the voltage, no Raman scattering from pyridine could be observed on the electrode but large signals from the pyridine were obtained after cycling. Initially, this result was believed to be due to an increase in the surface area when the surface was roughened, but it was later shown that there was enhancement per molecule by a factor of 10^6 over normal Raman scattering (Albrecht and Creighton, 1977; Jeanmaire and Van Duyne, 1977). This huge enhancement provides a form of Raman spectroscopy that is sensitive and selective, and this has led to a great interest in SERS.

SERS is now carried out using a wide range of substrates to which the analyte is adsorbed including aggregated colloidal suspensions, engineered solid surfaces, and electrodes. Alternatively, for solid samples, a rough coat of metal can be applied to the sample surface by techniques such as cold metal deposition. In practice, the most effective metals appear to be silver and gold, although many other metals do have a SERS effect including copper, iron, cobalt, nickel, ruthenium, rhodium, palladium, platinum, sodium, and lithium. The method will only work well if the analyte is effectively adsorbed on the surface during the time required to obtain a spectrum. Further, there is a possibility that the interaction with the surface may alter the analyte, and selectivity to SERS as well as to Raman scattering makes some analytes much more effective than others. Thus, the measurement can be relatively simple and the advantages of sensitivity and selectivity can make the method uniquely effective. However, it works best for problems where the analyte is a good SERS scatterer and naturally adheres to the surface or where the analyst has control over the chemistry and can promote surface adhesion.

Figure 3.1 shows a typical arrangement using nanoparticles of silver as the roughened surface. The exciting source is usually a visible or near-infrared laser and will have a wavelength between about 400 and 1064 nm. This is much bigger than the roughness features created here by the interaction of the 30 nm colloidal particles and the adsorbed analyte is smaller again by more than an order of magnitude.

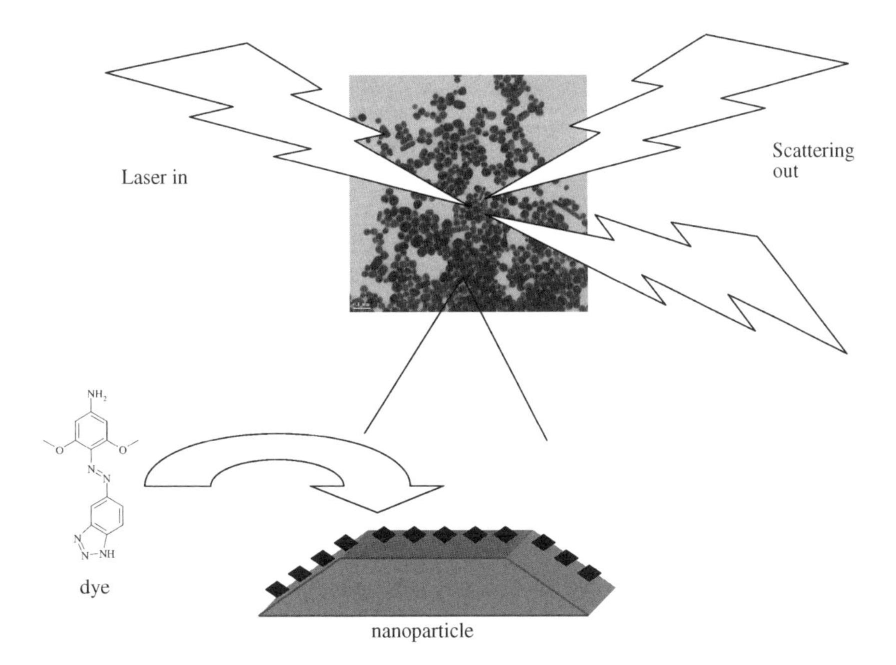

FIGURE 3.1 A schematic diagram of the SERS process. The roughened surface consists of 30 nm silver chloride particles immobilized on a surface. The exciting radiation to create the scattering can be focused down to an area about the size shown and each nanoparticle can adsorb a number of molecules as shown at the foot.

3.1 THEORY

One problem with SERS is that because the effect was discovered experimentally, a large number of theories quickly sprouted. There is still no clear defined theory. Most papers now refer to the theory as a combination of two effects, electromagnetic enhancement and charge transfer or chemical enhancement, and there is a growing consensus that the electromagnetic mechanism is the dominant effect (Kneipp et al., 2006).

In the concept of electromagnetic enhancement, the initial step is the creation of a surface plasmon when the excitation source used to create the Raman scattering interacts with the SERS-active surface. A surface plasmon is a collective oscillation of electrons that are bound at the metal surface. It is created when light interacts with the surface electrons and the frequency range over which this occurs efficiently is determined by the metal used and the nature of the surface. For example, the most efficient or resonant frequency of light to cause a plasmon to occur on a 30 nm silver particle is about 400–410 nm, and as the particle size increases the resonance frequency decreases. One of the principal reasons gold and silver are the preferred substrate for SERS is that they have resonant plasmon frequencies in the visible region. Another reason, often not stated, is that gold and silver are relatively stable, making it possible to create effective substrates that can be used relatively easily in the laboratory.

The behavior of the plasmon on the surface depends on the nature of the surface. On a smooth surface, the electrons are bound to the surface and oscillate across it. If the surfaces are roughened, then the oscillation contains a component perpendicular to the surface. This enables the light to be re-emitted from the particle by scattering. In a sense, the roughened features act like "lightening rods" from which the light is scattered. This process occurs in both plasmon resonance scattering and SERS. It is dependent on nanoscale surface roughness features. However, the Raman process is a molecular event and consequently for SERS, energy transfer between the plasmon and the molecules adsorbed on the surface is required. Thus, in the SERS process, on irradiation of a roughened surface with the analyte adsorbed on it, energy is transferred from the plasmon to the molecule so that the Raman process can occur. The energy minus one vibrational unit is then transferred back to the surface plasmon and scattered.

The position and orientation of the analyte on the surface is critical. The most effective SERS is obtained from molecules directly attached to the surface in the first monolayer and the greatest enhancement from these molecules is from those adsorbed at positions where there are sharp field gradients, for example, at particle–particle junctions in a colloidal aggregate. Too little is understood of these nanoscale events to give a definitive explanation of why this is, but the basic reason is that when the Raman process occurs there is a stronger field at these points, increasing the degree of polarization induced in the molecule by the incident excitation.

Thus, the magnitude of the enhancement is dependent on the strength of the field and the field gradient and therefore on the nature of the surface roughness created. Most authors now agree that the electromagnetic enhancement mechanism is important for SERS. It is likely to play a bigger part in the enhancement process than charge transfer or chemical enhancement (Kneipp et al., 2006).

The concept of charge transfer or chemical enhancement is that the analyte bonds covalently to the metal surface. The initial interaction is for the incident light to absorb into the metal close to the surface, creating an electron–hole pair in it. The energy is transferred via the hole to the adsorbed molecule, the Raman process occurs and the energy is transferred back to the metal from which it reradiated. This theory is still widely discussed. Although it appears less important in the enhancement process than electromagnetic enhancement, there are differences in the enhancement factors between similar molecules, and the enhancement is very high for molecules directly adsorbed onto the surface. It may be that the formation of a strong bond with the surface does have a part to play in the enhancement process.

It is very likely that the SERS process is one event and that the electromagnetic and charge transfer concepts are merely partial descriptions of that event. Cooperative oscillations of electrons across the surface (plasmons) are best described by band theory using delocalized wave functions to describe periodic behavior across the entire surface. However, the Raman process is a molecular process and the enhancement is much stronger from molecules directly attached to the surface. This type of bonding is often better described by localized bonding to a specific atom or atoms on the surface as described by charge transfer theory. On a metal surface with significant roughness, the likely truth is that the bonding experienced by a molecule on the surface covers many atoms but is not completely periodic, so a description of the surface between the single-atom–single-molecule approach used in charge transfer theory and the delocalized band theory approach used for electromagnetic enhancement would be required. The evidence for the involvement of the plasmon in the SERS process is convincing but very little is understood about the nature of the interaction of the molecule with the metal electrons and much more requires to be done to create a unified SERS theory. This continuing debate does not alter the ability to use SERS in a practical analysis.

3.1.1 Surface-Enhanced Resonance Raman Scattering (SERRS)

The surface-enhanced process can be made more selective and more intense when a coloured molecule of the correct absorption frequency is used as an analyte. The effect of the addition of a dye can be dramatic. While enhancement factors for SERS are quoted at about 10^6, enhancement factors for SERRS are quoted at about 10^{14} or 10^{15}. In practice, the additional enhancement provided by a dye depends on the relative frequencies of the plasmon resonance, the electronic transition(s) of the dye molecule, and the laser used to excite the sample. In theory, all should have the same frequency. In many cases, this will not be the case, and factors of about 100–1000 are more common in practical assays. Nevertheless, SERRS provides for Raman spectroscopy a sensitivity that can rival or surpass fluorescence, and single-molecule detection has been claimed (Kneipp et al., 1997; Nie and Emory, 1997).

As an example of the sensitivity that can be achieved, Fig. 3.2 shows the spectrum obtained from two nanoparticles immobilized on a surface that has been treated with an azo dye designed to adsorb onto silver. Figure 3.2 illustrates the sensitivity of the method and the nature of the particle–particle interactions that create "hot spots"—areas of intense scattering activity which contribute a significant proportion of the total scattering activity—in clusters of particles.

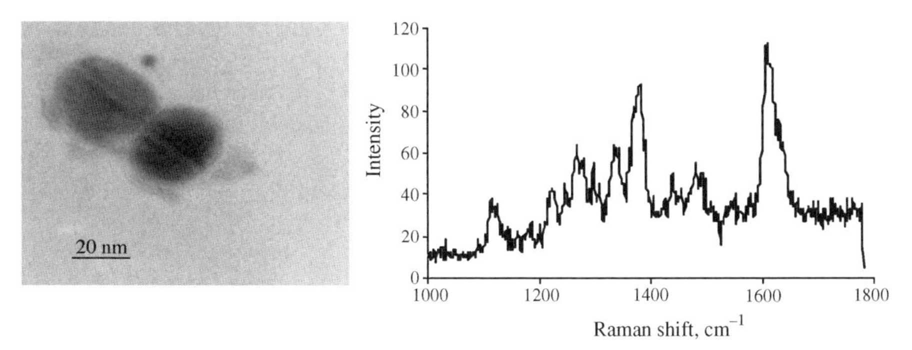

FIGURE 3.2 A pair of nanoparticles coated with dye and their SERRS spectrum. The bands at about 1600 and 1400 cm^{-1} are due to a stretch of a phenyl ring and the azo stretch, respectively. (Courtesy of Dr I. Khan Imperial College.)

3.1.2 Practical Application of SERS and SERRS

The ready availability of analytical quality and reliable instrumentation for Raman spectroscopy provides an opportunity for SERS to be used in practical applications. SERS retains some of the advantages of Raman spectroscopy. The measurement can be carried out *in situ* in aqueous solution and the signals obtained are sharp and molecularly specific. SERS has additional advantages. As discussed, it is much more sensitive than Raman scattering, making the detection of small quantities of material in solution down to about 10^{-6} M relatively simple to achieve. Relatively short measurement times of less than 1 s to 1 min are possible. In addition, the presence of the substrate required for SERS quenches fluorescence from any molecule present on the surface, reducing, but not eliminating, fluorescence interference that is one of the major problems of Raman spectroscopy. One remaining problem is that any material that is not adsorbed on the surface may still fluoresce, and this can compete with the SERS signal. The additional intensity of SERS can be used to reduce the problem by diluting the sample before addition to the colloid. This will reduce the fluorescence and the analyte concentration, but if sufficient analyte is present to cover most of the active surface, the ratio of SERS to fluorescence may still be more favourable.

The major disadvantage of SERS is the need to carry out the measurement in the presence of a substrate. Firstly, this interferes with the sample in a way that Raman spectroscopy does not; and secondly, it makes the technique selective for those molecules that adsorb readily onto the surface of the chosen substrate. This can be an advantage. If the molecule of interest naturally adheres to the substrate, the selectivity is increased since the molecule mitigates against other species and, in addition, the requirement to adsorb on a surface can mean that a sample can be concentrated on a small area, making it easier to carry out Raman measurements with equipment such as Raman microscopes where high power densities in a small area is an inherent advantage of the equipment. Thus, whether SERS should or should not be used is very much a decision for the analyst on a specific problem. One feature of SERS compared to other techniques such as mass spectroscopy is that SERS is much more analyte dependent. Some analytes adsorb strongly and are good SERS scatterers and others either do not adsorb or are weak SERS scatterers. Therefore, SERS is better used in cases where a specific analyte is selectively enhanced over interferents in the system through a combination of high SERS cross section and effective surface adsorption.

A further disadvantage of SERS for general analysis is that the orientation of the analyte on the surface can have an effect on the SERS signal. The surface creates a selection rule (Creighton, 1998), which, in very basic form, is that a component of the polarization induced in the molecule has to be perpendicular to the surface. For example, pyridine at low concentration adsorbs on the surface with the aromatic ring parallel to the plane of the surface. Since the largest polarization changes that occur are parallel to the ring plane, the SERS from pyridine in these conditions is weak. If the concentration of pyridine on the surface is increased, the molecules stack up with the aromatic ring perpendicular to the surface plane. When this happens, the signal per molecule increases very substantially because of the greater perpendicular polarization component.

For surface scientists, this is an interesting phenomenon. It allows studies of suitable molecules in solution in water in which the orientation of the molecule to the surface can be ascertained to some extent. From a general analytical point of view, this is a real disadvantage. It can make the intensity of the scattering sensitive to small changes in experimental conditions due to changes in the orientation of the molecule on the surface. Additionally, depending on the selection rules, new peaks not present or weak in the Raman spectrum can appear in the SERS spectrum. The formation of these new peaks is also a disadvantage. The very high enhancement factors mean that if a low level contaminant adheres strongly to the surface, it could provide an effective spectrum. It is therefore essential that the analyte be recognized on the surface by comparison with the Raman spectrum, and this may not be straightforward because of the additional peaks.

Figure 3.3 shows an example where the SERS spectrum and the Raman spectrum from a polymer sample were not the same (McAnally et al., 2003). In addition, it illustrates the involvement of the metal plasmon in SERS. The spectrum was obtained by coating a layer of polyethylene terphthalate (PET) with a thin (5 nm) layer of silver. Excitation and collection were from the silver side away from the polymer. In this arrangement, the silver acts as an optical filter. The excitation energy has to pass through the silver and induce scattering and the much weaker scattering has to pass back through the silver before it is detected. Even if sufficient excitation energy passes through the silver, no Raman or other scattering signal will be strong enough to pass back though it. However, SERS is obtained because, with such a thin layer of metal, the plasmon is coupled across the silver. Given that SERS gives only surface information, this is a simple way of obtaining only surface information from a solid.

SERRS is being more widely used as an analytical technique because it overcomes some of the disadvantages of SERS. Firstly, the use of a chromophore greatly

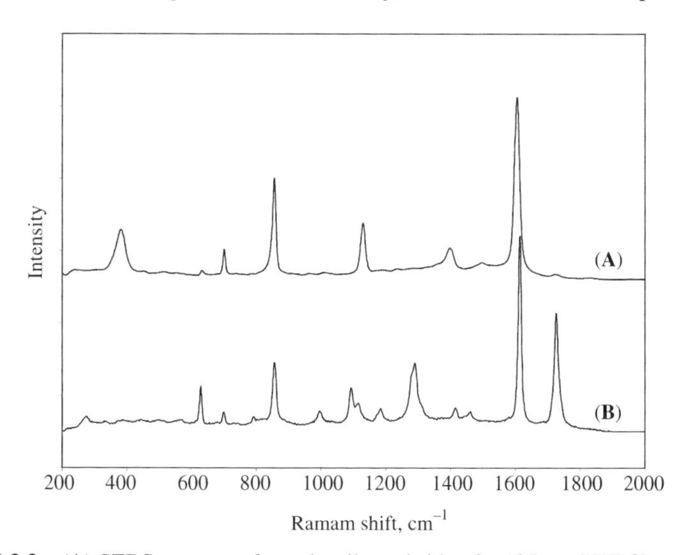

FIGURE 3.3 (**A**) SERS spectrum from the silvered side of a 125 μm PET film coated with 5 nm of silver (cold-deposited on to the PET at 0.02 nm s^{-1} at 1×10^{-5} mbar). Shown for comparison is the (**B**) Raman spectrum of PET at the same excitation wavelength (632.8 nm).

enhances the scattering. Since only a species with a chromophore can compete with the analyte in terms of sensitivity, this prevents interference from almost all interferents. Secondly, for most SERRS-active analytes, where excitation of a frequency close to the molecular absorbance peak frequency is used, there are much smaller changes than with SERS in the peak intensities when the orientation of the analyte to the surface is altered. Usually, a SERRS analyte can be readily identified from the equivalent resonance spectrum. This positive identification and the reduced dependence on the orientation of the analyte to the surface makes SERRS a much more robust and reliable technique for the identification of a species *in situ*. The additional sensitivity gives more scope to dilute the sample to reduce any interference from fluorescence from nonadsorbing molecules. Thus, SERRS provides selective signals that can be identified in water at very low concentrations and the molecular selectivity means that specific analytes can be picked up in complex mixtures.

With DNA, a study has been carried out of mixtures of oligonucleotides each labeled with different dyes. Many of these dyes are fluorescent, but efficient adsorption on the surface quenches any fluorescence at the low concentrations at which this type of measurement is usually carried out. An example with three different dyes is given in Fig. 3.4.

Currently, up to six components can be identified over a range of concentrations, but this number will increase with time. This degree of multiplexing is greater than can be obtained from fluorescence in a comparative manner, indicating the potential for this type of analysis.

Thus, the use or otherwise of SERS/SERRS is very much a matter of choice for the analyst. It cannot be seen as a general technique for all molecules, but for specific molecules where it is effective, it could easily be the technique of choice as it can be simple, selective, and fast. Examples of some types of analysis are given after the description of the experimental setup.

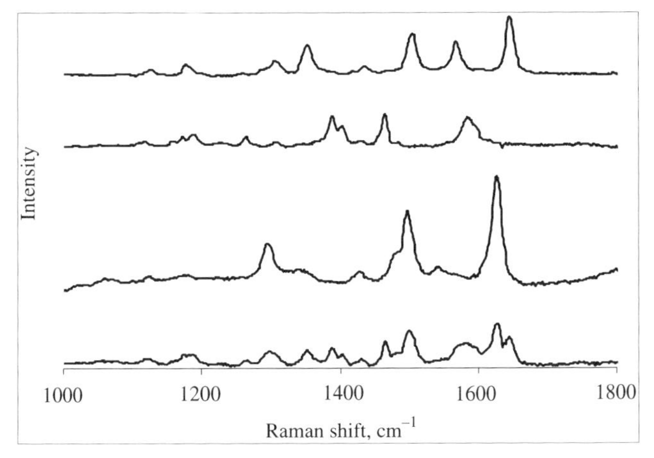

FIGURE 3.4 SERRS from three separate oligonucleotides labeled with different dyes and the spectrum of a 1:1:1 mixture.

3.2 THE EXPERIMENTAL SETUP

A key choice in SERS is the choice of the enhancing substrate, and there are a very wide range of these. As already discussed, silver and gold are usually the preferred metals. They have plasmons in the visible region which can be shaped by manipulating the form of the surface, and they are sufficiently stable to be used in a chemical analysis procedure. Among the most effective forms of substrate are island films where the metal has been cold deposited onto a metal surface to create a roughened layer, electrodes, surfaces created with UV light, colloidal suspensions aggregated to form stable aggregates in suspension, colloidal surfaces deposited on substrates, and specially designed and engineered solid substrates. Two classes of substrate that are currently being widely used are colloidal suspensions and engineered substrates.

The colloidal suspension method is simple to use analytically. Stable silver or gold colloids are readily made by reduction of salts with organic agents such as citrate. These suspensions can have lifetimes of many years, if the correct method is chosen and care is taken in their preparation. To achieve the correct degree of surface roughness, it is usual to aggregate the suspension with agents such as sodium chloride. As discussed in Section 3.1, this creates very strong fields at the interstices between particles and produces intense SERS. Thus, in a typical analysis, a colloidal substrate is aggregated with sodium chloride and the analyte added. A cuvette containing the aggregated suspension is then interrogated using a standard laser Raman system. In principle, gold would be preferred over silver as a substrate because it is more inert. However, with excitation frequencies in the visible region, the scattering/absorption ratio of silver colloidal particles is more favorable than that for gold and there is a wider range of attachment chemistries available. Consequently, rather better sensitivities can usually be obtained.

Figure 3.5 shows the UVvisible spectrum of silver colloid unaggregated and aggregated with sodium chloride. The spectrum is caused by the absorption of light to produce the surface plasmon, and, therefore, it should be a good guide to the most appropriate frequency to use to excite SERS. Both the resonance frequency of absorption and the resonance frequency of scattering are important, but to the accuracy we require for the design of experiments, they are usually not sufficiently different to affect the result. The colloid before aggregation has an absorbance maximum at about 406 nm. After aggregation, a more complex and broader spectrum is obtained, the exact shape of which is dependent on the state of aggregation. Therefore, the laser frequency should be in a region where there is plasmon activity. This means that a laser frequency in the middle of the visible region is often effective.

For SERS, we require scattering as well as absorption, and the ratio of the intensity of scattering and absorption depends not only on the nature of the surface but also on the metal used. Gold has a greater absorption-to-scattering ratio than silver and consequently for gold surfaces, lasers with wavelengths longer than the absorption edge of the gold plasmon, usually just above 500 nm, are more effective. Thus,

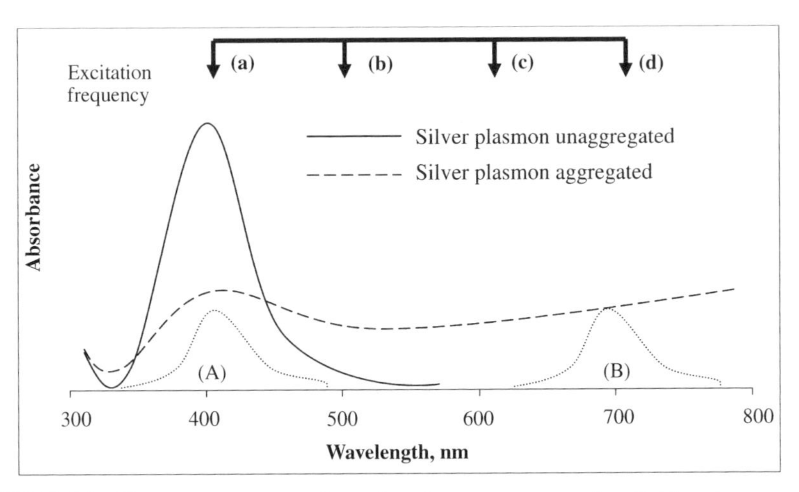

FIGURE 3.5 Plasmon resonance absorption from unaggregated colloid and aggregated colloid showing the increased breadth of the aggregated colloidal spectrum. Underneath are bands that represent the absorption profiles for two possible dyes A and B. For unaggregated colloid, the ideal condition should be with dye A and excitation (a), but with aggregated colloid the choice is broader.

wavelengths across the visible region can be used to produce SERS on silver, but to take account of the plasmon resonance wavelengths for aggregated colloid, they are usually 514 and 532 nm, and further toward the red. Wavelengths used for gold are usually 632 nm and above.

With SERRS, the choice of laser excitation frequency is a bit more complex. Ideally, it should be close to the frequency of the plasmon resonance and the frequency of the molecular adsorption. If this condition is required to be met precisely, SERRS would be very difficult to apply to real analysis procedures. Fortunately, we have discovered in practice that in aggregated systems some separation in frequency between the plasmon resonance and the molecular chromophore can be tolerated and that a fairly broad range of frequencies are effective. For example, using a dye that had an absorption maximum at 442 nm and using aggregated silver colloid as the substrate, the signal intensity with 514 nm excitation was a factor of 4 greater than with 632 nm excitation. Given the huge SERRS enhancement factors, this makes both the wavelengths useful for the development of practical analysis procedures.

Figure 3.5 shows the different arrangements (Cunningham et al., 2006). Four excitation frequencies are considered. For A, the frequency of the plasmon resonance and the molecular resonance coincide with the frequency of the laser excitation. For unaggregated colloid, the plasmon (solid line) is quite sharp. SERRS can be obtained from unaggregated colloid and from single particles, and the signal strength is strongly dependent on the frequency of the molecular absorption and the surface plasmon. However, a stronger signal is obtained from the aggregated particles due to the extra enhancement caused by high local fields at the particle–particle junctions. The dotted line shows the plasmon from aggregated colloid. At this point, the plasmon is very broad across the visible region. This is because a number of clusters of different sizes have been formed and each has a different plasmon resonance frequency. This means that lasers with frequencies (a)–(d) can all be used and the frequency of the chromophore can be anywhere between A and B. In practice, in silver colloid, we have found that with visible-range dyes, 514, 532, and 632 nm laser excitation frequencies have been very effective.

In a different approach, specific solid surfaces have been engineered to give a structure that locates the plasmon in certain wells and areas in a very reproducible fashion. These provide a surface that gives a relatively even enhancement across the whole surface. Such materials can be kept for reasonable lengths of time and used when required. They give intense SERS and for some analytes can be used repeatedly. The problem currently is that they are relatively expensive and require a method of applying the sample evenly to the surface such as a fluid or gas flow system if quantitative results are be achieved. Spotting a sample onto such a surface may be effective in simply detecting a material, but the way in which it dries out gives an uneven signal across the surface not because the surface will not produce an even excitation but because of the rate of drying on the surface makes for variable layer thicknesses. To overcome this, a flow device or some other method of ensuring even distribution of the sample of the surface to be interrogated is required. One big advantage of such solids is that they have the same advantages as a colloid but do not require the preparation of a colloidal suspension. Although colloidal suspensions give reliable and reproducible answers, this is usually interbatch and reproducibility between batches is poorer. Of course, this can be overcome by using a standard, but the ability to engineer the surface does provide a good alternative method for obtaining SERS/SERRS effectively (Perney, 2006).

Extensive studies have been carried out to detect SERS/SERRS signals from single nanoparticles or clusters of nanoparticles. These are often particles obtained from dried out colloid but interesting results can also be obtained from engineered surfaces. Much of this work is done with a view to achieving a better understanding of the SERS/SERRS process. Using this method, it is possible to isolate single particles, dimers, and small clusters and larger clusters of nanoparticles so that each can be studied separately under a microscope. This immediately creates a location problem in that it is necessary to define a nanoscale feature under an optical microscope. One way to do this is to combine an atomic force microscope with a Raman spectrometer or use near field optical microscopy. In another approach, lettered TEM grids have been treated with micron-scale beads so that the nanoparticles can be accurately located both in the Raman microscope and in a high resolution TEM. Single-particle SERRS has been demonstrated. In one study, 2% of particles could be detected as active at the limit of detection of the spectrometer and the signal per entity increased with cluster size. All clusters of over 15 particles were active (Khan et al., 2005). The larger signals from clusters agree with the previous work discussed on colloid. It was earlier shown that this enhancement is not even but there are hot spots in the clusters which give intense activity (Makel et al., 1999).

There are a number of reports of the detection of SERS/SERRS from single molecules (Kneipp et al., 1997; Nie and Emory, 1997), opening up the possibility of the use of SERS/SERRS for single-molecule detection. In practice, a problem arises in proving the presence of a single molecule. A very low concentration of analyte has to be added to silver or gold substrates so that only one or a few isolated molecules occur on the surface. It is very difficult to avoid contamination and to measure concentrations at these levels. A particular worry for single-molecule SERS/SERRS is that the largest signals come from interstices between particles following aggregation. On a solid surface, it is known that not all interstices are equally active and that "hot spots" create the maximum intensity. Thus, even if it is believed that only a few molecules are present on any one particle, it is also essential that the position of the molecule on the particle is known. Often, experiments are carried out on solid surfaces where particle aggregates are formed. The single-molecule detection is adduced from properties such as blinking of the signal. However, the SERS/SERRS signals are produced on a surface of silver or gold which could well be photo active. Under these circumstances, fluctuation in the signal from a single particle with time could be not only due to single molecule blinking effects but also due to particle surface alterations. Nevertheless, there are some convincing papers that suggest that single-molecule SERRS can be achieved. For example, Kneipp carried out a convincing study using a flow system in which the number of events was calculated with time (Nie and Emory, 1997). It was clear that single molecules could be detected.

In suspensions of colloid, it is possible to calculate the volume where the focused laser beam creates the maximum photon density and work out the number of molecules present in the beam at any one time. Since it is possible to use a series of dilutions to obtain a calibration curve, the drop in signal with concentration can be used to conform that the result is due to the sample added and not due to contamination. In addition, the collection volume is much larger than that of any one nanoparticle, and it is likely that thousands of nanoparticles are present in the detection volume at any one time. Further, the Raman event is very fast and many Raman events per particle could occur in an accumulation time of about 1 s which is usually the minimum used. As a result, the measurement is averaged over many events smoothing out particularly intense signals from any one hot spot. This work clearly demonstrates that it is possible to detect SERRS down to the level equivalent to that of a single molecule. This type of averaging is not possible with single-molecule devices such as single-particle *in-vivo* sensors but where larger surface areas of engineered surface or where colloidal suspensions are used, it is possible to obtain quantitative results as described in some applications in the next section.

3.3 EXAMPLES OF SERS/SERRS ASSAYS

3.3.1 Drugs of Abuse

Drugs of abuse such as cocaine and amphetamine can be interrogated using SERS (Faulds et al., 2002). These molecules give definitive Raman scattering and can be recognized in solution from the Raman spectra even in mixtures. This has led to the use of hand-held Raman systems as white powder detectors. One problem with this type of analysis is that street samples often contain materials that are fluorescent and this completely obscures the Raman signal. Another problem is that it is sometimes required to detect trace amounts of material, and the limited sensitivity of Raman spectroscopy can make this difficult. The sort of application where this may be important would be to examine traces on the outside of a suitcase or samples extracted by filtering air. For this type of more sensitive analysis on less pure samples, SERS can be effective. Addition of either cocaine or amphetamine onto a silver or gold colloid in suspension enables detection the Raman spectra. Detection limits between 10^{-5} and 10^{-6} M can easily be achieved, and the effect is quantitative over about 2 orders of magnitude. In practice, gold proved to be the superior substrate, but the sensitivity is limited by the fact that neither gold nor silver adsorbs the drugs efficiently. It appears that somewhat more adsorption is obtained on the gold surface, giving a somewhat better detection limit. However, if a surface could be obtained in which the drug is adsorbed efficiently, these detection limits could be much lower.

3.3.2 Glucose

SERS has considerable advantages for the sensitive and continuous monitoring of glucose from small samples. It could be used as a very good technique for detection of glucose *in vivo*. However, as in the previous section, the effective adsorption of glucose onto the SERS-active surface is a key requirement. Van Dyne and his colleagues have given careful consideration as to how this could be done in a reliable and efficient manner (Lyandres et al., 2005). They use a very interesting substrate to achieve this. An array of polystyrene nanospheres is created by deposition of this species onto a supporting substrate. Silver is then deposited on the top of these spheres. This creates a very regular surface pattern with clear points of interaction between the spheres to provide the high fields required for SERS. In general, these surfaces have been found to be very effective. They are stable and give reproducible answers over a wide area of the surface.

Glucose does not absorb effectively on bare silver. To achieve glucose adsorption, various long-chain thiols were attached to the surface. A particularly effective surface was created by using a combination of decanethiol and mercaptohexanol. Decanethiol on its own will produce a hydrophobic surface that is not suitable for glucose capture. It is difficult to be certain at a molecular level how these mixed films of SAMs form on a surface, but the computer modeling suggests that there are pockets of the hydrophilic mercaptohexanol present on the surface which will capture the glucose. Glucose adheres very well to the surface and the combination of the effective capture surface and the well-engineered surface layer enables sensitive and reliable detection.

A key advantage of this approach is that the selectivity of SERS can be used to create an effective implantable glucose sensor. If the sensor is to work continuously, the adsorption of the glucose to the surface requires to be reversible. When between 0 and 100 nM of aqueous glucose was added in cycles to the censor without flushing between measurements to simulate real time sensing, reversibility of the SERS glucose signal was found to occur over a period of time.

A further requirement for an *in vivo* sensor is that it must be capable of detecting the analyte, in this case glucose, in the physiological relevant concentration of 100–450 mg/dl at the physiological pH and in the presence of interfering analytes. Using a calibration model constructed with partial least squares leave-one-out (PLS-LOO) analysis with 46 randomly chosen independently spectral measurements of known glucose concentrations, it was possible to demonstrate that the SERS sensor is able to operate efficiently in the physiologically relevant range. The sensor appears to have significant potential for use for *in vivo* glucose measurement.

3.3.3 Mitoxantrone and Other Drugs

Mitoxantrone, a drug that has been used for the treatment of breast cancer, adheres well to silver colloid, is colured, and gives excellent SERRS (McLaughlin et al., 1992). The standard methods for the determination of mitoxantrone from blood require separation steps and are slow, so there is a need for an effective and fast analysis procedure. When a drop of serum was added directly to silver colloid and SERRS recorded with a standard Raman spectrometer, very intense signals were obtained from the mitoxantrone. Serum is not a good medium for Raman spectroscopy since there is a high fluorescence background. However, with the SERRS method used, the serum was diluted in the colloidal suspension. Thus, fluorescence from proteins and other fluorescent material adhering to the silver surface is quenched and fluorescence from non-adsorbing material is substantially reduced by the dilution process. Mitoxantrone adheres to the surface sufficiently strongly to compete with other adsorbing material with the result that a SERRS measurement from a sample from a patient can be taken with an accumulation time of 10 s.

Figure 3.6 shows a graph of the intensity of one of the peaks from the mitoxantrone SERRS spectrum against the concentration of mitoxantrone added to a serum sample. It is linear over a relatively long range that encompasses the clinically relevant range. The reason the graph flattens off at high concentrations is that this is equivalent to the complete coverage of the surface of the silver by mitoxantrone. SERS/SERRS gives the largest enhancement close to the surface and mostly from the first layer and so mitoxantrone adsorbed onto a second layer provides much less effective scattering. Thus, this is an example in which SERRS could be used to provide effective and fast determinations of the concentrations of a clinical drug directly from serum and it could well be the ideal choice of detection technology. However, this assay relies on the fact that mitoxantrone adsorbs efficiently on the silver surface and is colored. It is therefore not a general technique that can be applied simply to all clinical drugs. Related drugs such as doxorubricin and epirubricin were also studied and SERS can be obtained but not at clinically relevant concentrations. Here a simple preconcentration step would be required, but the molecularly specific signals give a definitive identification of the analyte and therefore the technique may still have advantages. An obvious direction to follow is to develop the surface engineering used to produce effective glucose determination to other targets.

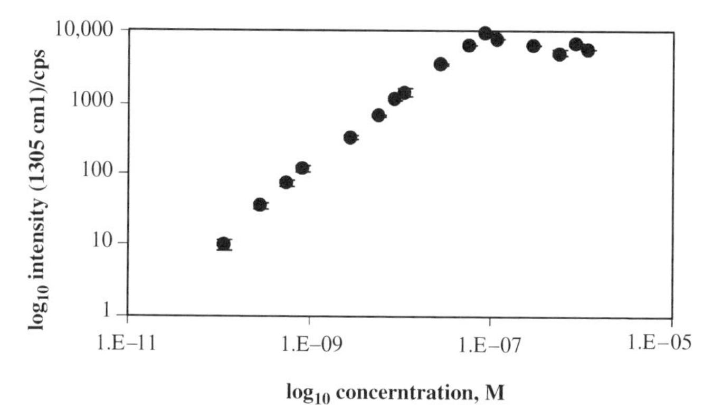

FIGURE 3.6 A SERRS intensity vs. concentration graph for mitoxantrone added to serum that spans the range of clinical interest.

3.3.4 Proteins

SERS from proteins tends to provide weak spectra, but there are notable exceptions that are more effective because of the lack of signals from the other material present. The natural selectivity of Raman spectroscopy for some molecules over others is clearly demonstrated by the results for proteins and this can be improved by using resonance particularly in the UV region where aromatic groups and peptide bonds can be picked out by using the correct excitation frequency. However, with the current state of development of SERS, the need to create effective plasmon resonance coupling means that most experiments are restricted to the visible and near-infrared regions. Good SERS has been obtained for a number of proteins including small proteins like lysozyme, but the enhancement is not large. As a general tool to determine protein structure, this approach has problems. The protein on adsorption on the metal surface may well deform, giving misleading information, and the part of the protein surface directly in contact with the metal is the most strongly enhanced. This of course provides surface information that is difficult to obtain by other techniques.

Some proteins contain a natural chromophore and for them SERRS scattering can give some very impressive results. Cytochrome C is the most widely studied of these proteins (Hildebrant, 1992). The heme group in it has a visible chromophore. It can readily be identified with little to no spectra from the rest of the protein being observed. This is despite the fact that part of the protein surface rather than the heme group is directly in contact with the metal surface. It is therefore possible to obtain *in situ* the detailed spectra of the heme group and to use the spectra to assess aspects of the heme environment. These results can be obtained at lower concentrations than with resonance Raman scattering, there are far fewer problems with fluorescence and there are less problems with sample damage since lower powers can be used.

Marker bands for oxidation state and spin state of the heme have been determined and other heme proteins such as myoglobin and hemoglobin have been studied. One particular enzyme that has potential value to the pharmaceutical

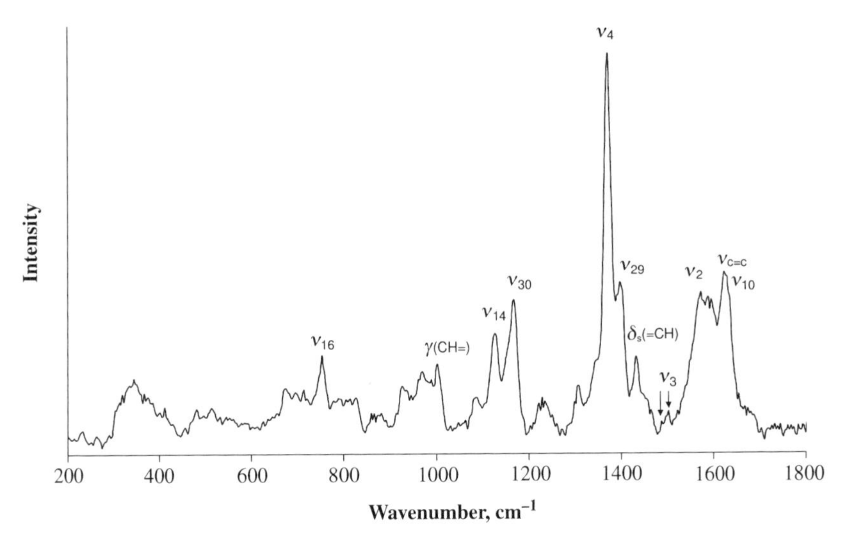

FIGURE 3.7 SERRS spectrum of P-450 BM3 aggregated with 20 μl 0.1% L-ascorbic acid. Protein concentration ~3.2×10^{-8} M. 10 accumulations of 5 s static scans, centered at 1100 cm^{-1} recorded from plastic cuvettes using 514.5 nm excitation immediately after aggregation.

industry is the enzyme P-450. For this enzyme, marker bands for oxidation state and spin state have been identified and bands assigned to the vinyl groups that extend into the protein from the periphery of the ring have also been studied (Smith et al., 2003) (Fig. 3.7). These vinyl groups stick to the protein surface in the active pocket. On distortion of the protein, the planarity of these groups and therefore the resonance interaction with the heme chromophore changes and consequently intensities change markedly. Further, it has been discovered that when nitric oxide is added to these proteins in the presence of oxygen, nitration of a tyrozine in the active pocket is achieved (Quaroni et al., 1996). Nitration of this tyrozine and of those outside of the pocket containing the active site give different Raman scattering since the environment within the pocket and outside it have different dielectric constants. The effect of some substrates and inhibitors has been studied and changes in the spectrum shown to occur. The SERRS of P-450 could be used in pharmacological screening of potentially active compounds.

3.3.5 DNA

A very effective method of using SERRS is the one in which the chemistry can be controlled by the operators. For this reason, the use of SERRS in biological assays where labeling is carried out on an individual molecule have been successful. In this form, SERRS is essentially an alternative to fluorescence. With SERRS, DNA analysis can be carried out with a sensitivity equivalent to or better than fluorescence, and with an inherent advantage in the discrimination of a large number of labels without separation (multiplexing). Figure 3.3 shows a diagram of a mixture of oligonucleotides labeled with different dyes. A significant number of labels can be discriminated by eye, but the sharp spectra also provide a wealth of information with which the data processing software could increase the numbers that can be discriminated significantly. Long linear ranges and good detection limits have been obtained (Faulds et al., 2004a, b). Thus, in assays of DNA that use labels, SERRS has significant potential. In addition to the sensitivity and multiplex advantages over fluorescence, both fluorophores and nonfluorophores can be used as labels so that a much wider labeling chemistry can be employed and fluorescence is quenched removing many interference. This methodology has been progressed to demonstrate the detection of a point defect in a cystic fibrosis gene. The two SERRS label were detected without separation.

For some applications, the ultimate sensitivity and selectivity created by a label may not be necessary. Molecular diagnostics that are label free is an attractive target. For DNA, the four bases, all give SERS, but they are not equally intense per molecule. It is possible to detect ratios of the bases such as the A/C ratio. Any assay work in this direction has to take into account the way in which the basis are adsorbed onto the metal surface. They can, for example, be adsorbed directly by breaking up the DNA into its individual bases and allowing them to adsorb onto a suitably SERS-active surface. However, it is often more desirable to use the distance from the metal to detect changes in the ratio of the bases from the surface.

3.3.6 Other Applications

The high sensitivity of SERRS lends itself to the development of Raman scattering techniques that can work on a very small scale. One leading technique is tip-enhanced Raman scattering (TERS) (Hayazawa et al., 2002; Stockle et al., 2000; Pettinger, 2006). This technique uses AFM tips that have been coated with silver. When the silver tip is brought close to or onto the surface being interrogated, only the area of the tip of the AFM that contacts the analyte on the surface will experience the enhanced field. The result is that Raman spectra can be taken from a surface with very high resolution. Large enhancements have been obtained from these tips. A major feature of the size of the enhancement is the nature of the silver or gold coating used on the AFM tip. Particular degrees of roughness are suitable to form plasmons that have a resonance in the visible region and the ability to achieve high spacial resolution depends on the ability to shape the tip of the probe. The ability to obtain intense SERS depends on the surface, the roughness on the tip, and the distance between the two. Figure 3.8 shows a schematic of a very simple arrangement, but there are now a number of designs all with there own advantages.

Another example of the use of SERRS on a small scale is when it is combined with microfluidics. The molecular specificity of SERRS is very effective here since it enables the detection of a specific analyte *in situ* within the chip (Fig. 3.9). It is possible, for example, to make colloid in the chip by mixing silver nitrite and borohydride when the stream of particles formed is then mixed *in situ* in the chip with an analyte in a flow in a controlled regime, quantitative answers can be obtained with high sensitivity (Docherty et al., 2004).

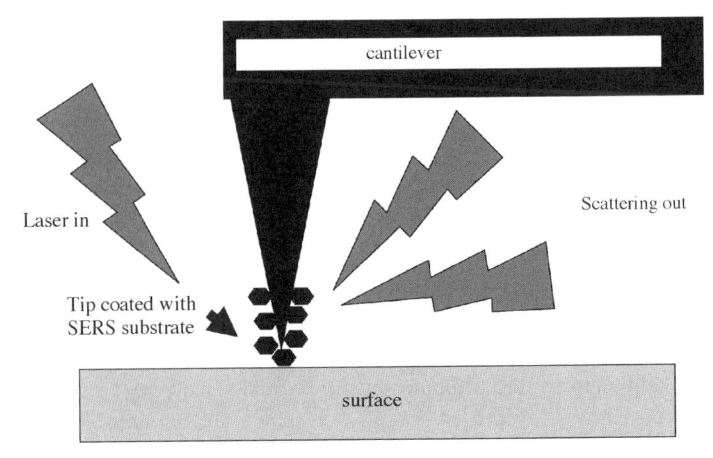

FIGURE 3.8 Diagram of a simple layout for TERS.

Immunoassays are extensively used for biodetection and labels are often used to get sensitivity and selectivity. In many cases, it is desirable that several different targets with different labels are detected at the same time. Of course, this can be done using chip-based approaches. However, one very efficient approach would be to use the additional multiplex ability of the SERRS signal so that a number of antibodies could be detected in suspension without separation. There are a

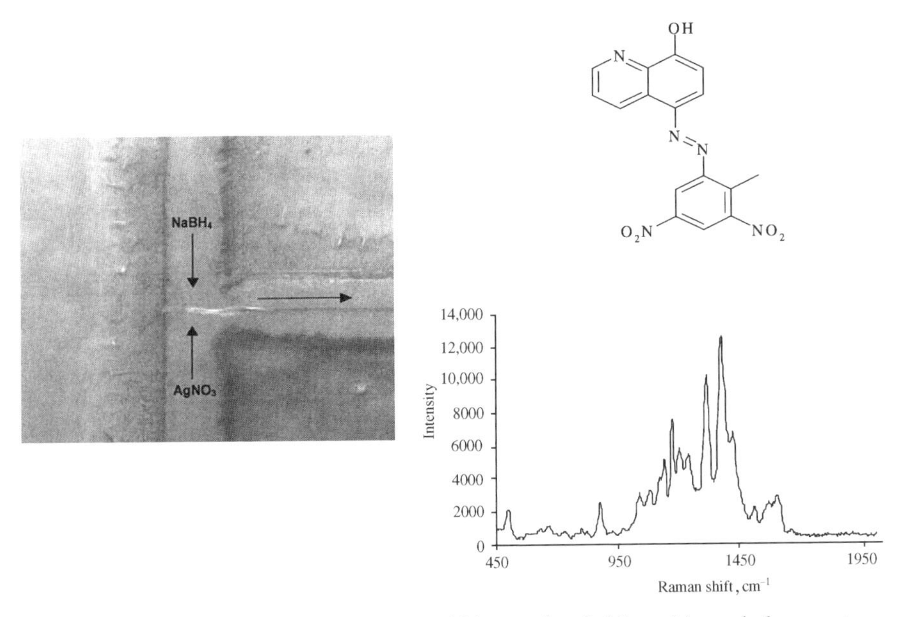

FIGURE 3.9 Colloid produced *in situ* within a microfluidics chip and the spectrum recorded after mixing in a dye. The spectrum is obtained by focusing the microscope onto the channel.

number of conventional techniques that employ this technology, including scintillation counting, fluorescence and absorption, electrochemistry, chemiluminescence, and Rayleigh scattering. The sensitivity and selectivity of SERRS in principle should outweigh these. This field is in the early stage. A number of reports of successful assays are now available, indicating that the technique is feasible (Cao et al., 2003; Dong and Channon, 2000; Gong et al., 2004, Xu et al., 2004).

There are many possible formats for such assays. A standard one is to immobilize a capture antibody onto a suitable surface. The analyte or antigen is then added to the surface and captured onto the antibody. A second antibody containing a SERRS label is then added to the solution and this antibody recognizes the antigen and complexes to the surface. At this point, colloid can be added to the suspension so that the SERRS label is coated with the colloid. Determination of the antibody antigen complex is then carried out using standard Raman spectroscopy equipment.

Thus, although SERS/SERRS is not widely used currently in the pharmaceutical industry, it may well have advantages in niche areas. Some molecules will give sensitive analysis in aqueous solution, others will give very effective analysis in clinical environments. As more effective and more targeted labels and substrates are produced, a significant potential for use in molecular diagnostics will develop.

REFERENCES

Albrecht MG, Creighton JA, 1977. Anomalously intense Raman spectra of pyridine at a silver electrode. *J. Am. Chem. Soc.*, **99**, 5215–5217.

Cao YC, Jin R, Nam J-M, Thaxton CS, Mirkin CA, 2003. Raman dye-labeled nanoparticle probes for proteins, *J. Am. Chem. Soc.*, **125**, 14676–14677.

Creighton JA, 1998. *Spectroscopy of Surfaces*, Clark RHJ and Hester RE (eds.) Wiley: New York p. 37

Cunningham D, Littleford RE, Smith WE, Lundahl PJ, Khan I, McComb DW, Graham D, Laforest N, 2006. *Faraday Discuss.*, **132**, 135–145.

Docherty FT, Monaghan PB, Graham D, Smith WE, Cooper JM, 2004. The first SERRS multiplexing from labelled oligonucleotides in a microfluidics chip. *Chem. Commun.*, 118–119.

Dong Y, Channon C, 2000. Heterogeneous immunosensing using antigen and antibody monolayers on gold surfaces with electrochemical and scanning probe detection. *Anal. Chem.* **72**, 2371.

Faulds K, Smith WE, Graham D, Lacey RJ, 2002. Assessment of silver and gold substates for the detection of amphetamine sulfate by surface enhanced Raman scattering. *Analyst*, **127**, 282–286.

Faulds K, Graham D, Smith WE, 2004a. Evaluation of surface-enhanced resonance Raman scattering for quantitative DNA analysis. *Anal. Chem.*, **76**, 412–417.

Faulds K, Barbagallo RP, Keer JT, Smith WE, Graham D, 2004b. SERRS as a more sensitive technique for the detection of labelled oligonucleotides compared to fluorescence, *Analyst*, **129**, 567–568.

Fleischman M, Hendra PJ, McQuillan AJ, 1974. *Chem. Phys. Lett.*, **26**, 163–166.

Gong J-L, Liang Y, Huang Y, Chen W-W, Jiang J-H, Shen G-L, Yu R-Q, 2006. Ag-SiO_2 core-shell nanoparticle-based surface-enhanced Raman probes for immunoassay of cancer marker using silica-coated magnetic nanoparticles as separation tools, *Biosen. Bioelectron.*

Hayazawa N, Inouye Y, Zouheir S, Satoshi K, 2002. Near field Raman imaging of organic molecules by an apertureless metallic probe scanning optical microscope, *J. Chem. Phys.*, **117**, 1296.

Hildebrandt P, 1992. *Relationships Between Structure and Function of Cytochrome P450—Experiments, Calculations, Models.* Academie Verlag.

Jeanmaire DL, Van Duyne RP, 1977. *J. Electroanal. Chem.*, **84**, 1–20.

Khan I, Cunningham D, Littleford RE, Graham D, Smith WE, and McComb DW 2005. From micro to nano: Analysis of surface enhanced resonance Raman spectroscopy (SERRS) active sites via multiscale correlations, *Anal. Chem*, **78**, 224–230.

Kneipp K, Wang Y, Kneipp H, Perelman LT, Itzkan I, Dasari RR, Feld MS, 1997. Single molecule detection using surface enhanced Raman scattering (SERS), *Phys. Rev. Lett.*, **78**, 1667–1670.

Kneipp K, Moskovits M, Kneipp H, Eds, 2006. Surface-Enhanced Raman Scattering—Physics and Applications Topics. *Appl. Phys.* 103.

Lyandres O, Shah NC, Yonzon CR, Walsh JT, Glucksberg MR, Van Duyne RP, 2005. *Anal. Chem.*, **77**(19), 6134.

Markel VA, Shalaev VM, Zhang P, Huynh W, Tay L, Haslett TL, Moskovits M. 1999. Near-field optical spectroscopy of individual surface-plasmon modes in colloid clusters *Phys Rev B*, **59**, 10903.

McAnally GD, Everall NJ, Chalmers JM and Smith WE, 2003. Analysis of thin film coatings on poly(ethylene terephthalate) by confocal Raman microscopy and surface enhanced Raman scattering, *Appl. Spectrosc.*, **57**(1), 44–50.

McLaughlin C, MacMillan D, McCardle C, Smith WE, 2002. Quantitative analysis of mitoxantrone by surface-enhanced resonance Raman scattering. *Anal. Chem.*, **74**, 3160.

Nie S, Emory SR, 1997. *Science*, **275**, 1102–1106.

Perney NMB, 2006. Tuning local plasmons in nanostructured substrates for surface enhanced Raman scattering. *Optics Express* **14**(2), 847–857.

Pettinger B, 2006. *Tip Enhanced Raman Spectroscopy in Topics in Applied Physics* Vol. 103 Kniepp K, Moskovits M, and Kneipp H, (eds.) Berlin, pp. 217, Springer–Verlag.

Quaroni LG, Reglinski J, Wolf R, Smith WE, 1996. Interaction of nitrogen monoxide with cytochrome P 450 monitored by surface enhanced resonance Raman scattering. *Biochim. Biophys. Acta Protein Struct. Mol. Enzymol.*, **1296**(1), 5–8.

Smith SJ, Munro AW, Smith WE, 2003. Resonance Raman scattering of cytochrome P450BM3 and the effect of imidazole inhibitors. *Biopolymers* 620–627.

Stockle RM, Suh YD, Deckert V, Zenobi R, 2000. Nanoscale chemical analysis by tip-enhanced Raman spectrosc. *Chem. Phys. Lett.*, **318**, 131.

Xu S, Ji X, Xu W, Li X, Wang L, Bai Y, Zhao B, Ozaki Y, 2004. Immunoassay using probe-labelling immunogold nanoparticles with silver staining enhancement *via* surface-enhanced Raman scattering, *Analyst*, **129**, 63–68.

4

RAMAN SPECTROSCOPY FOR IDENTIFYING POLYMORPHS

FRED LAPLANT
3M Corporate Analytical, Minneapolis, MN, USA
ANNE DE PAEPE
Materials Science Department, Pfizer Global R&D, Ramsgate Road, Sandwich, UK

4.1 INTRODUCTION OF POLYMORPHISM

Although Raman spectroscopy has made inroads into the pharmaceutical industry in a variety of important areas described elsewhere in this book, the analysis of polymorphs was the application that made Raman spectroscopy a standard part of the instrumental toolbox (Is et al., 1986; Raghavan et al., 1993). Polymorphs are simply different crystal forms that arise from a single molecular entity. While most practitioners in pharmaceutical formulation development will be very well acquainted with the nature and impact of polymorphism, it may be valuable here to review some basic definitions, how *polymorphs* arise, and why they are important, so that the impact of Raman spectroscopy on the characterization, quantitation, and control of polymorphs can be better understood in context with many other techniques that are also available for polymorph analysis.

The classical definition of polymorphs (from the inestimable Walter McCrone) is "a solid crystalline phase of a given compound resulting from the possibility of at least two crystalline arrangements of the molecules of that compound in the solid state" (McCrone, 1965). While strictly correct, a more inclusive view today recognizes that a compound may exist in a range of different energetic solid states, from the most stable crystalline form (having minimum ΔG) to fully amorphous state (having maximum ΔG). Depending on the desired attributes of the final drug product, any one

Pharmaceutical Applications of Raman Spectroscopy, Edited by Slobodan Šašić

of these states may be of very significant analytical interest. This also implies inclusion of a variety of other materials such as hydrates, solvates, and cocrystals (often referred to as pseudopolymorphs). Cocrystals in particular have drawn a great deal of attention recently because the range of cospecies is essentially limitless; this should allow great flexibility in controlling the functionality of the resulting drug product (Fleischman et al., 2003). Although not strictly covered in the old definition, these provide many of the same differences in solubility, stability, that make polymorphism that make polymorphism an important aspect of drug development. While this chapter will focus on analysis of "true" polymorphs, the technical approach used to characterize pseudopolymorphs is in most cases very similar.

As in any chemical process, while ΔG predicts the most energetically favored state, kinetics often plays an important role in determining what crystalline state will be formed. The rate of crystallization, level of environmental seeding, solvent system, and temperature are all key variables in determining which polymorph will be formed. Mannitol, for instance, can be selectively crystallized in its α or δ form using the appropriate solvents, rather than the favored β form (Campbell Roberts et al., 2002). The ΔG may also have a strong temperature dependence, such that a different crystal form will be favored above a critical temperature. This type of polymorph is known as an enantiotrope, as opposed to a monotrope in which a single polymorph is the most stable form regardless of temperature. Any of these factors can be used to induce a kinetically favored crystal form; however, this says nothing about the long-term stability of that form, or under what conditions it may remain stable (Maurin et al., 2002).

Why this is so important in the *pharmaceutical industry* is that it allows a single drug molecule to exhibit a variety of different physical characteristics. Depending on where the polymorph fits into the drug product strategy, each of these differences may be seen as either a curse or a useful formulation tool. More importantly, because these differences may impart novel and nonobvious properties, polymorphs can make valuable and defensible intellectual property.

The most important difference between polymorphs is almost always solubility. This is because solubility will directly alter the release characteristics of the drug which from a formulation standpoint is the most critical variable. Also, unlike the other properties outlined here, which may or may not differ between polymorphs, different ΔGs essentially guarantee that the solubility will be different. This may be inconsequential; for instance, if the drug is introduced such that a sink condition is always maintained, that is, the solubility limit is not approached. This may be the case for low dose or highly soluble drugs, so that the polymorphic forms are bioequivalent. Alternately, one form may be so insoluble that it does not completely dissolve before being excreted. More likely is some intermediate state where the release rate of the more stable form is slower, and the maximum concentration of the drug in plasma is both diminished and delayed in time. The potential impact on efficacy and toxicity continues to make this the principal driver behind polymorph analysis (Lu et al., 2006).

In addition to solubility, there are a variety of other variables that can be affected by crystal form. First, the density and hardness can differ. This can cause multiple effects on the performance of the drug product. The compressibility of the material can therefore be different, leading to differences in tablet hardness or friability (and subsequent dissolution rates). These variables, along with surface morphology, can also affect flow and mixing properties, leading to insufficient mixing, desegregation due to overmixing, and agglomeration. Crystal forms may also exhibit varying chemical stability. Vulnerability to attack by water and oxygen in particular can be drastically different, especially between amorphous and crystalline forms. Because each could in theory lead to unacceptable product variance, the validated processes and methods surrounding one polymorph may very well be invalid for another form. These issues have not gone unnoticed by the *FDA*, which follows the ICH Q6A guidance in order to determine what level of scrutiny is required of any process capable of producing multiple crystalline entities (Q6A International Conference on Harmonisation, 2000).

The FDA approaches this from two angles: the type of information that should be acquired during drug development and how this information relates to issues of drug quality. Throughout the drug development process, the pharmaceutical industry is asked to demonstrate different degrees of awareness of the possibility that other crystalline forms may occur, ranging from just establishing this awareness in preclinical development to performing full polymorph quantitation at later stages depending on whether this new form affects product safety, performance or efficacy. The burden on the analyst is then to appropriately characterize potential polymorphs at every stage of the development process.

These fall under three general categories: polymorph screening, where studies are performed in an attempt to find the most stable forms before formulation begins; process control, where analytical tools are implemented to optimize manufactured material; and product quantitation, where the polymorphic purity of drug substance and product is ascertained. Each of these areas poses its own unique measurement challenges; a brief overview of the common techniques available for polymorph characterization will follow so that the relative merits of Raman spectroscopy can be determined with regard to the specific goals and limitations of each application.

4.2 INSTRUMENTAL METHODS OF POLYMORPH CHARACTERIZATION

A variety of studies have compared multiple modern techniques to characterize specific polymorphs (Sheikhzadeh et al., 2006; Van Hoof et al., 2002). These techniques are summarized in Table 4.1. Although sophisticated analytical instrumentation and software can now be used to give a wide variety of information about the properties of polymorphs, simple physical observation of crystal form should never be overlooked. *Microscopy* has a long history in crystal form identification (Haleblian and McCrone, 1969; Krc, 1977; Kuhnert-Brandstatter, 1974), and although it is rare that the modern analyst would need to separate tartaric acid crystals from each other by hand as Pasteur did, this does not mean that the technique is no longer valuable. Often underutilized because it can be time consuming and requires a skilled practitioner, microscopic inspection of API can still be an unequivocal and information-rich method to

TABLE 4.1 Summary of Analytical Techniques for Polymorph Characterization.

Technique	Advantages	Disadvantages
Thermal methods	Cheap; fast; easy to implement	Nonspecific; not appropriate for drug product; can be difficult to quantify
PXRD	Direct measurement of crystal form	sensitive to crystal orientation
ssNMR	Can measure full dosage form with minimal sample preparation; assess amorphous content; provide detailed structural information	slow; costly
Near-infrared	Flexible sampling; strong signal; sensitivity to hydrated states; high depth of penetration	Reliant on complex modeling; very sensitive to physical state of sample
Mid-infrared	Ubiquitous instruments; fast	Limited sampling flexibility; low depth of penetration; low selectivity for API
Raman Spectroscopy	Flexible sampling; generally selective for API, polymorphs; insensitive to water	Fluorescent interference, signal can be weak

characterize polymorphic content and morphology. For instance, a great deal can be learned about the crystallization kinetics from observation of crystal quality, a parameter which can be difficult to measure with any other method. The uniformity of the crystal physical presentation can be a predictor for both polymorphic as well as chemical solid-state stability. But however useful microscopy is as a tool for fundamental understanding, the qualitative results and the difficulty of making fast, statistically relevant measurements limit its general application.

There are a variety of other nonspectroscopic methods that are routinely used. Because of the inherent difference in melting temperature, *thermal methods* such as DSC (Giron and Goldbronn, 1997; Sacchetti, 2001; Wu and Liao, 2000) are commonly used for API. TGA can also be useful when the crystal form is a hydrate or solvate, in which the crystal loses mass during thermal challenge. These measurements can be quantitative, and sensitive to multiple states that may be present in the sample. However, difficulties can arise from interstate conversion; for instance, as the temperature is increased in an enantiomeric system, conversion from one form to the other is possible, making quantitation complicated. This can be ameliorated by very fast temperature slew rates, but some uncertainty always exists (in the absence of supporting data from other techniques) that the heated sample in the DSC may not be fully representative of the sample as it exists at room temperature. Polymorphs whose melting temperatures do not differ by a significant amount can also be difficult to analyze thermally. Other methods such as gravimetric water sorption can be quite sensitive, especially for amorphous material. However, these methods are generally limited to API and have very limited sampling flexibility, so their implementation across the entire development platform is unlikely.

Although the thermal techniques measure an obvious inherent difference between polymorphs (melting temperature), it may be less obvious why there should be any difference between the *vibrational spectra* of different crystal forms. If we look at the technique naively, we could expect the vibrational spectrum of a solid to look identical to that of a liquid based on the vibrational assignments of the different functional groups in the molecule. However, the spectrum of a material in the solid state is not determined solely by the symmetry of the molecule but by the additional symmetry rules brought into play by variations in the crystal packing structure and nonisolated movement of molecules in these crystals. More elaborate explanations of occurrence of these lattice vibrations on a range of compounds can be found in a number of articles (Cruickshank, 1958; Loudon, 1964; Prasad, 1979; Turrell, 1972). This is why Raman spectroscopy and infrared spectroscopy can also be used to detect degrees of crystallinity within a given material (Okumura and Otsuka, 2005). One must take into consideration that the range of order that is probed by techniques such as PXRD, ssNMR and vibrational spectroscopy is very different.

One of the simplest compounds to demonstrate the effect of placing a molecule in a lattice is the ubiquitous polymer polyethylene (PE). If we took the chain of molecules as a separate entity, the Raman spectrum would look very simple, comprised of a couple of CH stretches and bends. The lowest expected frequency would be around $1060\,\mathrm{cm}^{-1}$ for the C–C stretch. However, when we place the PE chains in the lattice, there will be a reduction in symmetry compared to the isolated PE chains. For PE, the unit cell comprises two chains and is indicative of an orthorhombic crystal structure. This will cause crystal field splitting of the methylene bending and can be seen in the Raman spectrum of PE by a doublet at 1416 and $1440\,\mathrm{cm}^{-1}$. Because these motions occur predominantly perpendicular to the chain direction, these vibrations are thought to result in a more intense interchain interaction in the crystal (Lagaron, 2002; Strobl and Hagedorn, 1978).

In practice, the vibrational spectrum of true polymorphs will have a wide variety of very similar vibrational modes, many of which will likely be identical. Certain modes, however, can shift in both intensity and frequency according to the constraints put on the molecule by the crystal lattice, and studies have shown that these differences are generally statistically large (Mehrens et al., 2005). The solid *amorphous form* of the material will have features quite similar features to that exhibited by the liquid, and in general all of the allowable modes will be presented, often as broad bands because the range of populated physical orientations is large. The spectra of any polymorphic crystalline states will express bands present in the amorphous/liquid spectrum, although certain bands may be suppressed (to virtual invisibility) or shifted, while others are dramatically expanded. Occasionally, the vibrational spectrum of two polymorphs will be so different that it is not clear that they even come from the same parent molecule. However, the liquid/ amorphous spectrum often acts as a spectral bridge revealing the vibrational modes from which all of the polymorphic spectral features arise.

The spectroscopic methods are usually much more amenable to application across the entire pharmaceutical development life cycle. Mid-IR is the most common solid-state technology and has been applied successfully in a variety of polymorph analyses. By far the most mature technology, MIR's long standing as a compendial technique (the USP allows the use of MIR as a completely specific, stand-alone test for material ID) means that it is ubiquitous in pharmaceutical laboratories, and at least one validated system is typically present in laboratories performing GMP work. Additional benefits include being relatively cheap and fast compared to other spectroscopies. Sensitivity to water is also an advantage when looking for pseudopolymorphs; associated with this, MIR is sensitive to most hydrogen-bonded species, such as alcohols and amines, so that polymorphs exhibiting changes in these moieties will be more easily detected. Although it is true that Raman spectroscopy in general will be more sensitive to changes in crystal structure (e.g., the spectra will show more variation), MIR will almost always show some variance between polymorphs. With the widespread use of multivariate tools, the more subtle and less well-resolved spectral differences in MIR (and near-IR) can be effectively used as polymorphic indicators. Attenuated total reflectance modules (ATR) can be used for rapid and quantitative determination of polymorphic purity (Head and Rydzak, 2003) but one must avoid pressure sensitive polymorphs.

However, much like the nonspectroscopic methods, there are several critical limitations that make MIR a less appealing technology for most applications other than simple API qualification. Oversensitivity to water can also be a significant problem; for instance, it may be difficult to deconvolve the effect due to waters of hydration that may be present in excipients when analyzing drug product. Sampling in aqueous media, while not strictly impossible, typically yields very limited spectral information. Measuring material in nonaqueous media may also be difficult; although reaction monitoring can easily be accomplished when the reactant is in solution, obtaining useful spectra from slurries (the typical form of polymorphic conversions) is problematic. Additional disadvantages when analyzing drug product are that the method is not particularly selective for API; many excipients have very strong IR absorptions, which tend to mask the effect of the drug substance; and the depth of penetration into the sample is very low (on the order of a few microns when using an ATR crystal), making it difficult to ensure representative sampling. Furthermore, remote sampling with MIR is still limited. While great improvements in IR-transmissive fibers have been made over the past decades, the fundamental physics of coupling and maintaining photons in a waveguide will always vastly favor the visible and near-infrared (NIR) regions of the spectrum, where losses are measured in dB per kilometer rather than dB per meter. *In situ* ATR has also been implemented both for fiber optics and as direct beam coupled to a spectrometer, although this is typically practical only for liquids or dissolved species. Because of sampling flexibility and sample presentation issues, MIR will likely continue to make its largest impact in polymorph characterization at the benchtop rather than online.

The near-IR spectrum arises from essentially the same vibrational absorptions as mid-IR, so that any polymorph that is differentiable by MIR should also be differentiable in the NIR. But rather than absorbing at the fundamental frequencies of the vibrations, the NIR spectrum is made up of overtones and combination bands, yielding a mixture of broad, unresolved absorptions. NIR is similarly sensitive to water, and is thus well suited for hydrates; identification of bound and free water within a crystal lattice is also facile. Because of a relatively weak absorbance intensity and a high scattering efficiency, the depth of penetration of NIR radiation into the sample is much larger. This allows for a much broader range of sampling approaches; for instance, diffuse reflectance, where the sample is simply illuminated with NIR light and resulting scattered radiation is collected, is very common in NIR. Because of the large amount of radiation available for detection, NIR data collection also is often the fastest measurement of any of the techniques described here (Blanco et al., 2005; Luner and Patel, 2005). These advantages have made NIR the technique of choice for many online applications (Davis et al., 2004).

The principal disadvantages of NIR are not related to the ability to make a measurement, but rather to the reliability of the data that are obtained. The spectra obtained from solid samples using diffuse reflectance are very dependent on the physical state of the sample, especially particle size. While this sensitivity can be diminished by appropriate preprocessing of the data, the degree to which NIR relies on a highly representative calibration set may make NIR a less attractive selection in many cases. This will be discussed further in the section on quantitation of drug product. In a manufacturing setting, the amount of time saved running large numbers of routine tests using NIR versus other tests (XRD, for instance) will outweigh the cost of developing extensive calibration sets. This does not hold true in drug development, where the speed and ease of development of a calibration are much more important.

A couple of other techniques that should be mentioned here are *Terahertz (THz) spectroscopy and solid-state NMR*. THz (traditionally known as far-IR, because it probes the area between the midinfrared and microwave regions of the EM spectrum) has received a fair amount of attention in the pharmaceutical industry recently, at least in part due to work done in collaboration with the FDA. This spectroscopy probes low frequency molecular rotations and vibrations, and hence can be quite sensitive to changes in crystal structure (Taday et al., 2003). Modern instruments can also be used in time-resolved mode to determine coating thickness, among other things, using the time-domain reflectometry signal arising from the differences in refractive index at layer interfaces. However, from a practical spectroscopic standpoint, it has much in common with NIR. It is highly penetrating, the bands are not well resolved, and it is quite sensitive to the presence of water and the physical state of the analyte. Although awareness of the far-IR spectral region is not new, assignment of spectral features does not begin to approach the maturity of the other IR regions, and significant theoretical and analytical development has yet to be done. Often many of the higher frequency features (300–50 cm^{-1}) can also be accessed by Raman spectroscopy instruments. Because of its relatively recent introduction to the market, lack of predictive experience in pharmaceuticals, and price—about an order of magnitude more than a typical NIR spectrometer for a time-resolved unit—it is probably not a good choice at this time except for very specialized applications. ssNMR has a variety of advantages, which make it an appealing alternative, including being a much more mature technology than THz. Because it is specifically sensitive to the physicochemical environment of the probed nuclei, allowing interpretation of the structure including the assignment of proton bonded nuclei, which can be difficult to place in the crystal lattice by XRD, it is often used in conjunction with XRD for the complete solution to the crystal form (^{13}C NMR has also been used successfully for quantitation and characterization of polymorphs in API (Giordano et al., 2001; Moynihan and O'Hare, 2002). It has higher sensitivity than PXRD, as well as better selectivity for amorphous material. In terms of limit of detection, for specific nuclei it can compete with dispersive microscope Raman spectroscopy mapping and detect compounds well below 1% (w/w). However, the lack of speed, need for larger sample size, and expense will severely limit its application for general analysis.

Whilst all of these techniques have sensitivity to some aspect of polymorphism, only X-ray diffraction detects the fundamental crystal lattice, and hence it must be the gold standard by which other methods are judged (Ashizawa, 1989; Takahashi et al., 1962). Modern powder XRD instruments are relatively competitive with most other techniques in terms of speed and sensitivity. The principal weakness of XRD is that it can be quite sensitive to the size and orientation of the analyte crystals. Although grinding the sample and performing capillary spinning or other sample translational techniques can be very useful, this can induce form conversion and other variance in the sample that is often difficult to control repeatably. In addition, the X-ray spectrum is equally sensitive to any crystalline material present, so excipients are particularly interfering. Quantitation of the amorphous state is also difficult, because the spectral response is nonspecific and can be complicated by the presence of other noncrystalline material (polymers, for instance).

Raman spectroscopy is not a fundamental measurement the way XRD is; although virtually every true polymorph will have significant spectral differences, pseudopolymorphs may not, and Raman spectroscopy would need to be judged on a case-by-case basis based on spectral selectivity. Even so, Raman spectroscopy has a variety of advantages that often make it a very strong candidate. First, the intensity of the Raman spectroscopy spectra for the typical API is stronger than that of potentially interfering species, such as water or excipients. This provides an inherent selectivity for the analyte of interest. Secondly, Raman spectroscopy is more sensitive to symmetric stretching associated with the carbon backbone of a molecule. This usually imparts significantly more spectral selectivity between polymorphic forms (including amorphous or glassy states (Vassou et al., 2005) than other vibrational spectroscopies. Frequently low frequency modes ($< 400\,\mathrm{cm}^{-1}$) are extremely selective for crystal forms to which Raman spectroscopy has relatively easy access. Raman spectroscopy bands are also more highly resolved, making quantitation more straightforward. Thirdly, Raman spectroscopy combined with a microscope set-up allows for very high sensitivity. Finally, the sample presentation is very easy, and little or no sample preparation is necessary.

Occasionally, the drug itself or trace impurities can induce background fluorescence which makes any measurement difficult or impossible. Although this drawback is often highlighted, in our experience it is rare that a measurement in either drug substance or product is impossible due to fluorescence. Raman spectroscopy may also be sensitive to orientation of the molecule because of polarization effects. However, these can be eliminated by taking care to scramble the polarization of the incident laser. In short, assuming that fluorescence is low and the spectral differ-

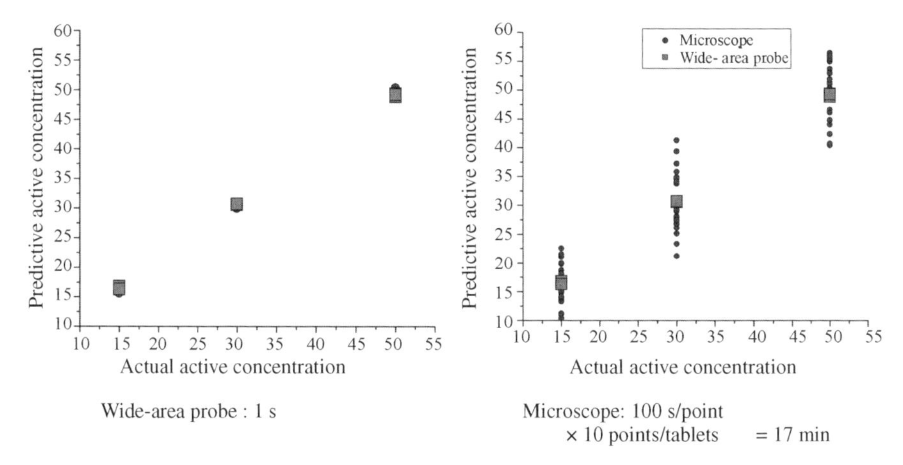

FIGURE 4.1 Comparison of microscope performance with that of a widearea probe for a typical pharmaceutical tablet active assay.

ences are sufficient to allow characterization, Raman spectroscopy is a very powerful technique to study polymorph conversions as API and in drug product (Deeley et al., 1991; Findlay and Bugay, 1998; Paul et al., 1990; Taylor and Langkilde, 2000). One of the classical difficulties against Raman spectroscopy is undersampling—it is experimentally difficult to measure (microscopy system) or a few hundreds (FT-Raman spectroscopy system) a statistically representative amount of the analyte when the focus of the laser is tens of microns. Typical solutions for this involve rotating or wriggling the sample, or translating combined with multiple replicates, techniques also implemented to reduce sample heating (De Paepe et al., 1997; Taylor and Langkilde, 2000). Application of a wide-area Raman spectroscopy sampling probes using large fiber-bundle collection has vastly improved the statistical sampling. Spot sizes for these types of probes can be greater than 4 mm, an obvious advantage. A large working distance (~4–6 in.) also yields a much increased depth of field. This depth of field also equates into higher sample penetration and less spatial rejection, allowing deeper probing of the sample (several millimeters in a typical tablet formulation) combined with decreased sensitivity to sample presentation. For multilayer systems, this sort of sampling has shown to be useful for subsurface sampling as well (Morris, 2007). This improved sampling volume leads to very high signal levels compared to traditional optical configurations, in spite of the fact that the probe *f*-number calculation would not appear to be very favorable. The advantage of this kind of system can be seen graphically in Fig. 4.1 (these data represent a potency measurement; the expected variability for a polymorph measurement could potentially be significantly larger). The wide-area data are compared using a standard microscope-based system with approximately 20-μm focus. The laser power and signal to noise parameters were comparable between the two systems; the principal difference was in the acquisition time. The wide-area probe data show three measurements at each point using a 1-s acquisition time with a single replicate. Although the average data from the microscope were essentially equivalent to the wide-area result, because of lower throughput and sample inhomogeneity, the microscope analysis required approximately 17 min of acquisition time.

Another consideration is the choice of Raman spectrometer. Both dispersive and FT instruments have seen wide application in polymorph analysis. Although it is outside the scope of this chapter to go into the fundamental differences between the two instruments, certain key features tend to make one of the instruments preferable based on application. FT-Raman spectroscopy has traditionally been used as the choice for quantitation because of its superior frequency accuracy and better fluorescence suppression stemming from the use of higher excitation wavelength (Auer et al., 2003; Taylor and Langkilde, 2000). Dispersive instruments are easier to adapt to fiber optic and microscope-based applications and are more amenable to miniaturization, although they have been used extensively in quantitative studies as well (Vehring, 2005; Wang et al., 2000), and offer greatly enhanced sensitivity compared to FT-Raman systems. Although FT instruments will likely always have important niche applications, with advances in technology yielding improved calibrations, lower noise, smaller packages, and decreased cost, it is likely that the current trend toward dispersive instruments will continue in the foreseeable future.

Some examples from the literature will highlight that while Raman spectroscopy is a valuable stand-alone spectroscopic tool, it is often used in conjunction with other techniques described above for full understanding of polymorphic structure and function. Ayala et al. (2006) used Raman spectroscopy and infrared (mid and near) spectroscopy to characterize olanzapine polymorphs and to understand their structural relationship. This paper nicely demonstrates that by correlating different structural techniques (references to ssNMR and X-ray diffraction data) and DFT calculations, useful information can be obtained from the vibrational spectra that will aid in structural understanding. After assigning vibrational bands, they determine the larger entropy form based on NH stretch frequencies (in accordance with Burger and Ramberger rules (1979a,b). This is consistent with previous reports on the stability of these polymorphs. When assessing the stretching vibrations of the double bonds, they notice a splitting in one of the bands for one of the forms. They rule out the presence of multiple molecules per asymmetric unit from previous ssNMR data and the fact that this splitting is not consistent throughout the whole spectrum. DFT calculations are then performed to try and attribute this splitting to conformational changes between the forms. From the fact that there is no good correlation between the DFT plots with the spectra obtained, this theory is then discarded and leaves one last option, namely, differences in energy of atoms directly or indirectly involved with NH-hydrogen bonding. The low frequency region (below $400\,cm^{-1}$) of the Raman spectra are compared. Above $100\,cm^{-1}$, there are relatively small differences between the spectra, again confirming that there are no large conformational differences between the molecules in the two forms. The region below $100\,cm^{-1}$, which encompasses lattice and librational modes, does show changes between the forms. Lastly, the NIR spectra are assessed and show differences in the hydrogen bonding in different forms.

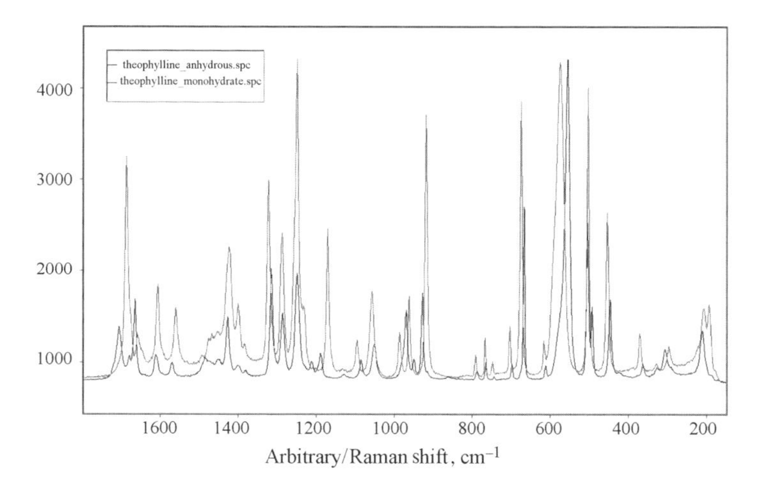

FIGURE 4.2 FT-Raman spectra of theophylline anhydrous and monohydrate.

Taylor and Langkilde (2000) describe how they use FT-Raman spectroscopy to assess polymorphic form of an API in tablets and capsules. For one of the solid dosage form studied, they were even able to detect the API polymorphic form at 1.25% (w/w). They describe how sample rotation is used to reduce subsampling and laser heating. They also looked at a case of pseudopolymorphism. The asthma drug theophylline forms hydrates when exposed to water. The Raman spectra of the two forms are displayed in Fig. 4.2. The area of the spectrum where the interaction with water is most pronounced is that between 1750 and 1500 cm^{-1}. In their paper,

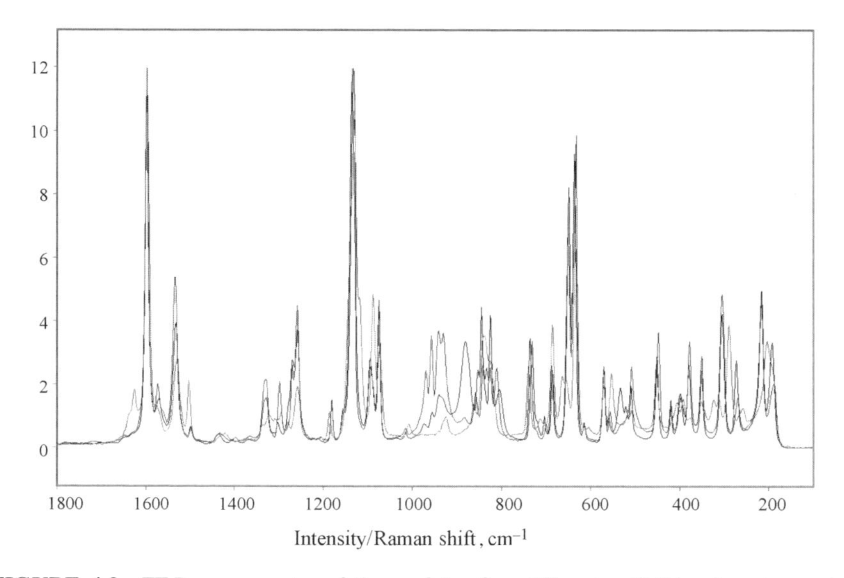

FIGURE 4.3 FT-Raman spectra of three of the five different sulfathiazole polymorphs. Impurity levels are variable.

Taylor and Langkilde studied several commercial solid formulations identifying monohydrate in one and anhydrous form in the other.

The polymorphic forms of the antibiotic drug sulphathiazole have been studied extensively. There are several studies where different analytical techniques have been used to study different polymorphic forms (Apperley et al., 1999; Blagden et al., 1998; Hughes et al., 1999). Anderson et al. (2001) published a study where they determined the onset of crystallization of sulfathiazole by UV–Vis and real-time calorimetry. Raman spectroscopy, DSC, and FTIR are used for subsequent polymorphic characterization. They found that UV–Vis monitoring of the onset of crystallization was much more sensitive than the real-time calorimetry experiment. By comparing DSC data with vibrational spectroscopy data, they found that Raman spectroscopy and FTIR were consistent in naming four forms, whereas the DSC experiments almost always yielded multiple endotherms for this compound. This is attributed to the ability of sulfathiazole to go through multiple transformations as it is thermally cycled. Spectra of three of the five sulfathiazole polymorphs are shown in Fig. 4.3. Zeitler et al. (2006) used THz pulsed spectroscopy and low frequency Raman spectroscopy and showed that the five polymorphs can easily be distinguished from one another. Firstly, the polymorphs were characterized using different techniques. Due to the stacking of molecular layers in the different forms, Form III appears to be similar to the sum or an intermediate of Forms IV and V. However, in the PXRD patterns and the far-Raman spectra, peaks from Form III lie in between those of Forms IV and V and hence allow for distinction between all the forms. However, due to the closeness of peaks and possible overlap of other forms, they suggest multiple techniques to assign forms or confirm the presence of more polymorphs. They then monitored the phase transitions using THz pulsed spectroscopy. As mentioned above, a drawback of this technique remains that assignment of the spectra is still in its infancy.

4.3 POLYMORPH SCREENING

Recent pharmaceutical history is littered with unfortunate products that suffered the appearance of a new crystalline form late in the development process. The most notorious, certainly, is Abbott's ritonavir (Chemburkar et al., 2000) in which the appearance of the new polymorph did not happen until the drug had been on the market for 18 months. How a crystalline state can escape discovery throughout the drug development process is a lesson in statistical thermodynamics. Because early lots of drug substance tend to be more impure than subsequent lots, kinetically stable polymorphs often dominate. As the production lots become more pure, crystal growth can slow down and thermodynamically stable forms appear. The appearance of a certain crystalline template may be kinetically unfavored, so that it may be statistically rare even though it is energetically more stable. But once the energetically most favored state has been discovered, it is essentially impossible to completely exclude this state from any subsequent drug lots. Although a variety of routines exist for generating unfavored polymorphs and subsequently inhibiting conversion, in some cases it can prove virtually impossible to produce the previous polymorph at an acceptable level. This proved to be the case with ritonavir; the drug was subsequently withdrawn from the market and reformulated, but the development costs and lost revenue were substantial. The lesson from this episode has been well learned by the entire industry, and improved methods are now generally in place to ensure that all polymorphs are characterized before the development campaign is launched. Only time will tell if the current methods will prove sufficient to avoid further polymorphic unpleasantness.

The challenge is that as the number of compounds in early development continues to increase, automated polymorph and hydrate screens are becoming more and more essential. This is compounded by the use of multiple *experimental variables*, such as solvent/antisolvent type (e.g., polar, nonpolar), temperature, speed of addition of solvents, stirring rate, and concentration. The classic first step in polymorph screening involves crystallizing samples from various solvents and then measuring the compound by XRD (or some other technique of choice), usually in small reaction vessels. Screening now more typically takes place in high-throughput-type 96-well plates with solvents of varying polarity, along with various cosolvent mixtures. Salt selection and cocrystal screening can use similar technology (Kojima et al., 2006). Further steps to understand API stability include measurements of a slurry of the known polymorph in various sparingly soluble solvents (Giordano et al., 2001). Given sufficient time, thermodynamically more stable species are given a better opportunity to form. Future directions include the use of polymer templates as crystallization masks; this may allow the use of a set of generic templates to reduce the kinetic barrier in the formation of crystals (Lang et al., 2002).

The challenge for any of these methods is to rapidly and unequivocally *characterize the resulting species*. Features of interest in these experiments include the following: are there crystals in the reaction mixture; where are these crystals located; how do these crystals differ from the starting material, if at all; and are these crystals solvates, cocrystals, or polymorphs? In addition, the analysis method would ideally be made *in situ*, with minimal interference from solvent or reaction vessel. Although other methods could theoretically be used in these applications, in practice the ability of Raman spectroscopy to sample very small amounts of material, in any solvent, and with complete geometric flexibility, from microscope objective to fiber optic dip probe to wide-area collection, makes it the strongest choice.

The determination of the presence and location of crystal is probably the most challenging aspect of these experiments. In slurries, this is a nonissue, since the sample is translated continuously through the measurement zone by mixing. The assumption is that the product of any polymorphic transformation will remain in the suspended state where it will be accessible to the Raman probe. But for static crystallization experiments, such as those carried out in 96-well plates, the position and indeed the presence of crystals is by no means guaranteed. Three approaches are currently implemented for this purpose.

The first method is simply hunting for the crystals manually with a polarized light microscope. This is complicated by the three-dimensional nature of the sample, since crystals could have formed anywhere within the volume of the well. Any crystal found then needs to have its Raman spectrum obtained. Although this may also be performed manually, Raman systems will allow multiple measurement sites to be selected, and then measured in sequence automatically based on stored stage positions (Lowery et al., 2006).

The obvious extension of this method is to automate the Raman microscope to search for particles all by itself. This is certainly an appealing proposition, especially when performing high throughput screening, and a variety of automated image analysis packages exist for this purpose. Once the visible light image analysis has identified crystalline areas in the wells, the Raman spectrum of the material is obtained using the same microscope. The principal technical difficulty is again that potential samples may be present in an unspecified spatial location. The rate of false positives from optical or physical discontinuities in the well, as well as missing crystals that may be present due to low contrast or poor spatial presentation, can be high. Improvements in machine vision mechanisms will almost certainly improve the capability of automated systems, but finding crystalline material still highly depends on appropriate lighting and rejecting spurious images. Alternately, the Raman microscope can be used in a sort of volumetric imaging mode without regard for the visible image, where the microscope is used to probe every position in the sample volume, one point at a time. While generally more robust, even this technique can be prone to errors, especially because crystals are commonly associated with the walls of the vessel. The wall material can overwhelm the signal from the crystals unless the focus is very accurately adjusted. This technique is also extraordinarily time-consuming.

The third route is to measure the entire well simultaneously using a wide-area fiber-bundle probe as described previously. The depth of field for this type of probe can be adjusted to match the required sample volume by adjusting the probe optics. The probe can theoretically analyze most, if not all, of a typical well volume in a single measurement. While this technique has a variety of very appealing advantages, such as speed, large sample coverage, and no requirement for viewing the sample, the large sampling volume also brings with it several drawbacks. The contribution from the well wall-material will be substantial, because no spatial discrimination is taking place. Although this can be accounted for by standard multivariate techniques, the interference that this adds to the signal can make measurement of very small signals difficult. The limiting size of the detectable sample is therefore larger as that of a purely microscope-based system.

Another possible solution to this problem involves altering the size of the solution, rather than changing the method of probing. By using very small volumes of liquid for crystallization, the area over which the crystals may form is decreased greatly, and the subsequent analysis of the crystallization volume becomes much easier (Anquetil et al., 2003).

The complete solution for polymorph screening will almost certainly employ a combination of the methods described above, with macro–micro Raman spectroscopy measurements coupled with automated visible detection of crystals. It is likely that this is the only approach that can lead to complete, zero-fault detection of crystalline material in a high throughput fashion.

4.4 PROCESS CONTROL

Raman spectroscopy is an invaluable tool in the area of online process monitoring and control. Batch reactions, process streams, feed stock, and distillates can all be monitored using Raman spectroscopy. The ease of coupling signal into fiber optics makes Raman spectroscopy the easiest and most versatile technique when adapting a monitoring technology to an existing manufacturing process. The application of online Raman spectroscopy to liquid and suspended samples is widespread, but will almost certainly become equally widespread for solid samples with the introduction of wide-area, large depth of field probes. Within the pharmaceutical industry, the FDA's process analytical technology (PAT) initiative may also help drive acceptance and implementation of online spectroscopic tools (Andrews and Dallin, 2004).

The area of industrial process control is covered elsewhere in this volume at length, and in general the application of this technique to polymorph conversion monitoring and control does not vary significantly from other crystallization applications. The spectrometer hardware, fiber optic probe geometries, sample coupling, probe placement, software, and safety restrictions are all equivalent and will not be discussed here. The quantification is generally more straightforward than the drug substance/product case discussed in the following section, and similar mathematical schemes can be employed. This section will therefore not cover the different aspects of process control in any great detail, but highlight the aspects important to polymorph control specifically along with a few examples.

As with any crystallization process, the control scheme must be founded on the phase diagram for the material of interest. The effects of temperature and solvent composition are critical in "normal" crystallizations for optimizing yield, purity, crystal size, and efficiency; they are equally critical for polymorphic control. Kinetic factors such as the rate of cosolvent addition, stirring rates, and reactor volume also play important roles. These highlight the necessity of having good online monitors present during every scale of the process. Conceptually, if the full phase diagram of the crystal is known, then control can be maintained by knowing the temperature and chemical composition within the reactor. In practice, this is not possible for a number of reasons. First, the phase diagram is typically developed at very small scale using a small subset of solvent variations. Second, the kinetic factors mentioned above are almost impossible to predict during the scale-up process; the solvent mixing properties, temperature profile, or crystal settling rate, for instance, are certain to vary as the reaction size increases. This means that continuous *in situ* monitoring of crystallization is not just a good idea, but essential for efficient crystallization process development (Starbuck et al., 2002). An online Raman spectrometer will typically be coupled with instruments to track particle size, shape, and distribution to give a complete analysis of the reaction (Calderon De Anda et al., 2005, O'Sullivan et al., 2003).

Quantification of the process should be much less restrictive than drug product. The principal difference is that the process characterization is not a release method, and should not be held to the same level of accuracy or scrutiny as that placed on drug product release methods. Certainly, this was part of the motivation behind much of the FDA's PAT initiative, in which fit-for-purpose qualification of methods, risk-based analysis, and process understanding are meant to replace knee-jerk development of full GMP-level validation of every single quantified method. Whether this in practice will become the norm is too topical to discuss here; as always, the degree of validation required in any pharmaceutical application must be the result of collaboration between scientists, quality agents, and regulators.

But whatever the eventual practice, the truth of the matter is that most crystallization experiments can be monitored reasonably accurately using simple two-point models. Assuming a crystallization of Form A with the possibility of formation of other Forms (B, C, D, etc.), a spectrum obtained from a slurry mixture of pure component A, B, C, and D, respectively, can be used to generate a reasonably accurate calibration model for the real reaction. Alternately, solid calibration samples can be measured, and the spectra assumed to correspond to the those obtained in the reaction vessel (Ono et al., 2004). This case differs from the solid-state case described in the next section in the sense that the variables requiring a very accurate calibration set (excipient particle size interactions, packing density, etc.) are essentially nonexistent during crystallization. The principal variable, particle size, which will affect the path length of the incident radiation, is minimized because the calibration value is always presented as fractional quantity relative to the total API crystal population, rather than an absolute quantity. Other variables, such as crystal shape and orientation, are also minimized because the presentation of the crystals to the probe should be generally uniform and reproducible.

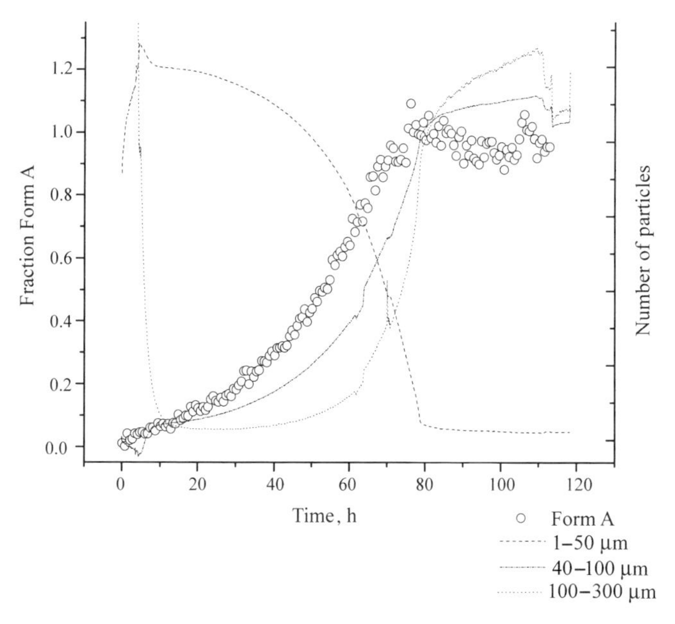

FIGURE 4.4 *In situ* polymorph conversion monitoring with Raman spectroscopy and crystal size data.

An example of how Raman spectroscopy fits into a reaction monitoring scheme is shown in Fig. 4.4. Raman spectroscopy must be used in conjunction with other monitoring and control mechanisms in order to maximize its utility; in this case, particle size using Mettler–Toledo Lasentec FBRM is used in conjunction with a Kaiser-Optical fiber probe to monitor the conversion of polymorph A to polymorph B in slurry. A variety of information can be obtained from this plot. First, by monitoring the shape of the conversion curve (coupled with curves taken at other temperatures) the conversion kinetics can be determined. The effect of conversion on crystal size is also shown; the size of the large crystal fraction increases with time, while the small fraction decreases. Also of interest, while the very large crystals increase as well, the discontinuity towards the end of the reaction indicates that these crystals are no longer being presented to the probe; for example, that they are likely dropping out of suspension and no longer providing the same surface area for continued crystal deposition. Although this is a relatively simple example, this illustrates the sort of information that is obtainable using standard online tools. By implementing these tools early in the development process, key process variables can be defined, leading to fine control of the desired crystal morphology by adjustment of the appropriate reaction inputs.

A variety of studies have shown the utility of Raman spectroscopy in improving process understanding in pharmaceutical polymorphs. The asthma drug theophylline has been used in several papers to study how pharmaceutical processing and excipient choice affect polymorph stability. Jørgensen et al.(2002) studied both theophylline and caffeine during wet granulation. Samples were analyzed at-line using NIR spectroscopy and off-line by Raman spectroscopy. PXRD was used to confirm the results. From this study, they concluded that the hydrate formation of theophylline starts at an lower water concentration than caffeine. Caffeine hydrate formation was not completed with the amounts of water used whereas theophylline almost completely transferred into the hydrate. Both compounds form channel hydrates with caffeine having a larger tunnel cross-sectional area. This and the fact that caffeine forms a 4/5 hydrate, compared to theophylline monohydrate, make these observations unexpected. An explanation can be found in the NIR spectra which indicate a difference in water bonding in the two compounds. NIR was shown to be capable of distinguishing between free water and bound water. Raman spectroscopy and NIR methods coupled with PCA were then used to visualize the changes in the state of water during process monitoring. Caffeine dehydrates more readily than theophylline. As expected, small caffeine crystals dehydrate more readily than large crystals. When vigorous mixing occurs, caffeine is not expected to form long, regular crystals. This accounts for the fact that even though there is enough water for all the caffeine to be transformed into the hydrate, dehydrating still occurs. The mixing time after water addition, was found to have an influence on the hydrate formation showing that when the mass was mixed for a shorter period after each water addition, anhydrous theophylline transformed to the hydrate only at higher moisture content.

Raman spectroscopy studies have also indicated the effect of excipients on form transformations during processing. Wikström et al. (2005) used Raman spectroscopy for in-line monitoring of the theophylline hydration process during highshear wet granulation. The effects of mixing speed, changes to the API including monohydrate seeding, concentration variations, and ball-milling, in addition to binder solution variations on the transformation were also assessed. They also show that a simple solvent-mediated transformation model can be used for estimation of the timescale for hydration during highshear wet granulation.

They found that the transformation to the hydrate was faster with increasing mixing speed within the range 100–400 rpm. At higher mixing speeds (600+ rpm), the transformation time was increased. This is probably due to a decrease in supersaturation caused by the temperature increase of the granules at these speeds. Both seeding monohydrate at 5% (mol/mol) and in a second experiment increasing the theophylline concentration from 30% to 45% (w/w) in the formulation (keeping the ratio of mannitol to microcrystalline cellulose constant) did not result in changes in the transformation rate. When unprocessed theophylline was substituted with ball-milled theophylline, a marked increase in the transformation rate was observed. No significant changes in the transformation rate were observed when the binder was added as a dry powder compared to when added as an aqueous solution. Airaksinen et al. (2003) studied the effect of α-lactose monohydrate and silicified microcrystalline cellulose (SMCC) on hydrate formation in wet masses containing theophylline. PXRD, NIR, and Raman spectroscopy were used to characterize the hydrate formation. They confirmed earlier results that the presence of α-lactose monohydrate enhances hydrate formation of theophylline during wet granulation. When comparing this to the results when having SMCC present, it is clear to see that the onset of hydrate formation is retarded in the latter case. Apart from at the beginning of the granulation, SMCC could not protect theophylline from hydrate formation.

4.5 POLYMORPH QUANTITATION

It is always preferable not to have to quantify polymorphs. The previous sections described the process by which the most stable polymorph may (hopefully) be found, and the sort of controls that can be put in place to help manufacture consistent crystalline material. However, in some cases the crystal state of the API cannot be guaranteed. Following the ICH Q6A guidance, if the bioavailability of the two (or more) forms is shown to be equivalent, there should be no need to quantify their relative amounts. One may argue based on cGMP requirements that consistent crystal state must be maintained simply to demonstrate appropriate process controls. In this case, it is up to the interpretation of the individual organization as to what level of crystal characterization is sufficient to provide regulatory satisfaction. However, if the crystal states are not bioequivalent, and there is cause to believe that multiple crystal states may be formed during processing, then quantification of API and subsequent drug product becomes an absolute necessity.

There are several different scenarios of varying severity depending on the relative energetics of the crystal forms and under what circumstances they were formed. The most benign situation is when the most stable polymorph is the predominant species during production of the API, although some level of another crystal form is present. Depending on the level of polymorphic impurity, typically the impurity can either be ignored, or reprocessed to enrich the desired state. In either case, the only quantitation is likely a limit test of the API. Somewhat more complicated is the same situation after the crystal is formulated into drug product. It is not uncommon for some processing steps (e.g., spray coating, milling) to generate changes in crystallinity, especially amorphous material. Quantitation of drug product is significantly more difficult, and will be addressed in detail below. Finally, the worst-case scenario is when the unstable polymorph is actually the target form. The most common reason for this is that the most stable form would otherwise have insufficient solubility, or that the other form is patent protected. In either case, the presence of the more stable form must be minimized, since even in the solid-state polymorphic transformation can occur at a rate based on the initial level of seed crystals. This means full quantitative characterization will need to take place at every step in the manufacture of the drug product, from the crystallization of the API, through any intermediate processing steps, to the production of the dosage form, and even beyond into the API and drug product stability protocols.

The only factor in ameliorating the difficulty of quantifying polymorphs is that the measurement is virtually always a relative one. For instance, a potency assay is always reported in percent of target (e.g., milligrams), which is an absolute measurement of the amount of drug substance in a given aliquot. This value is ensured through the use of an internal standard, or a near-internal standard, such as bracketing standards during an HPLC run. However, a polymorph assay is always reported as a fraction of the total amount of API. Since the amount of API in any volume can always be considered as unity, the API itself acts as its own internal standard. Theoretically, the measurement can be made extremely accurately; in practice, this is complicated by the generation of a suitable calibration curve, as described in the following section, which will discuss the issues involved with developing a calibration model that can assess polymorphic purity. The focus will be on drug product, although in general the same procedures apply to drug substance. As discussed above, features of this process may also be applied to calibration of process control analyses.

4.6 CALIBRATION SET AND SAMPLE PREPARATION

The most critical task in any solid-state assay is assurance that the calibration set is representative of the samples that will be measured. Possible variance can arise from differences in particle size, packing, hardness, excipient distribution, absorbing impurities, as well as other factors. These are mentioned here as "other factors" not because there is insufficient space to list them all, but because there are often indefinable factors that affect model as well as product performance. Anyone involved in any form of manufacturing in which a raw material is used will be familiar with the mystery product failure, which eventually turns out to be some change in the supplier or supplier process. Although the incoming raw material still meets specifications (for instance, the USP test specifications), the resulting product is now detectably different. Analytical methods can also suffer from these issues, but appropriate care in developing a robust model can minimize these effects.

Solution state assays are generally much more straightforward, as the presentation of the sample to the analyzer/detector is much more uniform and reproducible. Somewhat more complicated is the standard spectroscopic potency assay (whether NIR, Raman spectroscopy, or other technique) in which a calibration model is developed based on an orthogonal reference technique such as HPLC. If subsequent samples do not match the calibration set, they can be tested by the reference method and then added to the model. The spectroscopic assay is then used as a quicker, easier, less polluting, and online or at-line replacement for the reference method.

The difficulty with a polymorph assay is that it is unlikely that there is a fully reliable reference technique available. As discussed in the section on alternative methods, every technique (even XRD) is vulnerable to inconsistencies in the physical state of the sample. To some extent, the application and agreement of multiple techniques will increase the confidence in a certain measurement. But these calibrations must eventually rely on gravimetric preparation of standards that must closely match the expected sample, either as drug substance or drug product. Alternatively, the sample can be manipulated in order to make it more like the calibration set. In either case, extreme caution must be exercised if an accurate, representative measurement is to be made.

Based on experience with XRD measurements, where controlling particle size and preferred orientation are crucial, a seemingly reasonable approach is to grind both the sample and calibration set until they are a uniform powder. It has been extensively reported that this technique can lead to the generation of amorphous material for API alone; although the addition of fillers to API is often recommended to inhibit formation of amorphous material, our experience has been that grinding becomes significantly more problematic for drug product. There are a variety of different pathways by which the crystal state of the sample can be altered. First, the amorphous forms of either excipients or API can be formed, which can be highly dependent on the formulation and the method of grinding. This amorphous material may have variable stability based on degree of grinding, presence of water, and so on, leading to further conversion over time. However, mixed crystalline states are possible between the API and excipients, or direct conversion to the more stable polymorph may occur. All of these transitions have been observed in our laboratory, with often only fairly minimal sample manipulation.

These experiences lead to the general conclusion that manipulation of the sample is to be used only as a last resort, and that it is generally easier to make the calibration set approximate the sample rather than vice versa. In addition, the relative insensitivity of Raman spectroscopy to solid-state variables makes sample manipulation much less of an issue than it is with other techniques. Consider the NIR and Raman (785 nm excitation) spectra shown in Fig. 4.5. Although there are other factors involved such as matrix absorption, the difference due to wavelength-dependent scattering will be significantly larger for the near-IR spectrum than for the Raman, simply because the wavelength covered by the Raman spectrum is so much narrower. This also means

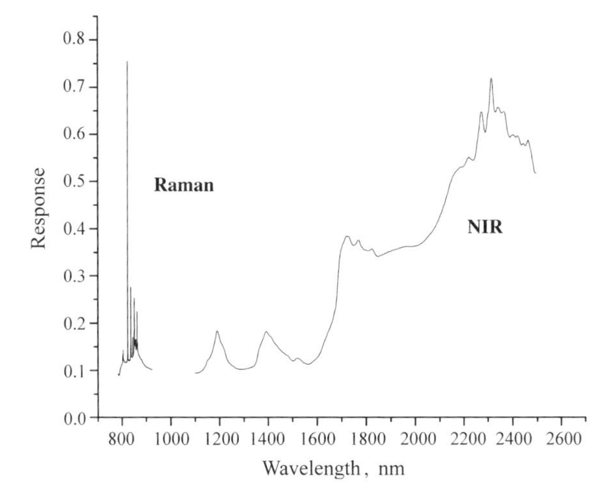

FIGURE 4.5 Comparison of spectral range for NIR versus Raman spectroscopy data with 785 nm excitation.

that variable absorptions should have much less effect on the Raman, since they will tend to be broad compared to the Raman bands. This may lead to absolute intensity variations, but is less likely to lead to the relative intensity changes that would cause quantitative variance. The normalized spectra shown in Fig. 4.6 are representative of this insensitivity. Although there is some small difference in the drug product central

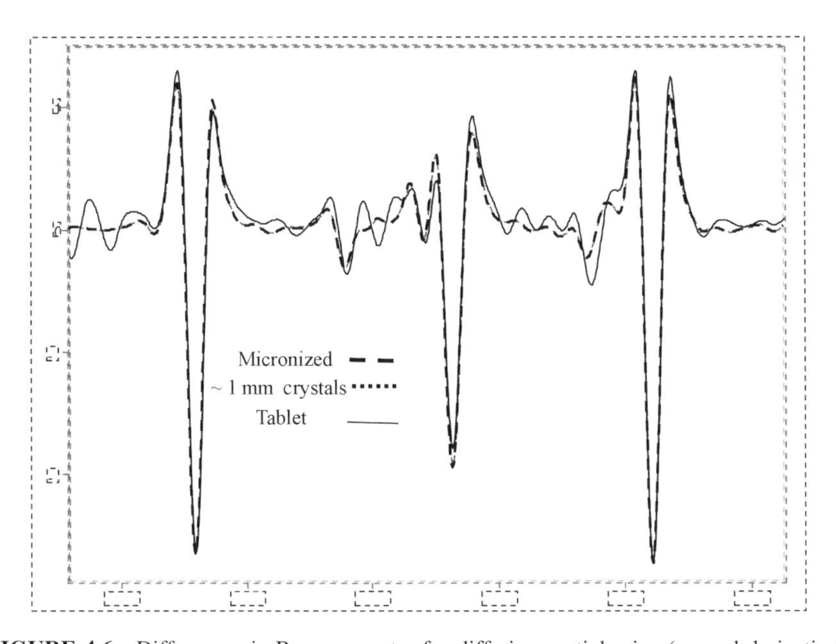

FIGURE 4.6 Differences in Raman spectra for differing particle size (second derivative).

frequency due to excipient contributions, the relative intensities remain quite constant from micronized API, to very large crystals, to formulated drug product. Although absolute intensity and depth of penetration are certainly affected by physical variations in the sample, in the case of polymorph quantitation the relative nature of the measurement tends to minimize the effect of these variables. The final judgment as to how important these effects are will depend on the degree of validation and the absolute accuracy required from the final measurement.

The most straightforward approach for calibration preparation is therefore to generate a set of standards as much like the sample as possible. If the dosage form is a tablet, then the calibration standards can be gravimetrically prepared, suitably mixed, and then pressed into tablets according to the formulation protocol. Capsules likewise can be prepared by mixing the appropriate components. Many more exotic formulations, such as sprayed coatings, extrusions, lyophilized cakes, and so on may induce variable amounts of polymorph during their manufacture. It may simply be impossible to generate reliable calibration standards that sufficiently mimic the drug product. In these cases, it may be necessary to rely on the analyst to create a reasonable justification for an imperfect calibration set, with the understanding that the quantitation provides answers that are proportional and precise, but with unverifiable accuracy.

4.7 QUANTITATION

Once an acceptable calibration set has been prepared, the analysis approach needs to be selected. This will depend on the degree of spectral difference of the polymorphs, as well as the desired limits of quantitation. Although often not explicitly stated, most experimental techniques have a fair amount of flexibility on the limit of quantitation obtained. Depending on what aspect of the polymorph needs to be monitored, the lowest LOQ may not be the desired result of the analysis. By selecting the acquisition time for the spectrum, the Raman bands for analysis, and the analysis method, it is fairly easy to select the limit of quantitation from a few tenths of a percent (under good conditions) to whatever may be desired. The very lowest limit of detection will likely involve the longest acquisitions and result in the least robust calibration model; alternatively, methods with a higher LOD can be made more robust and have shorter measurement times.

The selection of univariate versus multivariate analysis is also important, as the tools for development of multivariate analysis are becoming increasingly ubiquitous. If the polymorph spectra are sufficiently different, for example, there are well-resolved bands for either polymorph, then a univariate approach is an option. First and second derivative spectra are also commonly utilized (if sufficient signal to noise is present) not only as robust methods of background correction, but also to provide more narrow bands that may improve selectivity. Using appropriately selected band ratios should yield quite reasonable calibration results. The primary difficulty with a univariate approach is that, in spite of the previous arguments that Raman spectroscopy is less affected by particle size than other techniques, these

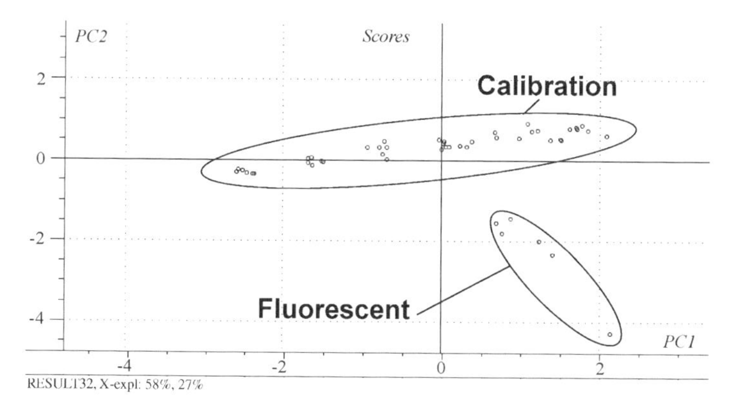

FIGURE 4.7 Differences in calibration model predication based on fluorescence interference.

variables cannot be completely removed from any solid-state measurement. The robustness of any univariate measurement must always be questioned, because no other parameters from the spectrum are being monitored. At the very least some sort of multivariate identity function (such as a Mahalanobis distance measurement) should be performed to confirm that the sample is sufficiently comparable to the calibration set that the univariate quantitation will be valid.

In almost every case, a multivariate approach will be superior to the univariate measurement in both robustness and limit of quantitation (Strachan et al., 2004). Numerous papers describe the methods for developing multivariate models (Dyrby et al., 2002; Niemczyk et al., 1998; Skoulika et al., 2001), and the controls necessary for obtaining the best results (for instance, low wavelength calibration drift) (Vogt et al., 2004). A good example of the necessity for multivariate analysis is given in Figs. 4.7 and 4.8. A multivariate calibration was prepared using various spectral regions intended to maximize polymorphic variance. However, as seen in Fig. 4.7, several samples were tested that had increased fluorescence compared

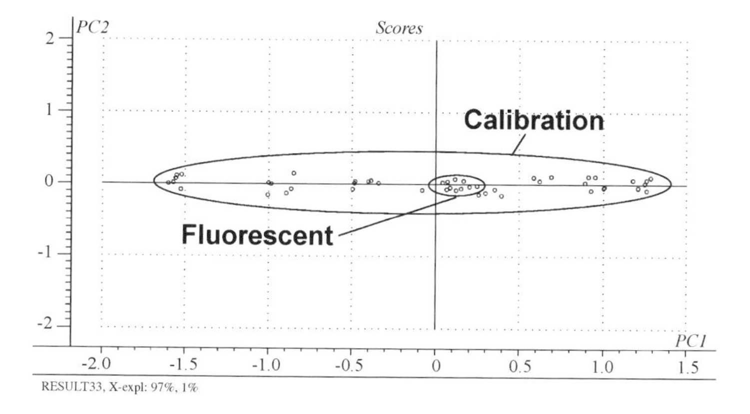

FIGURE 4.8 Calibration model adjusted to ignore influences of fluorescent interference.

to the calibration set. It is possible that the fluorescence is the primary cause of the spectral difference; however, it is also possible that other process parameters that gave rise to fluorescent impurities (such as increased temperature) caused these samples to have other physical/spectral differences. In either case, the samples should not be measured using this calibration model; if a univariate method alone was being used, this could have gone undetected and the results would be erroneous. By reducing the spectral region selected for the model, these fluorescent samples could be appropriately quantified (Fig. 4.8).

While the power of this sort of multivariate analysis is evident, widespread acceptance of it in the context of Raman spectroscopy analysis is still significantly lacking. For NIR, where multivariate methods are essentially required to obtain meaningful results, the application of these techniques is unquestioned, because there is no alternative. However, with Raman spectroscopy, there are often quite well-resolved bands that can give rise to HPLC-like quantitation routines, which are easy to calculate, easy to confirm, and very comforting for both the analyst and the reviewer. Because a choice is possible to not use multivariate techniques, it is much easier to not do so. Although theoretically all possible variables in the sample would be addressed during method validation, considering the calibration set constraints described above this is an unrealistic expectation for solid-state drug product polymorph analysis. Use of pure univariate methods, while feasible, provides no assurance of robustness or detection of outliers, and their indiscriminate application is strongly discouraged.

The quantitation schemes described above are generic for bulk measurements, where no spatial information is retained. Indeed, for this type of measurement uniformity of signal is paramount and unrepresentative sampling arising from sample inhomogeneity is unacceptable. Spectroscopic imaging, however, is premised on the fact that drug product is inhomogeneous; if the samples were homogeneous, there would be no reason to image them. From a quantitative standpoint, there are a variety of instances where mapping or imaging can provide better sensitivity and accuracy than traditional methods. Although we will focus on Raman spectroscopy similar advantages can be obtained using NIR or MIR imaging (within the context of the advantages/disadvantages discussed above). Since Raman imaging is covered in detail elsewhere in this book, this discussion will focus on the aspects particular to polymorph imaging.

The principal difference between macro-Raman and Raman mapping is the massively parallel amount of spectral data that is generated in an image. The spatial resolution is typically also very high for Raman mapping, and although single-point confocal measurements will have similar resolution, the typical macroquantitation aims to have the largest measurement area possible. This high spatial resolution means that the local spectral variance will be high; that is, individual pixels may contain highly pure spectra or combinations of only a small number of components (Henson and Zhang, 2006).

There are a variety of consequences for trace measurement. First, the limit of detection is completely altered from that of the bulk measurement. In a macromeasurement, the LOD is effectively limited by the noise in the measurement, which in turn is a function of the level of interference from the other species present in the sample. By spatially reducing the effect of interferents, the signal from a given region of analyte is maximized. Rather than looking for a spectral contribution of one part in 10,000, the quantitation becomes a matter of assessing the number of pixels containing pure (or nearly so) analyte spectra. The limit of quantitation is effectively limited only by the area which can be sampled, and the time available to perform the measurement. For a Raman spectroscopy measurement, this can be a fairly serious restriction. Mapping/imaging experiments can take days to perform if covering a wide area at high spatial resolution and reasonable signal to noise. Depending on the Raman cross section of the target polymorph and the expected particle size, NIR or MIR imaging may be the only practical approach.

Secondly, because the spectra should be more pure than those obtained in a macroexperiment, the restrictions on generating a very accurate calibration set may in some cases be relaxed. Especially, when using a line mapping system (see Chapter 6), which spatially discriminates in the *Z*-direction, the effect of particle size and multiple scattering is greatly decreased over a system that does not discriminate by depth (such as a global illumination mapping system). In this case, the calibration curve for the sample can be made virtually a priori by using the spectra of the pure components when an indication of sampling volume is known. Although this may not be as satisfactory from a regulatory standpoint, this form of direct measurement may in some cases yield superior results than relying on difficult-to-prepare and unreliable calibration samples.

It must also be noted that this technique is most powerful when the polymorphic impurity is expressed as a spatially localized particle. If the unique spectrum of the analyte can be measured as a single-seed crystal in the sample matrix, then astounding sensitivities can be achieved. Improved but not miraculous limits of detection can be obtained when the impurity is associated with the parent API. The spectrum of the analyte must be deconvoluted from that of the predominant API spectrum, but interference from the excipient matrix can be minimized. If the polymorphic impurity is distributed evenly throughout the sample and at a sufficiently small size so that its signal is mixed uniformly with the surrounding matrix, then imaging offers no improvement over bulk methods.

4.8 INTELLECTUAL PROPERTY

As mentioned in Section 4.1, it is essential to gain patent rights for as many polymorphs of a pharmaceutically active compound as possible. This can prevent competitors from potentially producing an alternate version of an otherwise protected drug product; it may also expand the range of possible development strategies. New solid-state forms may also aid in extending the period of exclusivity for a drug product. In recent years, several instances of patent litigation have involved crystal form issues.

Because intellectual property is such an important aspect of polymorphism, the use of Raman spectroscopy to characterize polymorphs also has important implications in this area. Although the use of Raman spectroscopy in the applications described above yields equal leverage for the IP owners as well as patent challengers, when discussing intellectual property specifically, Raman spectroscopy becomes a very powerful tool for staking out and protecting a property position rather than opening an area for competition.

For instance, the most basic tool for protection is the content of matter patent and associated crystal form patents. While a variety of techniques will be included in these patents to support the chemical and physical structure of the material, including MS, NMR, DSC, mid-IR, and ssNMR data and of course X-ray crystallography, Raman spectroscopy is not necessarily automatically included. This is likely due to the fact that Raman spectroscopy has historically been unnecessary for any sort of structural assignments, since infrared would be sufficient to assign general functional groups, and the crystallography would characterize the crystal structure. However, as described previously, it may be very difficult to use either of these techniques to unequivocally ascertain the presence or quantify the amount of a polymorph in drug product. The inclusion of the Raman spectrum in the initial patent greatly simplifies the use of this technique to support any defensive strategy showing infringement at a later date.

The scenario when this would be important is typically the following: Company X produces a product containing drug in crystal form Y, a protected crystal form. Company A produces the same drug using the less-stable crystal form B. With the production of any crystal of lower free energy, the likelihood of "contamination" with the more stable form is strong. Company X would like to determine if its protected polymorph is being produced in this other product, and to what extent. As described in Section 4.7, with the possible exception of ssNMR, Raman spectroscopy is very likely the most sensitive method for determining the presence of this crystal form in drug product, and infringement is simplified if the Raman spectrum is included in the patent documentation.

Several points need to be clarified however with regard to the use of Raman spectroscopy as supporting evidence for infringement. Raman spectroscopy is not equivalent to crystallography for determination of crystal structure. The X-ray data are conclusive, irrefutable, and specific characterization of physical form. Unfortunately, X-ray diffraction often does not work well in the complicated matrixes of a drug product, where most measurements detecting infringement must take place. While Raman spectroscopy, like mid-IR, can be used as conclusive ID for a pure chemical material, in the presence of potentially interfering materials specificity is harder to prove. Specificity in drug product may be supported by a variety of means: assignment of all other bands to known actives or excipients; correlation of multiple bands of the analyte polymorph; highly accurate frequency correlation for the analyte; and use of multivariate methods to characterize principal components. However, especially at very low levels where even weak bands from other materials may interfere, the supporting evidence for polymorphic specificity must be closely scrutinized to confirm that no other interpretation of the spectral signature could be supported.

The ability of Raman spectroscopy to quantify the amount of polymorph in a sample is also an important feature, although the amount of protected polymorph in a competing product that would indicate an infringement is still open to considerable debate. The absolute amount value may depend on a variety of factors. Most important is the difference in functionality between the two forms; for instance, does one form give a significant pharmacokinetic advantage versus the other form, and to what degree does the presence of the protected form affect this performance. The stability of the form also plays a role, in that the degree of infringement may increase if the principal polymorph converts to the protected form over time. The synthetic route is also important and may be reflected in the final product; even though a protected form may not be present at high levels in the final drug product, the presence of the protected form may indicate that it was used as part of an earlier crystallization or purification step, but subsequently recrystallized in another crystal form.

The many advantages of Raman spectroscopy for measuring drug product, its ability to measure materials without altering the sample, usually strong Raman intensities for API compared to excipients, the spatial selectivity, and limits of quantitation (often less than 1% of the drug loading; Henson and Zhang, 2006) make Raman spectroscopy a strong choice for litigation support.

REFERENCES

Airaksinen S, Luukkonen P, Jorgensen A, Karjalainen M, Rantanen J, Yliruusi J, 2003. Effects of excipients on hydrate formation in wet masses containing theophylline. *J. Pharm. Sci.*, **92**(3), 516–528.

Anderson JE, Moore S, Tarczynski F, Walker D, 2001. Determination of the onset of crystallization of N1-2-(thiazolyl)sulfanilamide (sulfathiazole) by UV–Vis and calorimetry using an automated reaction platform; subsequent characterization of polymorphic forms using dispersive Raman spectroscopy. *Spectrochi. Acta A Mol. Biomol. Spectrosc.*, **57A**(9), 1793–1808.

Andrews J, Dallin P, 2004. Heterogeneous mixtures and polymorphs—choose your weapon from the PAT armoury. *Spectrosc. Eur.*, **17**(2), 32–33.

Anquetil PA, Brenan CJH, Marcolli C, Hunter IW, 2003. Laser Raman spectroscopic analysis of polymorphic forms in microliter fluid volumes. *J. Pharm. Sci.*, **92**(1), 149–160.

Apperley DC, Fletton RA, Harris RK, Lancaster RW, Tavener S, Threlfall TL, 1999. Sulfathiazole polymorphism studied by magic-angle spinning NMR. *J. Pharm. Sci.*, **88**, 1275–1280.

Ashizawa K, 1989. Polymorphism and crystal structure of 2R,4S,6-fluoro-2-methyl-spiro[chroman-4,4′-imidazoline]-2′,5-dione (M79175)’. *J. Phar. Sci.*, **78**(3), 256–60.

Auer ME, Griesser UJ, Sawatzki J, 2003. Qualitative and quantitative study of polymorphic forms in drug formulations by near infrared FT-Raman spectroscopy, *J. Mol. Struct.*, 661–662 (1–3), 307–317.

Ayala AP, Siesler HW, Boese R, Hofffmann GG, Polla GI, Vega DR, 2006. Solid state characterization of olanzapine polymorphs using vibrational spectroscopy. *Int. J. Pharm.*, **326**, 69–79.

Blagden N, Davey RJ, Lieberman HF, Williams L, Payne R, Roberts R, Rowe R, Docherty R, 1998. Crystal chemistry and solvent effects in polymorphic systems sulfathiazole. *J. Chem. Soc. Faraday Trans.*, **94**(8), 1035–1044.

Blanco M, Valdes D, Llorente I, Bayod M, 2005. Application of NIR Spectroscopy in polymorphic analysis: study of pseudo-polymorphs stability. *J. Pharm. Sci.*, **94**(6), 1336–1342.

Burger A, Ramberger R, 1979. On the polymorphism of pharmaceuticals and other molecular crystals. I. *Mikrochim. Acta.*, **72**, 259–271.

Burger A, Ramberger R, 1979. On the polymorphism of pharmaceuticals and other molecular crystals. II. *Mikrochim. Acta.*, **72**, 273–316.

Calderon De Anda J, Wang XZ, Lai X, Roberts KJ, 2005. Classifying organic crystals via in-process image analysis and the use of monitoring charts to follow polymorphic and morphological changes. *J. Process Control*, **15**(7), 785–797.

Campbell Roberts SN, Williams AC, Grimsey IM, Booth SW, 2002. Quantitative analysis of mannitol polymorphs. FT-Raman spectroscopy. *J. Pharm. Biomed. Anal.*, **28**(6), 1135–1147.

Chemburkar SR, Bauer J, Deming K, Spiwek H, Patel K, Morris J, Henry R, Spanton S, Dziki W, Porter W, Quick J, Bauer P, Donaubauer J, Narayanan BA, Soldani M, Riley D, McFarland K, 2000. Dealing with the impact of ritonavir polymorphs on the late stages of bulk drug process development. *Org. Process Res. Dev.*, **4**(5), 413–417.

Cruickshank DWJ, 1958. Lattice vibrations of benzene, naphthalene, and anthracene. *Rev. Mod. Phys.*, **30**, 163–7.

Davis TD, Peck GE, Stowell JG, Morris KR, Byrn SR, 2004. Modeling and monitoring of polymorphic transformation during the drying phase of a wet granulation. *Pharm. Res.*, **21**(5) 860–866.

De Paepe AT, Dyke JM, Hendra PJ, Langkilde FW, 1997. Rotating samples in FT-Raman spectrometers. *Spectrochim. Acta A Mol. Biomol. Spectrosc.*, **53A**(13), 2261–2266.

Deeley CM, Spragg RA, Threlfall TL, 1991. A comparison of Fourier transform infrared and near-infrared Fourier transform Raman spectroscopy for quantitative measurements: an application in polymorphism. *Spectroch. Acta A Mol. Biomol. Spectros.*, **47A**(9–10), 1217–1223.

Dyrby M, Engelsen SB, Norgaard L, Bruhn M, Lundsberg-Nielsen L, 2002. Chemometric quantitation of the active substance (containing CN) in a pharmaceutical tablet using near-infrared (NIR) transmittance and NIR FT-Raman Spectra. *Appl. Spectrosc.*, **56**(5), 579–585.

Findlay WP, Bugay DE, 1998. Utilization of Fourier transform-Raman spectroscopy for the study of pharmaceutical crystal forms. *J. Pharm. Biomed. Anal.*, **16**(6), 921–930.

Fleischman S, Morales L, Moulton B, Rodríguez-Hornedo N, Zaworotko M, 2003. Crystal engineering of the composition of pharmaceutical phases. *Chem. Commun.*, **2**: 186–187.

Giordano F, Rossi A, Moyano JR, Gazzaniga A, Massarotti V, Bini M, Capsoni D, Zanol M, 2001. Polymorphism of Rac-5,6-diisobutyryloxy-2-methylamino-1,2,3,4-tetrahydronaphthalene hydrochloride (CHF 1035). I, Thermal. spectroscopic, and X-ray diffraction properties. *J. Pharm. Sci.*, **90**(8), 1154–1163.

Giordano F, Rossi A, Moyano JR, Gazzaniga A, Massarotti V, Bini M, Capsoni DD, Zanol M, (2007). Polymorphism of rac-5,6-diisobutyryloxy-2-methylamino-1,2,3,4-tetrahydronaphthalene hydrochloride (CHF 1035). I. thermal, spectroscopic, and X-ray diffraction properties. *J Pharm. Sci.*, **90**(8), 1154–1163.

Giron D, Goldbronn C, 1997. Use of sub-ambient DSC to complement conventional DSC and TG. The study of water adsorption of drug substances and excipients. *J. Therm. Anal.*, **49**(2), 907–912.

Haleblian JK, McCrone W, 1969. Pharmaceutical applications of polymorphism. *J. Pharm. Sci.*, **58**(8), 911–929.

Head T, Rydzak J, 2003. Chemometric models using diamond attenuated total Reflectance IR and Raman spectroscopy to characterize and quantitate polymorphs in pharmaceuticals. *Am. Pharm. Rev.*, **6**(1), 78–84.

Henson MJ, Zhang L, 2006. Drug characterization in low dosage pharmaceutical tablets using Raman microscopic mapping. *Appl. Spectrosc.*, **60**(11), 1247–1255.

Hughes DS, Hursthouse MB, Threlfall T, Tavener S, 1999. A new polymorph of sulfathiazole. *Acta Crystallogr. C Cryst. Struct. Commun.*, **C55**(11), 1831–1833.

Ip DP, Brenner GS, Stevenson JM, 1996. High resolution spectroscopic evidence and solution calorimetry studies on the polymorphs of enalapril maleate. *Int. J. Pharm.*, **28**(2–3), 183–191.

Jørgensen A, Rantanen J, Karjalainen M, Khriachtchev L, Raesaenen E, Yliruusi J, 2002. Hydrate formation during wet granulation studied by spectroscopic methods and multivariate analysis. *Pharm. Res.*, **19**(9), 1285–1291.

Kojima T, Onoue S, Murase N, Katoh F, Mano T, Matsuda Y, 2006. Crystalline form information from multiwell plate salt screening by use of Raman microscopy. *Pharm. Res.*, **23**(4), 806–812.

Krc J, Jr, 1977. Crystallographic properties of flufenamic acid. *Microscope.*, **25**(1), 31–45.

Kuhnert-Brandstatter M, 1974. Thermomicroscopy in the Analysis of Pharmaceuticals. Pergamon Press, Oxford.

Lagaron JM, 2002. On the use of a Raman spectroscopy band to assess the crystalline lateral packing in polyethylene. *J. Mater. Sci.*, **37**(19), 4101–4107.

Lang M, Grzesiak AL, Matzger AJ, 2002. The use of polymer heteronuclei for crystalline polymorph selection. *J. Am. Chem. Soc.*, 124, 14834–14835.

Loudon R, 1964. The Raman effect in crystals. *Adv. Phys.*, **13**(52), 423–82.

Lowry S, Dalrymple D, Song A, Rosso V, Pommier CJ, Venit J, 2006. Integrating a Raman microscope into the workflow of a high-throughput crystallization laboratory. *J. Assoc. Lab. Autom.*, **11**(2), 75–84.

Lu GW, Hawley M, Smith M, Geiger BM, Pfund W, 2006. Characterization of a novel polymorphic form of celecoxib. *J. Pharm. Sci.*, **95**(2), 305–317.

Luner PE, Patel AD, 2005. Quantifying crystal form content in physical mixtures of (±)-tartaric acid using near infrared reflectance spectroscopy. *AAPS PharmSciTech*, **6**(2) E245–E252.

Maurin MB, Vickery RD, Rabel SR, Rowe SM, Everlof JG, Nemeth GA, Campbell GC, Foris CM, 2002. Polymorphism of roxifiban. *J. Pharm. Sci.*, **91**(12), 2599–2604.

McCrone WC, *Physics and Chemistry of the Organic Solid State*, Vol. 2, Wiley Interscience, (1965), p. 725.

Mehrens SM, Kale UJ, Qu X, 2005. Statistical analysis of differences in the Raman spectra of polymorphs. *J. Pharm. Sci.*, **94**(6), 1354–1367.

Moynihan HA, O'Hare IP, 2002. Spectroscopic characterization of the monoclinic and orthorhombic forms of paracetamol. *Int. J. Pharm.*, **247**(1–2), 179–185.

Niemczyk TM, Delgado-Lopez MM, Allen FS, 1998. Quantitative determination of bucindolol concentration in intact gel capsules using Raman spectroscopy. *Anal. Chem.*, **70**, 2762–2765.

O'Sullivan B, Barrett P, Hsiao G, Carr A, Glennon B, 2003. *In situ* monitoring of polymorphic transitions. *Org. Process Res. Dev.*, **7**(6), 977–982.

Okumura T, Otsuka M, 2005. Evaluation of the microcrystallinity of a drug substance, indomethacin, in a pharmaceutical model tablet by chemometric FT-Raman spectroscopy. *Pharm. Res.*, **22**(8), 1350–1357.

Ono T, Ter Horst JH, Jansens PJ, 2004. Quantitative measurement of the polymorphic transformation of L-glutamic acid using *in situ* Raman spectroscopy. *Cryst. Growth Des.*, **4**(3), 465–469.

Paul SO, Schutte CJH, Hendra PJ, 1990. The Fourier transform Raman and infrared spectra of naphthazarin. *Spectroch. Acta A Mol. Biomol. Spectrosc.*, **46A**(2), 323–9.

Prasad PN, 1979. Raman study of molecular motions in organic solids. *Mol. Cryst. Liq. Cryst.*, **52**(1–4), 367–79.

Q6A International Conference on Harmonisation; Guidance on Q6A Specifications: Test Procedures and Acceptance Criteria for New Drug Substances and New Drug Products: Chemical Substances., Federal Register, 2000, **65**(251), 83041–83063.

Raghavan K, Dwivedi A, Campbell GC Jr, Johnston E, Levorse D, McCauley J, Hussain M, 1993. A spectroscopic investigation of losartan polymorphs. *Pharm. Res.*, **10**(6), 900–904.

Sacchetti M, 2001. Thermodynamic analysis of DSC data for acetaminophen polymorphs. *J Therm. Anal. Calorin.*, **63**(2), 345–350.

Sheikhzadeh M, Rohani S, Jutan A, Manifar T, Murthy K, Horne S, 2006. Solid-state characterization of buspirone hydrochloride polymorphs. *Pharm. Res.* **23**(5), 1043–1050.

Skoulika SG, Georgiou CA, 2001. Rapid quantitative determination of ciprofloxin in pharmaceuticals by use of solid-state FT-Raman spectroscopy. *Appl. Spectrosc.*, **55**(9) 1259–1265.

Starbuck C, Spartalis A, Wai L, Wang J, Fernandez P, Lindemann CM, Zhou GX, Ge Z, 2002. Process optimization of a complex pharmaceutical polymorphic system via *in situ* Raman spectroscopy. *Cryst. Growth Des.*, **2**(6), 515–522.

Strachan CJ, Pratiwi D, Gordon KC, Rades T, 2004. Quantitative analysis of polymorphic mixtures of carbamazepine by Raman spectroscopy and principal components analysis. *J. Raman Spectrosc.*, **35**(5), 347–352.

Strobl GR, Hagedorn W, 1978. Raman spectroscopic method for determining the crystallinity of polyethylene. *J Polym. Sci. Polym. Phys. Ed.*, **16**(7), 1181–93.

Taday PF, Bradley IV, Arnone DD, Pepper M, 2003. Using terahertz pulse spectroscopy to study the crystalline structure of a drug: A case study of the polymorphs of ranitidine hydrochloride. *J. Pharm. Sci.*, **92**(4), 831–838.

Takahashi H, Takenishi T, Nagashima N, 1962. Quantitative analysis of mixtures of L-glutamic acid polymorphs by x-ray diffraction. *Bull. Chem. Soc.*, **35** 923–926.

Taylor LS, Langkilde FW, 2000. Evaluation of solid-state forms present in tablets by Raman spectroscopy. *J. Pharm. Sci.*, **89**(10), 1342–1353.

Turrell G, 1972. Infrared and Raman Spectra of Crystals. Academic Press, New York. p. 60.

Van Hoof P, Lammers R, Puijenbroek RV, Schans MVD, Carlier P, Kellenbach E, 2002. Polymorphism of the CNS active drug org 13011: the application of high temperature analysis to detect new polymorphs. *Int. J. Pharm.*, **238**(1–2), 215–228.

Vassou D, Gionis V, Chryssikos GD, 2005. Glassy drugs: a Raman investigation of binary dihydropyridine systems. *Phys. Chem. Glasses*, **46**(2), 144–147.

Vehring R, 2005. Red-excitation dispersive Raman spectroscopy is a suitable technique for solid-state analysis of respirable pharmaceutical powders. *Appl. Spectrosc.*, **59**(3), 286–292.

Vogt F, Steiner H, Booksh K, Mizaikoff B, 2004. Chemometric correction of drift effects in optical spectra. *Appl. Spectrosc.*, **58**(6), 683–692.

Wang F, Wachter JA, Antosz FJ, Berglund KA, 2000. An investigation of solvent-mediated polymorphic transformation of progesterone using *in situ* Raman spectroscopy. *Org. Process Res. Dev.*, **4**(5), 391–395.

Wikström H, Marsac PJ, Taylor LS, 2005. In-line monitoring of hydrate formation during wet granulation using Raman spectroscopy. *J. Pharm. Sci.*, **94**(1), 209–219.

Wu, T-M, Liao, C-S, 2000. Polymorphism in nylon 6/clay nanocomposites. *Macromol. Chem. Phys.*, **201**(18), 2820–2825.

Zeitler JA, Newnham DA, Taday PF, Threlfall TL, Lancaster RW, Berg RW, Strachan CJ, Pepper M, Gordon KC, Rades T, 2006. Characterization of temperature-induced phase transitions in five polymorphic forms of sulfathiazole by terahertz pulsed spectroscopy and differential scanning calorimetry. *J. Pharma. Sci.*, **95**(11), 2486–2498.

5

RAMAN SPECTROSCOPY FOR MONITORING REAL-TIME PROCESSES IN THE PHARMACEUTICAL INDUSTRY

KEVIN L. DAVIS, MARK S. KEMPER and IAN R. LEWIS
Kaiser Optical Systems, Inc., 371 Parkland Plaza, Ann Arbor, MI 48103, USA

5.1 INTRODUCTION

This chapter reviews the application of Raman spectroscopy as a tool for process analytical technology (PAT). The initial sections outline the basics of Raman spectroscopy, its instrumentation, and provide an overview of Raman process analysis. In the later sections, coverage of select applications where Raman spectroscopy has proven to be an appropriate tool are provided. Section 5.9 incorporates specific case studies that illustrate to the new practitioner and analytical user of Raman spectroscopy the applicability and benefits of Raman spectroscopy for process understanding, analysis, monitoring, and control.

5.2 A BRIEF HISTORY OF RAMAN SPECTROSCOPY

In 1928, Professor Raman and, his student, Krishnan published their manuscript in Nature heralding the experimental discovery of "a new type of secondary radiation" (Raman and Krishnan, 1928). In 1930, Professor C.V. Raman was awarded the Nobel Prize for this discovery and the technique that would bear his name was born.

Pharmaceutical Applications of Raman Spectroscopy, Edited by Slobodan Šašić

From the time of its discovery until the Second World War, when the first commercial dispersive mid-infrared (MIR) spectrometer was built by Perkin-Elmer, Raman spectroscopy maintained a position as the predominant vibrational spectroscopic technique—due to the simplicity of the experimental instrumentation. Following the commercialization of infrared spectroscopy, MIR took that role. Since then, Raman spectroscopy has undergone two notable periods of user expansion both driven by instrumental developments; the first in the 1960s coincided with the development of the double monochromator, the laser, and electronic methods of signal detection. The second and current period started in 1986 and is still ongoing.

From 1944 until approximately 1988, Raman spectroscopy was used primarily for research purposes by experts trained in the technique. Large companies, PhD granting academic institutions, and elite government laboratories typically had one or perhaps two Raman experts on staff. Customized Raman instrumentation was built from components by these experts and optimized for specific applications. Laser laboratories were set up with precautionary warning signs and interlocks were affixed to the doors to avoid laser exposure to those who wandered by mistake into the laboratories. In addition, because instrumentation was not in an advanced state of commercialization, consistent Raman measurements were not readily accomplished. These factors contributed to the impression, even to the average colleague within a company, that Raman spectroscopy laboratories were mysterious places where the inner workings involved some form of "black art." Because of this, the cost of the instrument components, and the safety issues inherent with operating a Class IV laser in the open, Raman spectroscopy was not practiced to any great degree in the analytical or QC laboratory. Also many academic instructors in analytical instrumentation courses failed to mention Raman spectroscopy as a technique of any importance. It was considered an exotic curiosity practiced only by a few specialists and without any mainstream applications. Thus, new generations of scientists and engineers were not exposed to the potential benefits of Raman spectroscopy.

Over these last 20 years, four major developments have occured in the field of Raman spectroscopy; the advent of laser radiation blocking filters [particularly holographic notch filters (Owen, 2002)], the successful demonstration and quick acceptance of FT-Raman spectroscopy (Chase and Rabolt, 1994; Hendra et al., 1991; Hirschfeld and Chase, 1986), the coupling of spectroscopic grade charge coupled device (CCD) (array) detectors to dispersive Raman spectrometers (Bilhorn et al., 1987), and the development and adoption of compact lasers such as the NIR diode laser (Wang and McCreery, 1989; Williamson et al., 1989) and the diode-pumped solid state (Nd:YAG) laser (both at 1064 nm and frequency doubled to 532 nm).

These component developments coupled with advances in sampling (especially microscopy (Turrell and Corset, 1996), and fiber optic sampling (Lewis and Griffiths, 1996, 2002; Slater et al., 2000) have led, over the last 12 years, to an advanced state of commercialization and maturity of analytical Raman instrumentation. Consistent Raman measurements from the laboratory to the process line are now readily accomplished (Adar et al., 1997; Lewis, 2000; Everall et al., 2002; Jestel, 2005). A consequence of these advances has been a dramatic increase in the number of successful application areas for Raman spectroscopy and a broadening of the practitioners from university and government laboratories, to academics in nontraditional Raman fields, industrial problem solvers, analytical chemists, as well as process and control engineers, and quality control technicians.

5.3 BASIC THEORY OF RAMAN SPECTROSCOPY

Raman spectroscopy provides vibrational information about the chemical makeup of the sample under investigation. When a high energy photon interacts with a molecule (or crystal lattice), there is a small, but finite possibility of occurence of Raman scattering (Chase and Rabolt, 1994; Hendra et al., 1991; Lewis and Edwards, 2000). During this interaction, part of the photon energy may be retained by the molecule causing vibrational excitation of the molecule. Most commonly, a photon is emitted whose energy, compared to the incident photon, is reduced by this vibrational energy quantum and is thus shifted to a longer wavelength, that is, Stokes shifted. A lower probability event is that the emitted photon will have energy increased by the vibrational energy quantum and is shifted to shorter wavelength, that is, anti-Stokes shifted. Each molecular vibration is associated with a characteristic energy and therefore produces a characteristic frequency shift during a Raman interaction.

To observe a Raman spectrum, the sample must be excited with a monochromatic light source. The scattered light is then dispersed (separated by wavelength) to reveal a spectrum where each spectral peak or band corresponds to a specific molecular vibration (e.g., C–H stretch, CH_2 bend). Most commonly, only the long wavelength Stokes shifted radiation is studied.

Raman spectroscopy is a valuable technique for studying molecular vibrations, whether in the laboratory or in a manufacturing setting. There are several reasons for this. First, as with Fourier transform infrared (FTIR) spectroscopy, Raman detects fundamental molecular vibrations. The result is the production of sharp peaks that can often be unequivocally assigned to specific chemical species and can yield quantitative information. Thus, it is relatively simple to construct robust Raman models to derive quantitative compositional data. The ability to follow and interpret the loss or formation of chemical bonds provides valuable information to the synthetic organic chemist and scale-up scientist. This information helps them to understand mechanisms and pathways of synthetic processes or the appearance of by-products generated from ineffective stirring or other phenomena. Better process understanding leads to the potential for better process control. Similarly, for pharmaceutical scientists, Raman can provide useful information that leads to better understanding of secondary processes, such as granulation, blending, and coating, which ultimately produce final products. This chemical specificity is one of the key benefits that Raman spectroscopy has over "trending" process analysis tools such as near-infrared (NIR) spectroscopy.

5.4 GENERAL INSTRUMENTATION FOR RAMAN SPECTROSCOPY

A Raman analyzer is composed of four basic components: a laser, an optical sampling system, a wavelength separator, and a detector. The laser, spectrometer, and detector hardware are typically packed together into a "base unit." The optical sampling system includes a means for illuminating the sample with laser light and collecting the Raman scatter for input to the spectrometer.

5.4.1 Lasers

Table 5.1 identifies several lasers used for pharmaceutical applications. The most common lasers are the NIR type at 1064 and 785 nm, although visible lasers may be used for some API research applications.

UV lasers are also available. However, they are not generally used for analytical work. Their use generally results in low resolution spectra and a significant risk of damage for solid samples.

There are several considerations that must be accounted for when choosing a laser for a particular application. The first is that Raman efficiency increases by a fourth-order function as the laser frequency is decreased (Raman efficiency $\propto 1/v^4$). Hence, the lower the laser wavelength, the greater the Raman signal will be. This first consideration is mitigated by the second, however, in that lower wavelength lasers increase the risk of sample fluorescence. This is often an intractable problem limiting the utility of visible lasers. The third consideration is that of sample burning, and this is also more prevalent with visible than with NIR lasers. Ultimately, the choice of laser is tailored to the application.

TABLE 5.1 Typical Lasers Used in Pharmaceutical Applications.

Laser (λ, nm nearest whole number)	Type	Typical power at laser	Wavelength range, (nm; Stokes region, 100 to 3000 cm^{-1} shift)	Comments
NIR Lasers 1064	Solid state (Nd:YAG)	Up to 3 W	1075–1563	Commonly used in Fourier transform instruments
785	Diode	Up to 500 mW	791–1027	Most ubiquitous dispersive Raman laser
Vis Lasers 488–632.8	Ion gas and solid state frequency doubled lasers	Up to 1 W	488–781	Fluorescence risks

5.4.2 Optical Sampling

The sampling system may be optically interfaced to the base unit (containing the laser, spectrometer, and detection system) either by direct coupling or via fiber optics, the latter being particularly preferred for PAT manufacturing settings. Some of the more common sampling arrangements are microscopes, fiberoptic-based probes (either noncontact or immersion optics), and sample chambers (including a variety of sample holders and automated sample changers). These devices may be designed to permit observation of polarized Raman spectra. The use of polarizers can provide valuable information about the physical properties of the samples (McCreery, 2000).

5.4.3 Spectrometer

The function of the spectrometer is to separate the wavelengths that comprise the polychromatic Raman signal and present them to the detector. The most common classes of spectrometers are based on either dispersive spectrograph or Fourier transform (FT) spectrometer technologies (Chase and Rabolt, 1994; Hendra et al., 1991; Parker, 1994). Spectrographs typically using gratings to angularly separate the radiation into its individual wavelength components prior to imaging and, thus, measurement at the detector. In an FT-Raman spectrometer, the light to be analyzed is passed through a scanning interferometer, normally a Michelsen interferometer, to generate a temporal signal on a single detector. The Fourier transform of that temporal signal corresponds to the wavelength spectrum of the input light. A more detailed review of the theory of operation of FT-Raman spectrometers is beyond the scope of this chapter but can be found in the following references by Hendra et al., (1991), Chase and Rabolt (1994), and Parker (1994).

5.4.4 Detector

The choice of detector is coupled to the type of technology that is used to separate the wavelengths in the Raman spectrum (dispersive or FT). The CCD detector is nearly universal in dispersive spectrographs. It is used because of the fact that spectral data can be imaged to the detector. This permits instruments to be designed with a minimal number of moving parts. Composed of silicon, CCDs are sensitive to the detection of light in the visible and low-wavelength NIR region of the spectrum.

FT-Raman spectrometers, typically configured with a 1064-nm laser, are equipped with either a germanium (Ge) detector or an indium gallium arsenide (InGaAs) detector, both of which are optimized for measurement of NIR light. Ge detectors are more sensitive but have an innate inconvenience of requiring constant cooling with liquid N_2. InGaAs detectors perform adequately with good linearity but are not as sensitive as Ge detectors.

5.5 THE CHOICE—DISPERSIVE OR FT?

The availability of these two fundamentally different architectures poses a question to the user: Which analyzer should I choose?

The FT-Raman approach, when introduced, offered a number of advantages over dispersive instruments using visible excitation. These include the well-known FT-advantages of Jacqinot, Fellget, Connes as well as NIR excitation to minimize organic fluorescence. They also offered improved data analysis software and superior instrument packaging. FT-Raman spectrometers typically have no user adjustable parts with the exception of a sample positioning device while providing a Class 1 laser-safe turn-key bench-top package. However, FT-Raman spectrometers contain inherently more delicate moving hardware items, which are vibration sensitive, and are therefore most suited to the laboratory environment.

Dispersive analyzers have now largely resolved the packaging, software, and safety concerns mentioned above. 785 nm excitation also addresses a number (but not all) of the fluorescence issues. Comparisons between these architectures have been made by Bowie et al. (2000a,b) for laboratory settings and Gervasio and Pelletier, (1997) and Everall et al. (1995) for process applications. Given that many online or in-line process applications of Raman spectroscopy involve moving, flowing, turbulent and heterogeneous materials, multiplex noise is a significant concern for FT-based analyzers. Gervasio and Pelletier compared NIR-excited dispersive and FT-Raman analyzers for online production of phosphorus trichloride where the impact of bubbling and particulate movement is reported. This point coupled with the extensive range of fiber optic based sampling options have led to dispersive analyzers taking a lead role in online and *in situ* PAT applications.

5.6 PROCESS ANALYSIS AND PAT

In late 2002, the FDA, following numerous open discussions with the industry, published a draft PAT guidance document (US FDA PAT Draft Guidance, 2002). The draft guidance was followed by an enhanced guidance document that also encompassed an extension for biotechnology in 2004 (US FDA PAT Guidance, 2004). At the heart of this initiative for the pharmaceutical industry is a push for quality to be built in/by design rather than testing for quality. For the last 25 years (prior to the PAT initiative), pharmaceutical companies would develop a manufacturing process, run test batches, document the results as a recipe book, provide the FDA the book, and always follow the book. This approach would be fine if pharmaceutical yields were high and product consistency was excellent. However, pharmaceutical yields are typically in the range of 20–25%. Unfortunately, consistency is no better today than when the product was commercialized, and drug recalls can and do occur. In searching for a solution to the yield and quality problems, it was noted that the petrochemical and chemical businesses have struggled with their own business issues and have found a solution that increased production, quality, and allowed the cost per unit of product to decrease. This solution involved process analytics—and included steps to analyze, understand, monitor, and ultimately control the process. In addition, where regulatory obstacles could have been encountered (as was the case with the US EPA for gasoline refining), regulatory acceptance of process analytical tools and embracing a "Win–Win" approach have allowed significant cost reductions, reduced waste, reduced pollution, and increased yields. In view of this, the historical FDA approach to quality has now been recognized to stifle manufacturing innovation via regulatory "fear" and has lead to the mantra that quality cannot be tested into a product.

There are several objectives for the PAT initiative identified in the guidence document. These include:

1. to facilitate process understanding (manage variability),
2. to continuously improve or correct processes (time, automation, etc.),
3. to facilitate the development of risk mitigation strategies,
4. to build a knowledge database,
5. to identify where routine monitoring is needed,
6. to facilitate the possibility of continuous processing,
7. to facilitate mechanistic-based regulatory specifications,
8. to provide a framework for possible real-time release of pharmaceuticals,
9. to provide a mechanism to save money by eliminating nonoptimal practices that lead, in worst-case scenarios, to product rejection.

The pharmaceutical industry is currently focusing much attention on applying PAT tools and methods to current drug substance and drug product manufacturing in order to better understand and control production of today's drugs. In 2006, the FDA started to promote the widespread use of the term "Quality by Design" (acronym QbD) as the "rallying cry" for pharmaceutical manufacturing innovation and improvement. This change was made in order to better emphasis the control aspect of the ultimate embodiment of a PAT analyzer. PAT is best thought of as a subset, albeit an important one, of a QbD approach. The future of PAT is not, however, to apply PAT tools to a finalized production methodology (this would satisfy some of the PAT/QbD "take homes") but to utilize PAT tools from R&D, to optimization, to the Pilot plant and into manufacturing (Arrivo, 2003). PAT tools are used to better understand the process and, when understood, the design can both be optimized to yield the best process possible and controlled to ensure the best quality. As data from the full-scale manufacturing plant is generated, the process can be refined based on the information learned. The refinements to the process could, if appropriate, lead to a tightening of product specification. In doing this, the true spirit of the PAT guidance is captured.

There are two main locations where PAT analyzers are being evaluated, these locations are laboratory-based or process-based and are given in Table 5.2. The initial goal of implementing PAT analyzers is the same for both locations (laboratory and plant), that is, in both cases to provide information and thus understanding about the process of interest. However, the perspectives and requirements are often

TABLE 5.2 Possible PAT Locations.

Off-line	Laboratory (QC or Analytical)
At-line	In production area, analysis during production close to the manufacturing process
Online	Analysis on diverted stream connected to process
In-line	Process stream disturbed/nondisturbed
	Lab-Pilot–Plant scale-up
	Batch
	Continuous

different. In the laboratory, the measurement is typically not a real-time measurement and analyzers are selected because the time dependence is not critical. The results can be used as a process quality monitor on the past history of the process. In the case of production area measurements, the results are used in many cases to control processes through feedback in real time or to diagnose and explain process anomalies. These perspectives highlight two major differences between the two type of PAT analyzer locations. These differences are (1) sampling and (2) the rate of data collection.

5.7 WHY CHOOSE RAMAN AS A PAT TOOL? THE NEED FOR RAMAN

Emil Ciurczak (private communication, 2002) accurately assessed the past status of implementing new technology in the pharmaceutical industry:

> . . . when testing the final product, analysts are encouraged to fall back on USP/NF methods . . . despite the fact that (these) methods are often not specific and are usually more time-consuming than modern spectroscopic techniques... Despite these shortcomings, HPLC assays are clung to as if they were engraved in stone. . . All QA/QC departments. . .are familiar with the mantra of HPLC development and have become comfortable with what is expected. . .HPLC results are assumed to be perfect (when compared to new techniques). . . The reason we don't use newer technologies is that we fear that FDA won't accept them because of the agency's unfamiliarity with them. But the reason FDA is unfamiliar with newer technologies is that companies don't submit them. . .We owe it to our customers to use the best and fastest techniques to analyze our products.

This unwanted axiom that had become practice in the industry was acknowledged by the FDA stated in their draft guidelines (US FDA PAT Draft Guidance, 2002): ". . . the pharmaceutical industry generally has been hesitant to introduce new technologies and innovative systems into the manufacturing sector . . . one reason often cited is *regulatory uncertainty,* which *may* result from the perception that our existing regulatory system is rigid and unfavorable to the introduction of new technologies." The agency is attempting to alleviate concerns of this from hampering efforts with PAT. Though Raman represented "new" technology, the agency is more open to application of newer analytical devices than in the past (as can be surmised by the inclusion of Raman in the guidance as a possible PAT tool).

In spite of these assurances, one concern of industrial practitioners remains the validation of new technologies. However, a scheme has been proposed to circumvent major difficulties in this respect. This relates to appropriate perspective regarding the implementation of Raman and other newer techniques to QbD situations utilizing PAT tools. Industry and the FDA have agreed in principle to treat PAT methods used for-information-only, that is, not used to make regulatory decisions, differently than analytical methods used for release. This is a positive development for many situations in which Raman would be applied. In most cases, the information in a PAT application is used to help understand, optimize, or control the process. It is not used for release. Since only 2–5% of the cost of goods for pharmaceuticals is wrapped up in release testing, there is little incentive to use a PAT technique for regulatory decisions. In addition, Raman can be used in a noncontact mode, making validation of the technology for a PAT method more straightforward since it is not invasive to the process.

Pelletier made a particularly perceptive statement germane to the routine practice of Raman spectroscopy recently in a published article (Pelletier, 2003): "...improvements in instrumentation have allowed users to focus on their applications rather than on the operation and limitations of the instrument..." In past years, the state of Raman was such that only trained experts in the field of Raman could practice the technique. Instrumentation in the mid- to late-1990s advanced to the point at which the technique became available to a nonexpert. This has opened the door for the use of Raman in routine situations by those not specifically trained in the technique. Some of the benefits of applying Raman technology are driving its continued acceptance. These benefits can be summarized in three categories: (1) process understanding, (2) process optimization, and (3) process monitoring.

Perhaps the greatest benefit that Raman provides is the insight into the chemical processes, leading to a better understanding of the chemistry that occurs in each situation. This advantage of Raman is related to the copious information content in Raman spectra. Process understanding is a key goal of the PAT initiative. If used in development and scale-up operations, Raman has the potential to give workers valuable insight into the chemistry, measure reaction rates, percentage conversion to the product, and the presence of intermediates. In addition to insight, Raman has the potential to give proper controls for their individual processes. This allows beneficial management of variability when the process is implemented in production. Management of variability leads to the mitigation of the risk of producing product that is out-of-specification. This satisfies the regulatory requirement that pharmaceutical companies demonstrate control using validated processes.

Raman can also be used to *optimize* the timing of the process stages and the endpoint. When choices of various parameters must be made, Raman data can indicate which set of parameters help achieve the desired end-status for a process. For example, when choosing a catalyst for a hydrogenation reaction, Raman can potentially be used to assess which one facilitates the optimal kinetics and the optimal yield as well as to diagnose whether there are detrimental by-products formed during the process. Such knowledge is critical in choosing the ultimate conditions, reagents, ultimately implemented for production.

The obvious benefit derived from Raman for PAT applications is the potential for real-time process monitoring. Quantitative and qualitative trends can be assessed and adjustments can be made quickly during routine processing based on Raman analyses. Not every process needs to be monitored. In other words, just because they can be monitored does not mean that they should be monitored. The benefit of routine monitoring should be demonstrated. If processes are properly understood, adequate controls should allow management of the variability in many cases to the point that only simple parameters such as temperature and pH need be routinely monitored. However, in those cases in which process excursions are possible, qualitative, and quantitative chemical insight may be beneficial. Raman data can provide such insight. Because of the possibility for fiber optic coupling, Raman is a convenient tool for such a purpose.

The reason Raman spectroscopy has received the level of attention it currently has can be summarized in nine points.

1. The Raman spectrum contains information on the fundamental vibrational modes of the sample under investigation, thus allowing unequivocal chemical identification (similar to IR). Raman is a selective technique providing well-resolved information leading to better process understanding.
2. Homonuclear diatomic analytes, such as H_2, N_2, and O_2, can be measured. The capability of measuring headspace gases in reaction vessels or in product containers (Gilbert et al., 1994; Powell and Campion 1986), such as vials of lyophilized products, has been illustrated.
3. Raman can be effectively employed within an aqueous medium. This arises from the weakness of the Raman signature for water. This stands in stark contrast to techniques such as mid-IR and NIR in which water presents significant interference.
4. Measurement of various physical states (liquids, slurries, pastes, solids, powders, etc.) are possible.
5. In the analytical laboratory, sample preparation is generally not necessary; hence, samples can be measured as is. That is, samples can be measured nondestructively.
6. *In situ* sampling through glass and certain packaging containers/materials can be achieved through the confocal capability of Raman optics. It is possible to measure Raman spectra *in situ* because glass and many transparent polymers do not generate intense Raman signatures.
7. Raman allows sampling flexibility. Samples can be measured remotely using fiber optic interfaces.
8. Modern Raman instrumentation is easy to use. An "expert" is no longer required to perform Raman analyses. Fast measurements on the order of a few seconds to 2 or 3 min are possible.

9. Simple calibration models are viable due to the well-resolved information available in a Raman spectrum. Because of the high degree of spectral density (many peaks are assignable to specific functional groups), simple univariate models using peak area, peak height, or even peak position can be constructed (Shaver, 2002).

5.7.1 Raman PAT Analyzers

In Fig. 5.1, the basic pharmaceutical areas where Raman analyzers can be applied are highlighted. These areas can be served by two fundamentally different types of analyzers; standard laboratory or process optimized. The different types of analyzers are mapped onto the pharmaceutical areas highlighted in Fig. 5.1. Off-line and at-line, including QA/QC opportunities, for PAT are important areas of industrial interest, and these areas in general can be adequately addressed with standard laboratory Raman instruments including FT-Raman spectrometers and dispersive Raman microscopes. The remainder of the PAT opportunities in the industrial environment, including in-line and on-line opportunities, can only be implemented using a true process Raman analyzer.

Despite the different requirements imposed by the installation location, all PAT analyzers must meet requirements for laser safety, electrical emission, appropriate laser-specific site certification, and appropriate government regulations such as the EU ATEX regulations. Subsequent to the introduction of the draft PAT Guidance document FDA representatives clarified the requirement for PAT analyzers to meet the 21 CFR Part 11 guidelines. It was stated that, for PAT evaluation purposes, an analyzer that does not meet the 21 CFR Part 11 guideline could be used. However, for implementation of a routine PAT-based control method a compliant

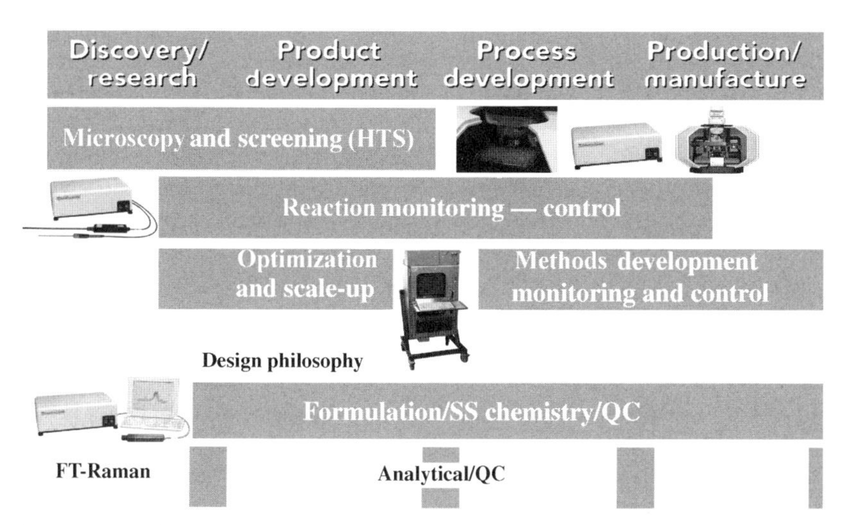

FIGURE 5.1 Schematic of where Raman analyzers can be applied to the pharmaceutical areas. (Reproduced with permission. © 2006 Kaiser Optical Systems Inc.)

analyzer should be used. Finally, the analyzers selected should be both validateable and "fit-for-purpose," that is, designed and tested to meet the environmental, longevity, and robustness requirements of the installation site.

For pharmaceutical applications, a stabilized NIR diode laser (e.g., 785 nm) is generally the preferred laser source, as this has been shown to virtually eliminate background fluorescence in many samples while still allowing efficient detection by the silicon-based CCD detector. For these reasons, the preferred Raman analyzer approach for online or in-line analysis is to use the dispersive approach, although it should be noted that a few older reports of the use of FT-Raman spectrometers for online analysis have been published (Bauer et al., 2000; Claybourn et al., 1994; Everall et al., 1995; Farquharson and Simpson, 1992; Feld et al., 1993,1994; Freeman et al., 1993; Garrisson et al., 1992,1993; Gervasio and Pelletier, 1997; Roberts et al., 1991; Sawatski et al., 1998).

5.7.2 Off-Line and At-Line Analyzers Based on Laboratory Instruments

Analyzers for these areas are generally selected by the customer to meet a number of important criteria including the analyzer package (size and aesthetics), the availability of sampling devices and accessories, the availability of after-sales support and the serviceability of the unit. These criteria are similar to those used for selection of other types of laboratory instruments. Instruments equipped with microscopes and standard sample stages for universal sample adaptation are the typical choices for laboratory instruments.

5.7.3 In-line and On-line Analyzers Based on Ruggedized Process Instrument

To a process engineer, Raman offers sampling flexibility, this flexibility comes about in two distinct parts. One part is the interface of the analyzer to the sampling location, while the second part is the interface of the analytical tool to the process For online and in-line applications of Raman spectroscopy, the analyzer base unit, containing the excitation and detection system is, interfaced to the sampling location using conventional telecommunication grade optical fiber cables (Lewis and Rosenblum, 2002). This enables remote sampling with a fiber-coupled spectrometer at tens or even hundreds of meters from the spectrometer.

The analyzer base unit must be packaged suitable with typically choices being a rack-mounted analyzer for nonhazardous and a purged cabinet design for classified locations (Fig. 5.2a). At classified locations, analyzers are typically housed in stainless steel enclosures that allow them to be cleaned using a standard hose down procedure. These analyzers, at minimum, feature a certified purge system (ATEX or Z- or X-type rated) when they are for installation in a classified location. However for installation in manufacturing the EU, compliance and marking to the ATEX directive is a requirement. Warman and Hammond (2005) noted that the lack of ATEX compliance could seriously limit the potential of Raman spectroscopy for installation in manufacturing. In late 2005 the first Raman analyzer that was

FIGURE 5.2 Picture of a RamanRxn3 ATEX-certified Raman analyzer and ATEX-certified Pilot-E online Raman probe. (Reproduced with permission. © 2006 Kaiser Optical Systems, Inc.)

certified to comply with ATEX appeared (Fig. 5.2). This analyzer design can be supplied certified for any of three configurations, for samples requiring the IS limit, for samples that are nonhazardous but whose headspace is classified, and for samples that are nonclassified but the base unit and cables are located in or traverse an explosive/classified environment.

There are two common types of commercial optics (Slater et al., 2000). High pressure immersion optics, where the collection lens is located behind a chemical-resistant Raman-transparent window, allows measurements involving sample contact with the sampling device. Remote "telescope" optics, by contrast, allow noncontact sampling. Together, these optics allow a wide variety of chemical systems and products to be analyzed *in situ*.

A typical Raman probe for online analysis is shown in Fig. 5.2. In Fig. 5.3, some typical locations for probe installation are shown. For pharmaceutical applications, interfacing through the reactor's side walls is generally not possible since the vessels do not have side ports and are glass-lined. Another option for installation is to place the probe in a slipstream. In this scenario, material is pumped from the bottom of the reactor through a stream and then fed back into the reactor. This approach is not recommended because historical data (from the ISA organization) suggest that the slipstream itself will become the major source of analyzer downtime and will add significantly to the analyzer maintenance expense. For example, in reactors with samples being held at about room temperature, a small temperature

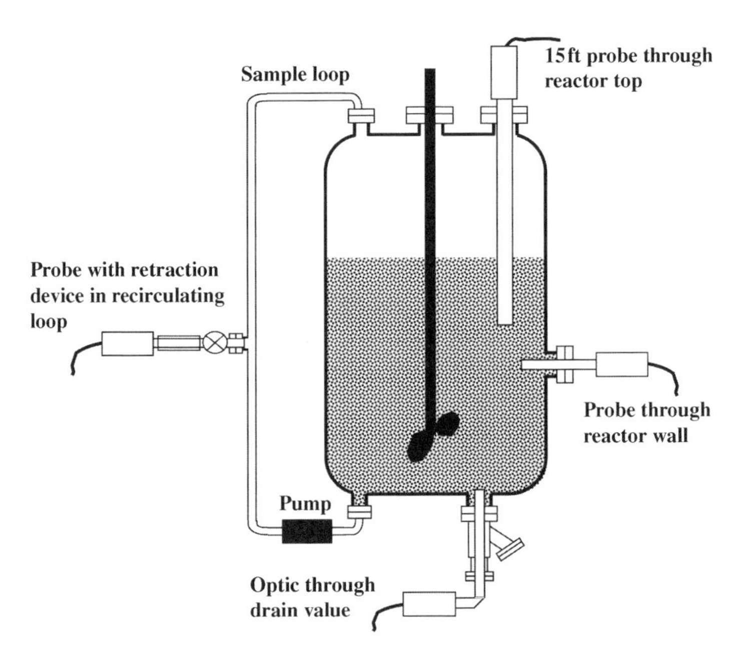

FIGURE 5.3 Diagrammatic representation of potential location for online Raman probe in an API reactor. (Reproduced with permission. © 2006 Kaiser Optical Systems, Inc.)

drop from the reactor to the slipstream can cause precipitation and clogging in the slipstream. Other problems with slipstreams include lack of representative sampling, maintenance, and safety concerns.

5.8 DATA ANALYSIS

The choice of appropriate qualitative or quantitative tool varies from project to project. Both pretreatment methods and analysis algorithms should be chosen wisely according to project goal. Raman libraries represent one common qualitative tool. Libraries have also been used for the purpose of raw material, intermediate, or formulation identity. Search algorithms are the same as those implemented for MIR. For quantitative online PAT work, analyzers that measure the entire spectral range of interest simultaneously are preferred to minimize sampling artifacts. High spectral resolution ($\sim 4\,cm^{-1}$) is also preferred for the majority of applications. This permits small changes in peak shape and position to be determined (e.g., polymorph analysis).

Two prominent measurement error sources in Raman are varying baseline intensity and shape, and changes in absolute Raman intensity. Raman is a single-beam technique that provides high information content that is often superimposed upon a broad featureless background (e.g., fluorescence). That background can be both sample dependent and variable over time. Changes in sample opacity, orientation, and effective laser power have a direct affect on Raman intensity. Therefore, baseline correction and intensity normalization are two key data pretreatments critical for Raman analysis.

The proper pretreatment sequence is baseline correction followed by normalization. There are several data pretreatments that are typically used for removal of baseline anomalies. These include multiplicative scatter correction (MSC) (Geladi et al., 1985; Hancewicz and Petty, 1995; Helland et al., 1995), standard normal variate (SNV) (Barnes et al., 1989) and derivatives (Savitsky and Golay 1964; Williams and Norris 1984). Pearson's method, an iterative approach to baseline de-curvature, works well for the correction of some sets of data. Beebe et al., (1998), Kramer et al. (1998), and Shaver (2000) provide excellent descriptions of the function of these pretreatment approaches. Reference peaks, internal and external standards, and total signal intensity represent some of the more common normalization approaches (Pelletier, 2003).

Isolated analyte and reference bands often make univariate models a good choice for quantitative work. If quantitative estimates are all that are necessary or the sample set is a closed set (no future predictions will be made), then MCR or PCA are good choices (Andrews & Hancewitz, 1998; Andrews et al., 1998; Beebe et al., 1998; Shaver, 2000). For reasonably extensive data sets with available reference data, MLR, PCR, or PLS can be used (Hancewicz and Petty, 1995; Seasholtz et al., 1989; Shaver, 2000; Svensson et al., 2000; Swienga et al., 1999; Williams and Everall, 1995). There are also other options that are less common but potentially effective.

5.9 APPLICATIONS

Despite the development of compact Raman analyzers and the myriad benefits of Raman spectroscopy as an analytical technique, the widespread adoption of Raman spectroscopy has only started to be demonstrated by the adoption of routine applications.

5.9.1 Select Nonpharmaceutical Raman Application Areas

There are several published reviews concerning the use of Raman in process applications (Adar et al., 1997; Everall et al., 2002; Jestel, 2005; Lewis, 2000). In the chemical industry, Raman has been used extensively to monitor the production of titanium dioxide (Besson et al., 1996; Clegg et al., 2001), phosphorous trichloride (Feld et al., 1993,1994; Freeman et al., 1993; Gervasio and Pelletier 1997), distillation of chlorosilanes [Lipp and Gross (1998)], polymerization studies and BTEX products (Cooper et al., 1995,1999; Marteau et al., 1994,1995). It has been mentioned that hydrogenation (Tumuluri et al., 2006) and Grignard (Kaiser, 2001) reactions have been accomplished.

Many needs of the polymer industry are similar to those of the pharmaceutical industry. In the most basic cases, the quantitative measurement of analytes is often of interest. However, in many cases, such as the production of polyolefin films, physical properties are also of interest. Raman has been used both off-line (Everall, 1998; Strobl and Hagedon, 1978) and online to examine crystallinity and orientation of polymers within the extruded product (Everall and King, 1999), dyeability of the polymer (Swierenga et al., 1999), and online measurement of polymer properties (Long and Marrow, 2003; Long et al., 2005).

These are only a selection of successful applications in order to demonstrated that Raman is a very effective tool to gather useful information about the chemical and in some cases the physical status of products during production.

5.9.2 Specific Application Areas of Raman Spectroscopy for PAT

Some broad categories of successes within the pharmaceutical industry are listed in Table 5.3. Table 5.3 represents a compilation of applications highlighted by Frank (1999), Williams (2000), Pivonka (2002), and Folestad and Johannson (2003). A brief review of the three largest segments of the Raman market, reaction monitoring (primary), crystallization (primary), and secondary manufacturing, are surveyed in more detail below. Pharmaceutical Raman applications are leading to successful process understanding, process optimization, and process monitoring.

5.9.3 Primary Manufacturing

Real-time reaction analysis can provide valuable information at all API development and primary manufacturing stages. As outlined by Wiss and Zilian (2003), online analytics can be considered to include three main application segments

TABLE 5.3 Demonstrated Raman Applications Opportunities.

API/drug substance manufacturing opportunities	Drug product manufacturing opportunities
Structure elucidation	Polymeric matrix in DP delivery
Polymorphic forms	DS in polymeric DP delivery
Crystallization/recrystallization	Lyophilization
Hanging drop crystallization	Drug distribution in transdermal patch
Grignard reactions	Blister pack analysis—shelf life
Hydrogenation	Blister pack analysis—adulteration
Purity of chiral materials	DP counterfeiting
Salt screening/salt form analysis	Polymer packaging laminate ID
Solvate form	DP tablet precoating
Hydrate form (DS, Excipient)	DP tablet postcoating
DS/excipient interactions	DP wet granulation
Aqueous solution measurements	Headspace analysis—shelf life
Solvent measurements	
HTS wellplate analysis	Raw materials ID of API and excipients
Polymerization	Formulations
Combinatorial chemistry	Blending
ID of TLC spots	Chemical imaging
Phase transformation	Polymorph transformations
Slurry/suspension analysis	Primary and secondary process monitoring
Reaction monitoring	Primary process monitoring of API and excipient synthesis
Chemical imaging	Individual dosage form uniformity
Drying	Drying

Reaction analysis for chemical R&D, reaction engineering for process R&D, and process monitoring for manufacturing efficiency.

Initial applications of Raman for PAT involved monitoring primary processes. Raman is particularly suitable for reaction monitoring. In terms of process understanding, the technique can be used to understand the kinetics and thermodynamics of a process. It can also help to identify the formation of intermediates. In a practical sense, it can be used to understand the effect of the deviation of process parameters on the course of the reaction. In this section, several applications are briefly reviewed that are of interest to PAT practitioners and demonstrate the potential of Raman spectroscopy as an *in situ* Quality by Design tool. These applications are broken down into those that are not specific to crystal form (reaction monitoring) and those that are specific to crystal form (crystallization).

5.9.3.1 Reaction Monitoring Among other properties, Raman spectroscopy has been utilized to provide mechanistic, pathway, intermediate, endpoint, conversion, and concentration information (Wiss and Zilian, 2003). Raman spectroscopy is therefore a well-positioned PAT tool, providing the opportunity to improve

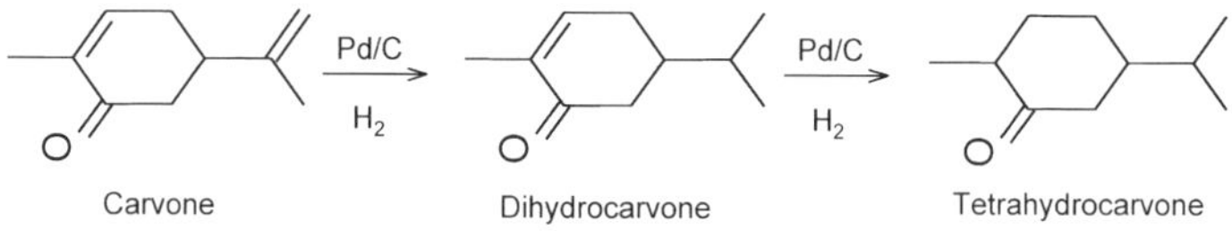

Carvone $\xrightarrow[H_2]{Pd/C}$ Dihydrocarvone $\xrightarrow[H_2]{Pd/C}$ Tetrahydrocarvone

FIGURE 5.4 Conversion of carvone to dihydrocarvone to tetrahydrocarvone. (Reproduced with permission. © 2006 Kaiser Optical Systems, Inc.)

processes, facilitate risk mitigation strategies, identify where routine monitoring is appropriate, and facilitate continuous processing.

Tumuluri et al. (2006) make use of model hydrogenation reactions to examine real-time analysis for a general class of organic processes commonly used in the pharmaceutical industry. Fig. 5.4 shows the mechanistic conversion of carvone to tetrahydrocarvone. In Fig. 5.5, the reaction profiles for the starting material carvone, the intermediate dihydrocarvone, and the product tetrahydrocarvone are provided.

The authors estimate that hydrogenation reactions represent 10–20% of all API manufacturer reactions. Olefinic to aliphatic bond conversion, nitro to amine functional group conversion, and functional group deprotection are examples of hydrogenation reactions commonly employed. Observation of reaction intermediates, kinetic profiles, and catalyst optimization are provided to illustrate the potential for real-time reaction analysis.

Radioactive labeling reactions have likewise been investigated with Raman spectroscopy. Heys et al. (2004) examine a range of tritium labeling reactions in real-time. Organoiridium-catalyzed heteroatom-directed exchange (HDE), olefin hydrogenation, and catalytic aryl dehalogenation reactions were investigated. Changes in peak intensities and frequencies attributed to labeling are observable at concentrations and reaction scales typical of tritium gas reactions. Raman bands sensitive to isotopic substitution provided real-time information over the course of these reactions.

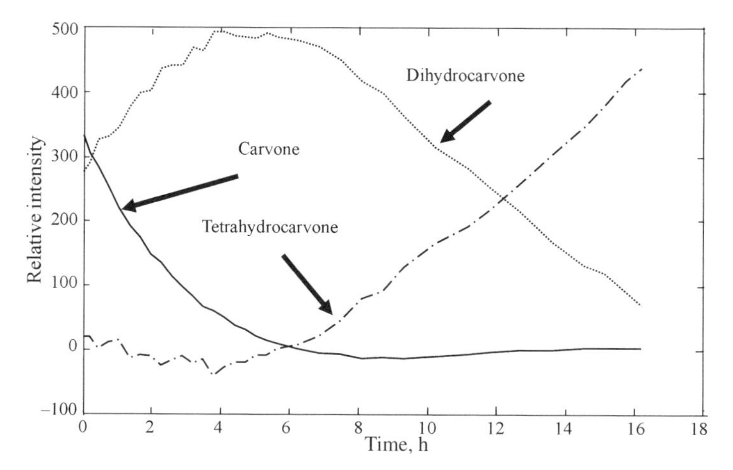

FIGURE 5.5 MCR data analysis profiles for carvone reduction. (Reproduced with permission. © 2006 Kaiser Optical Systems, Inc.)

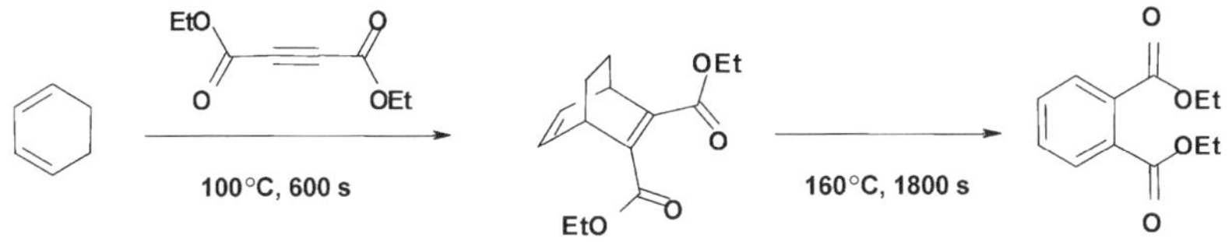

FIGURE 5.6 The Diels–Alder cycloaddition reaction. (Reproduced with permission. © 2006 Kaiser Optical Systems, Inc.)

Microwave-assisted organic reactions represent an expanding research area in discovery and optimization groups. Improved yield, accelerated kinetics, and no or minimal solvent consumption promote this interest. Pivonka and Empfield (2004) integrate a Raman probe into a commercial microwave synthesizer to provide real-time mechanism, intermediate, and reaction kinetic information. Amine and Knoevenagel coupling model reactions were investigated. Raman spectroscopy was shown to eliminate limitations present in traditional analytical techniques (IR, NMR, MS, HPLC, etc.).

Diels–Alder reactions are part of a family of highly important pericyclic reactions that yield stereospecific products. Stereospecificity can be an important property when developing drug-like compounds. Microwave-assisted Diels–Alder reactions monitored by Raman spectroscopy have been reported (Kaiser, 2006a). Figure 5.6 represents the Diels–Alder cyclo-addition reaction of 1,3-cyclohxadiene and diethyl

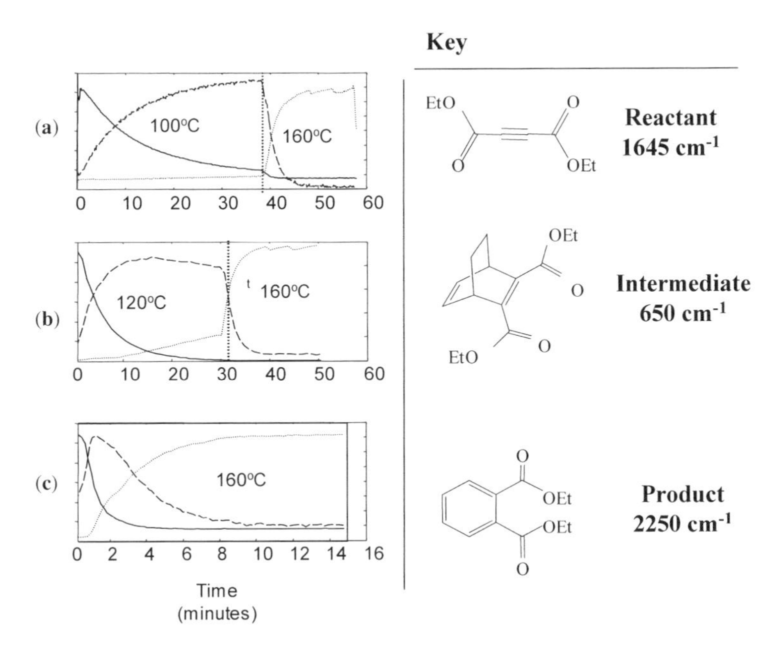

FIGURE 5.7 Profile plots of the three reactions: (**a**) 100 °C for 40 min, then 160 °C. (**b**) 120 °C for 30 min, then 160 °C. (**c**) 160 °C. The three traces in each plot correspond to the normalized intensities for the reactant (solid line), intermediate (dashed line), and product (dotted line). The dotted vertical lines indicate the times at which the temperature in the reactor was changed. (Reproduced with permission. © 2006 Kaiser Optical Systems, Inc.)

acetylene malonate. Reaction profiles are presented in Fig. 5.7. Each plot traces the appearance and disappearance of the reactant diethyl acetylene malonate, the intermediate, and the product using the intensities of a unique Raman band for each species. The Raman data give highly specific information that allows the state of the reaction at any time to be determined accurately. Lewis (2005) points out that Raman spectroscopy has the potential to address a significant roadblock to widespread acceptance of microwave-assisted reactors (i.e., lack of *in situ* analytics). *In situ* analysis provides the development chemist with both mechanistic and process understanding.

Reaction engineering directed toward process development and scale-up can benefit directly through reaction monitoring. Noninvasive sampling has proven to be key advantage for Raman spectroscopy. Typically used to produce new carbon–carbon bonds via nucleophilic addition, Grignard reactions (Kaiser, 2001) present an exothermic concern in scaled-up reactions. A common Grignard reagent (phenyl magnesium bromide) model reaction was used to illustrate real-time monitoring (Kaiser, 2001). The results show that Raman spectroscopy provides a viable methodology to evaluate and mitigate potential thermal runaway of Grignard reactions without the limitations of ATR-based IR sampling configurations.

Safety concerns are again illustrated by Wiss and Zilian (2003) in another reaction engineering exercise. In this example, two reactions were examined, the synthesis of tri-*n*-butylin azide along with the synthesis of 2-chloroanaline. In the former reaction, toxicity is the primary concern, in the second, an unstable hydroxylamine intermediate represents a potential safety hazard. While both ATR-IR and Raman spectroscopy provide potential monitoring options, a fiber optic based Raman approach is deemed more appropriate for industrial application due to the ease of installation of fiber optics and the ability to locate the analyzer upto 200 m from the reactor. The hydroxylamine intermediate reaction is considered further in an additional note (Kaiser, 2006b).

Figure 5.8 illustrates the reaction summary for the reduction of *ortho*-nitrochlorobenzene with hydrogen, at atmospheric pressure, in methanol with a platinum/

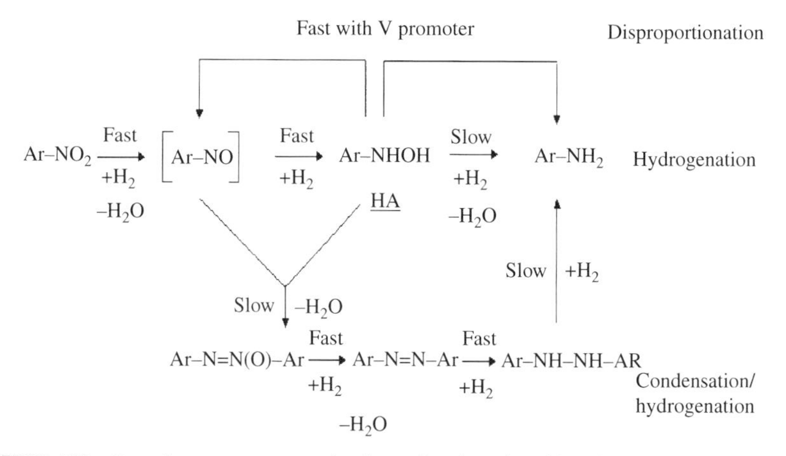

FIGURE 5.8 Reaction summary: reduction of ortho-nitrochlorobenzene with hydrogen. (Reproduced with permission. © 2006 Kaiser Optical Systems, Inc.)

carbon catalyst to form *ortho*-chloroaniline. The reaction is known to follow one of two pathways. At elevated pH, the reaction can be shown to follow the upper route, with the two intermediates predicted for the condensation/hydrogenation not being present. In this study, the impact of a vanadate-based catalyst can be demonstrated, where the reaction order changes and the formation of the predicted hydroxylamine intermediate no longer occurs. Reaction profiles generated by real-time Raman spectroscopy appear in Fig. 5.9. An important benefit of Raman, that can be capitalized upon by appropriate design, is that the optics and the measurement system do not limit the spectral range of the analyzer and a range from 150 to 3400 cm^{-1} can be observed with 785 nm excitation. Spectral information may be acquired throughout the entire vibrational spectrum, even down to low wavenumbers, without the optical limitations imposed by a FTIR spectrometer.

Batch process monitoring and control are discussed by Svensson et al. (2000). Metoprolol (a selective beta receptor blocker) synthesis is used as a model reaction. Aspects of multivariate data analysis and spectral preprocessing are evaluated with

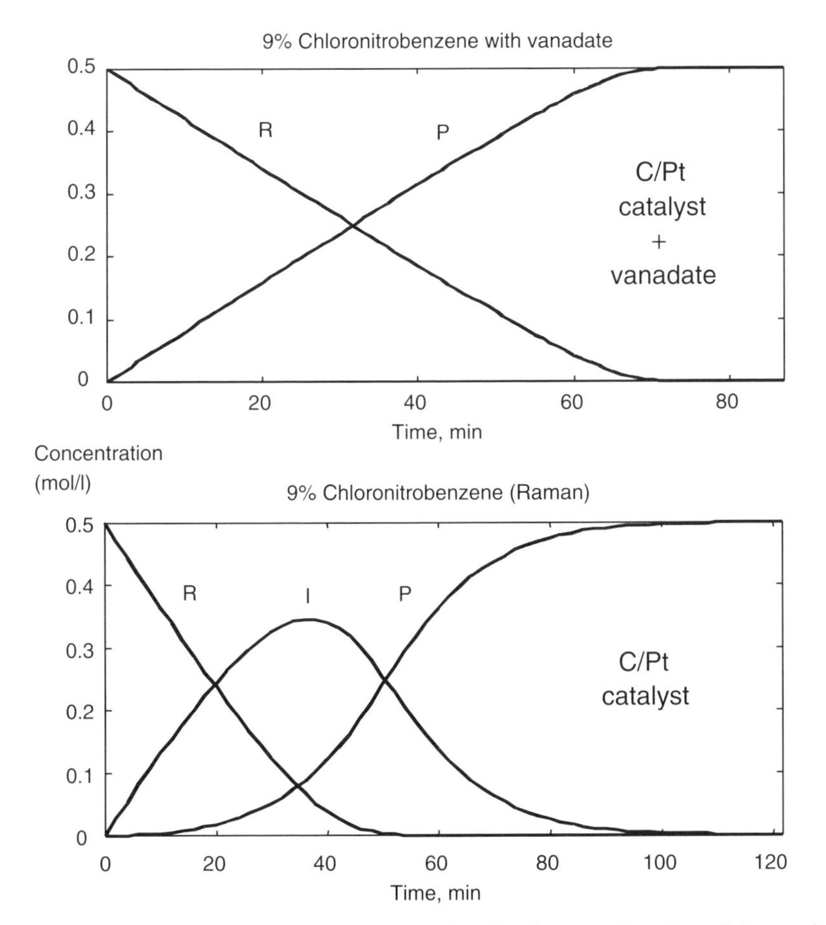

FIGURE 5.9 Changes of nitro reduction reaction kinetics as a function of the catalyst system. (Reproduced with permission. © 2006 Kaiser Optical Systems, Inc.)

respect to manufacturing efficiency. When several similar molecules are used in a reaction mixture, the Raman spectrum becomes more complex, making multivariate methods preferable. The monitoring results (PCA and PLS) show that through optimizing reagent composition and reaction thermal gradient, the time to produce the API can be reduced.

Ray et al., (2005) promote a streamlined approach for implementing PAT technology in a pilot plant facility. Unlike the manufacturing environment, the development activities that take place in pilot facilities often do not offer the opportunity for detailed, robust model development. The authors advocate developing quantitative models with appropriate limit tests to incorporate PAT technology into day-to-day operations where significant process changes are anticipated. Careful examination of the particular process step to be studied along with a detailed understanding of acceptable error constraints permit the implementation of PAT with minimal development time, resources, and batch data. Two examples are provided. In the first, FTIR is used to monitor a process waste stream. In the second, Raman spectroscopy is used to monitor the amount of ethyl acetate remaining during a solvent swap with a higher boiling solvent.

In a different publication, Wethman et al. (2005) provide an example of reaction monitoring for a key synthetic intermediate. *In situ* Raman spectroscopy is used to follow a Schiff's base reaction and subsequent hydrogen reduction. Here the authors focus is on technology transfer from development to manufacturing. Issues related to method transfer from the development environment to the manufacturing facility are discussed. Figure 5.10 compares Schiff base reaction data for the primary off-line (HPLC) method and the *in situ* Raman method. The methods are in excellent agreement as the reaction nears completion.

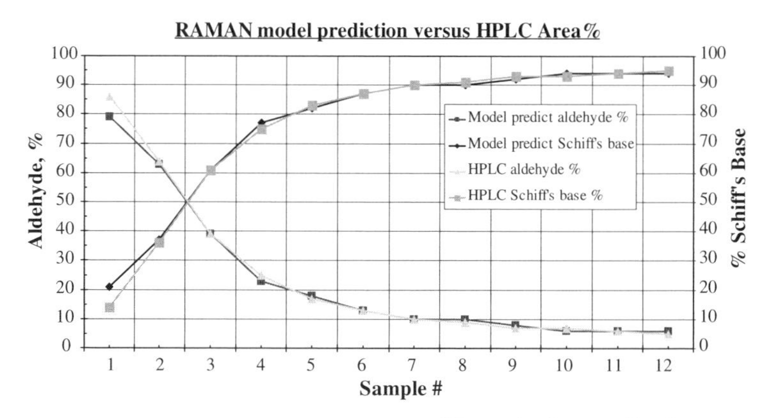

FIGURE 5.10 A comparison of Schiff base reaction data for the primary (HPLC) method and the Raman method. The methods are in excellent agreement as the reaction nears completion. (Reproduced with permission. © American Pharmaceutical Review, 2006).

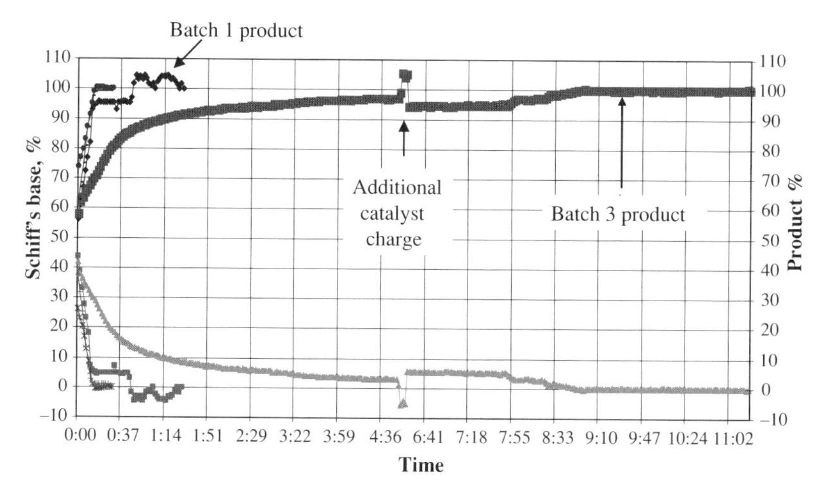

FIGURE 5.11 An overlay of all four pilot plant reduction runs demonstrating the slower reaction rate for batch 3. The PAT method immediately detected that the reaction rate was significantly slower than the other pilot plant runs and that at the expected call point (~30 min) a significant amount of starting material remained. An additional catalyst charge was made at ~5 h to drive the reaction to completion. The results of batch 3 highlight the value of a quantitative method when monitoring an atypical reaction. (Reproduced with permission. © American Pharmaceutical Review, 2006.)

Figure 5.11 provides an overlay of four pilot plant runs. The Shiff's base reaction requires greater than 92% completion to prevent downstream impurities that result in an increase in plant cycle time and cost. The PAT Raman method identified batch 3 as an outlier (Fig. 5.11). An additional catalyst charge drove the batch 3 reaction to completion, highlighting the use of Raman spectroscopy for monitoring atypical reactions. Method-specific PAT benefits for Raman spectroscopy are also provided by the authors. These include probe design flexibility, remote analyzer location (no explosion-proof instrumentation requirement), limiting the user's exposure to potential chemical and safety risks (required for HPLC analysis). Additional benefits include speed of analysis. This study suggest that if Raman spectroscopy was implemented for process control, the approach would offer enhanced analysis time, response time saving, reduction in the cost of disposables, similar product quality, and a reduction in the load on the QA/QC laboratory. Collaboration between plant operators, chemists, engineers, and analysts is emphasized as a key component to successful implementation of PAT strategies.

As can be seen, Raman spectroscopy offers significant benefits and potential for reaction analysis, monitoring, and control. The selection of Raman spectroscopy versus other potential reaction analysis techniques such as MIR, NIR, and UV-Vis will be dependent on the installation requirements, the chemistries being studied, and the functional groups of interest. Under these criteria, Raman spectroscopy is unique in being compatible with long fiber runs, is applicable to a wide range of chemistries (unlike NIR and UV-Vis), does not require a specific chromophore (unlike UK-vis), provides data that are interpretable to specific functional chemical groups and are applicable to both slurries, and homogeneous and heterogeneous processes.

These benefits must be weighed against the conflicting limitations of fluorescence and education. While fluorescence may limit the use of Raman spectroscopy from a universal applicability perspective, fluorescence is no more of a limitation to Raman spectroscopy than the inherent handicaps found with other reaction analysis techniques. Warman and Hammond (2005) noted for the API processes they studied 785-nm Raman spectroscopy has the potential to be applicable to approximately 70% of those processes. This ranks first equal with MIR when five potential reaction analysis techniques were surveyed. Interestingly, though, while MIR and Raman spectroscopies both are approximately 70% techniques, the combination of both tools allowed greater than 90% of all reaction to be studied. In summary, Raman spectroscopy has been shown to be appropriate for reaction analysis, monitoring, and control.

5.9.3.2 Crystallization Crystallization is a critical processing step in pharmaceutical active ingredient production. During this process, API's undergo crystal engineering in order to assure that the final desired crystalline form is produced. Physical-chemical issues such as particle size are important as is the assurance that the correct polymorphic form is produced. The FDA is keenly interested in the integrity of these processes. Hence, most pharmaceutical companies have groups of various sizes dedicated to crystallization expertise.

Raman is a very good means of distinguishing crystalline and amorphous materials as well as different types of crystalline forms of the same material, that is, polymorphic forms. The reason is that the molecules in crystalline materials are, by definition, ordered in a specific fashion. As a result, the Raman spectra for crystalline materials give rise to very sharp and intense bands. Amorphous materials, however, give rise to less intense bands that are generally very broad. Also, a particular moiety in one crystalline form may have a different environment than the same moiety in another form. This may lead to band shifts between polymorphs and provide the scientist with the ability to qualitatively distinguish between two forms and to apply a simple univariate model to quantitatively distinguish one polymorph from another.

Raman spectroscopy has found a place in monitoring crystallization phenomena and has become one of the preferred techniques for this purpose. Raman is often used during development so that crystallization processes are understood early on. As such, processes can be developed and optimized and, hence, can be controlled effectively on a routine basis. When processes are transferred to production, Raman is then often used as a routine monitoring tool. When employed in this fashion, Raman can be used (1) to monitor the progress of the crystallization process, (2) to assess the endpoint of the procedure, and (3) to assure that the correct polymorph has been formed.

In 1999, Everall et al. reported work that was a portent for the extension of Raman spectroscopy into online crystal form control in pharmaceuticals. In this report, Raman was used to control the rutile/anatase ratio of titanium dioxide. The use of Raman was critical to control the process and allow the production to

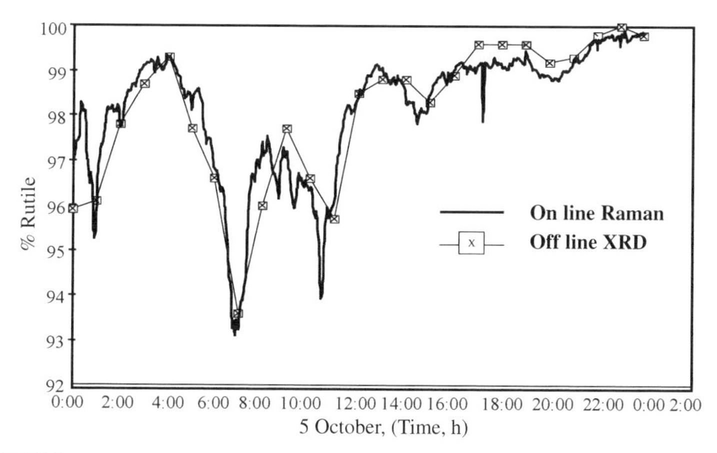

FIGURE 5.12 Process control diagram for titanium dioxide (TiO_2) production showing the percent Rutile content of the production batch. (Reproduced with permission. © 2006 Kaiser Optical Systems, Inc.)

efficiently produce material that was continuously in specification. The paper reported the use of a system that monitored four points simultaneously to maximize overall process control (see Fig. 5.12). The previous means of accomplishing the control on this process was off-line X-ray diffraction (XRD) crystallographic measurements. While accurate, XRD did not allow real-time control of the process. The figure shows that process excursions could be detected by Raman since it was employed online and allowed more frequent sampling. The comination of the results from the Raman analyzer and feedback controls resulted in improvements in batch quality and production efficiency. The study of Everall et al., (1999) represents a more complete disclosure of the use of Raman spectroscopy for titanium dioxide control than earlier publications (Beeson et al., 1996).

Several papers have been published in recent years that highlight the ability of Raman to convey valuable information in crystallization processes. Wang et al. (2000) described a very simple and effective application in which Raman was used to monitor a solvent-mediated transformation (SMT). In this process, Form II of progesterone was isolated after synthesis. Since Form I was the desired form for formulation, a process such as SMT was necessary. The Raman spectra of the two polymorphs indicated a $5\,cm^{-1}$ shift in the C=O band relative to one another (see Fig. 5.13). Hence, Raman could be used with a fiber optic interface to follow the process in real time. The relative percentages of Form I and Form II could be quantified based on the C=O band shift. The endpoint was clearly deciphered by following this shift while applying the calibration (see Fig. 5.14). Raman also proved to be a valuable tool in characterizing the thermodynamics of the process.

In 2002, Dell Orco et. al. (2002) showed the value of Raman for crystalline design. The technique proved to be valuable both *in situ* and for laboratory measurements in determining crystalline form. Process optimization was the key goal for these workers.

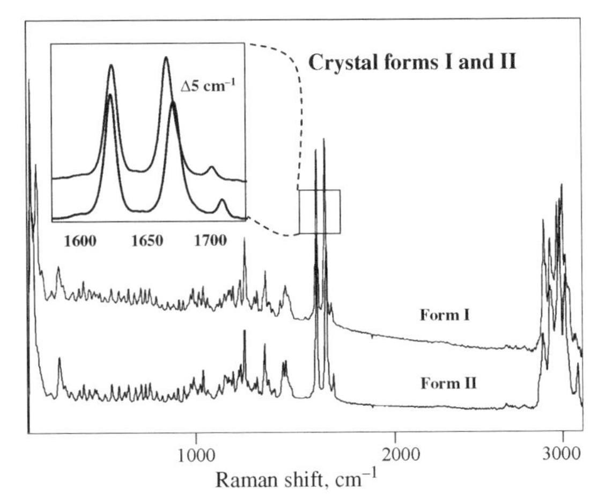

FIGURE 5.13 Spectra of Form A and Form B of progesterone showing the useful 5 cm^{-1} shift in the carbonyl region allowing the solvent-mediated transformation (SMT) to be monitored. (Reproduced with permission. © 2006 Kaiser Optical Systems, Inc.)

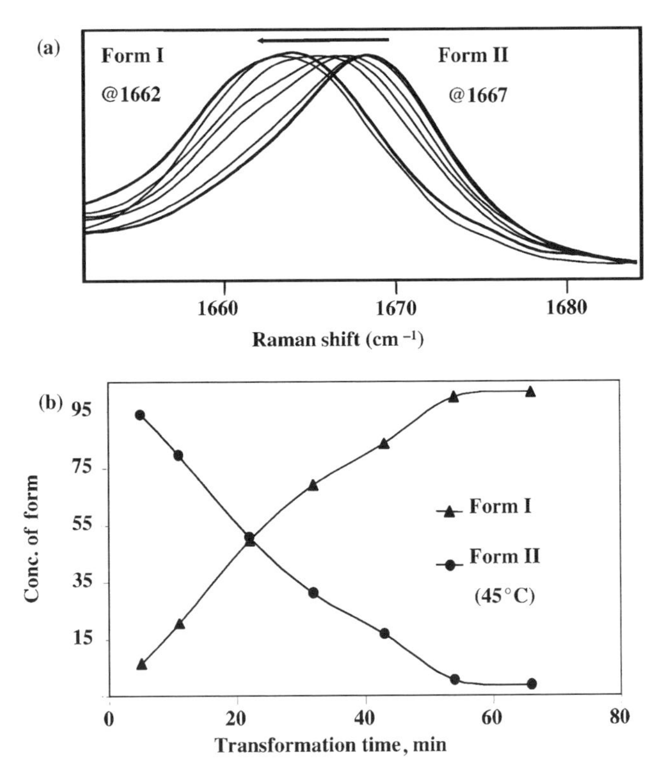

FIGURE 5.14 (a) Spectra of the carbonyl band shift at various times during the SMT process. (b) Trend plots showing the appearance of Form I and disappearance of Form II. (Reproduced with permission. © 2006 Kaiser Optical Systems, Inc.)

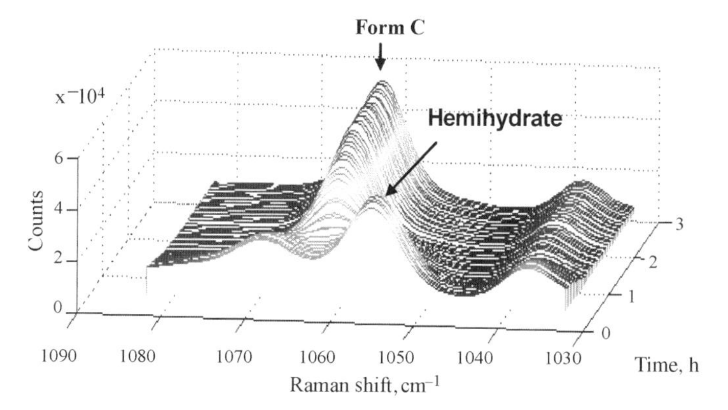

FIGURE 5.15 Three-dimensional plot of the Raman spectra acquired during the transformation of the hemihydrate to Form C for a Merck proprietary compound. (Reproduced with permission. © 2006 Kaiser Optical Systems, Inc.)

In 2002, Zhou et. al. (2002) investigated the use of Raman for monitoring an API crystallization process. During the process development and optimization stage, they were able to successfully follow the transformation of a proprietary compound from the hemihydrate to Form C and then further to Form A. They used characteristic bands at 1055, 1062 and 1060 cm^{-1} to diagnose the presence of hemihydrate, Form C, and Form A, respectively. Figure 5.15 shows the spectra that were collected during the transformation of the hemihydrate to Form C. They were able to use a second derivative pretreatment followed by a PLS regression to quantify the amount of Form C that was ultimately present during the transformation to Form A (see Fig. 5.16). These researchers noted that particle size was a limiting factor in obtaining accurate quantification.

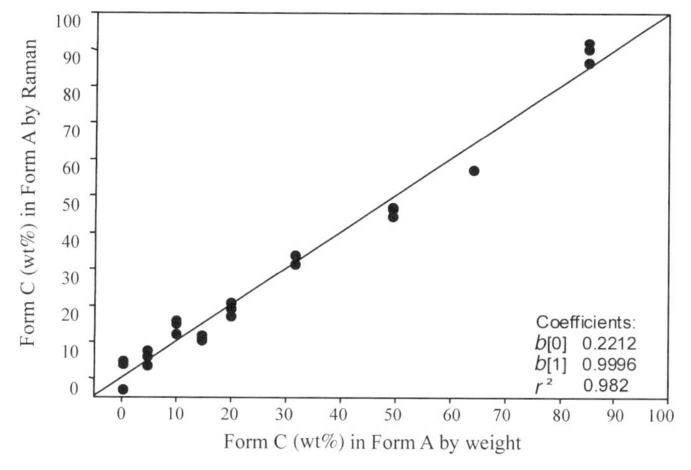

FIGURE 5.16 PLS calibration plot for the percentage of Form C in the presence of Form A. (Reproduced with permission. © 2006 Kaiser Optical Systems, Inc.)

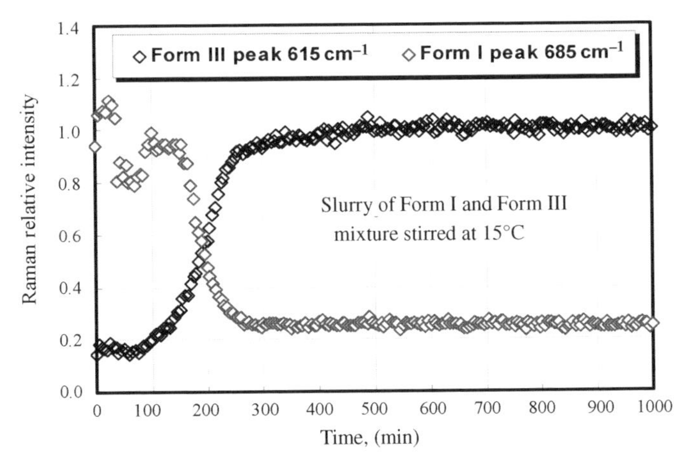

FIGURE 5.17 Figure showing the trend plot derived from the Raman spectra collected during the transformation of Form III to Form I. (Reproduced with permission. © 2006 Kaiser Optical Systems, Inc.)

The groups of Taylor and Myerson (Hu et al., 2005,2006) determined that wide spot illumination was invaluable in improving the accuracy in the quantification of actives in crystallization form transformation process, thus overcoming the limitations found by Zhou et al. (2002). Using flufenamic acid as a model compound in the development of a challenging crystallization process, they showed that the larger spot size allowed more accurate monitoring. This process required 10 h. Temperature reduction from 50 to 30°C showed no conversion. The crystallization required temperature reduction to 15°C and seeding with Form III was shown to accelerate the process such that it could be completed in less than 5 h. In this work, they discovered that there were characteristic peaks for Form I (685 cm^{-1}) and Form III (615 cm^{-1}) that allowed the researchers to follow the process successfully (see Fig. 5.17).

Falcon and Berglund (2003) used Raman spectroscopy to follow an antisolvent crystallization process. The crystallizations were carried out with cortisone acetate (CA) as the model compound. To effect crystal formation, water was added in a controlled fashion to a ternary solution of acetone (87.00%), water (10.50%), and CA (2.50%). The crystallization was performed with three different rates of water addition: 0.25, 0.50, and 1.00 ml/min. The Raman shift region of greatest interest was found between 1610 and 1680 cm^{-1}, which contains two peaks: one between 1600 and 1625 cm^{-1} and another between 1660 and 1680 cm^{-1}. The former was assigned to a double bond in the CA and the latter to the carbonyl. A significant aspect to this work was that the crystallization data were subjected to principal components analysis (PCA). The first two factors contained information that was obvious in the raw spectra. However, the third factor, when plotted versus time for the crystallization, increases to a maximum and then decreases steadily. It was determined that the maximum occurred when the solution was saturated. Hence, Raman combined with PCA allowed the determination of the onset of supersaturation. The third PCA factor appeared to correspond to solvent–solute

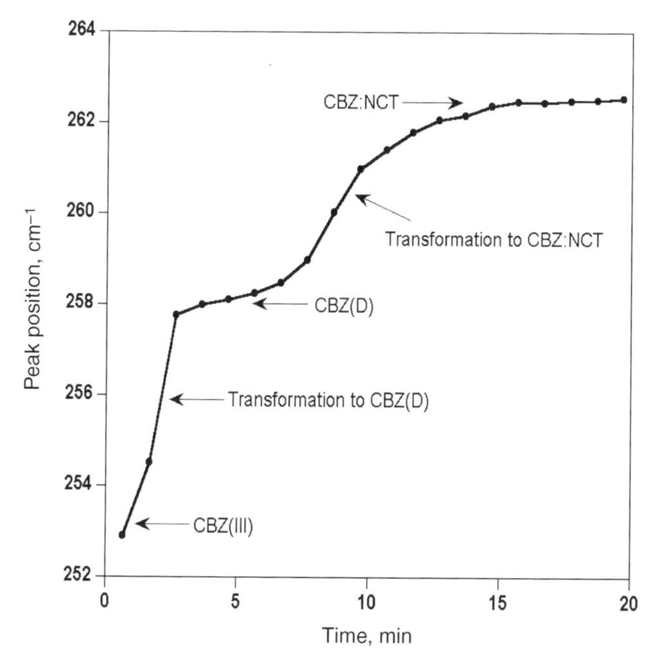

FIGURE 5.18 Raman peak position with respect to time showing the slurry conversion of anhydrous CBZ(III) to co-crystal CBZ-NCT at 23°C according to the pathway CBZ(III) to CBZ(D) to CBZ-NCT. (Reproduced with permission. © 2006 Kaiser Optical Systems, Inc.)

interactions during the addition of antisolvent, providing further information beyond what could be obtained from the raw spectra.

More recently, Rodriguez-Hornedo et al. (2006) have investigated the value of Raman spectroscopy in the engineering of co-crystals. In their seminal paper, they were able to follow the transformation of carbamazopine Form III [CBZ(III)] to the dihydrate form [CBZ(D)] and then to the co-crystal (CBZ-NIC) in the presence of dissolved nicotinamide. The process could be followed using the shift of a key fingerprint band from 254 cm^{-1} [CBZ(III)] to 259 cm^{-1} [CBZ(D)] and then to 264 cm^{-1} (CBZ-NIC) in a distinct three-step process (see Fig. 5.18). Follow-up studies were accomplished to determine the effect of excipients, such as povidone and fructose, on this process. These workers concluded that Raman analysis gave them key information that allowed them to monitor and understand these crystalline transformations in real time.

5.9.4 Secondary Manufacturing

Raman spectroscopy currently has found greater acceptance in chemical development/discovery and primary optimization and manufacturing than in pharmaceutical development and secondary manufacturing. However, much recent interest has been focused in these later areas, and publications outlining Raman analysis in the field of secondary processing are appearing.

Secondary manufacturing normally incorporates several unit operations; these may include dispensing (ID), granulation, drying, blending, compression, film coating, and QC. At each of these steps, it is possible to experience process variability and thus monitoring and controlling each critical step can be warranted. Ultimately, the goal is to develop tracking specifications and indicators for each critical quality parameter and control the process so the outcome of each step is within the required specifications.

In the dispensary (ID) stage, the incoming materials are verified at the dock and before the contents of the drum are loaded into the manufacturing equipment. Frank (1999) was one of the first to document the use of Raman with a library search routing for materials ID confirmation for this purpose.

In the next stage, during granulation, the particles of the active and excipients are mixed together. Granulation of the dry powders is termed dry granulation and when granulation is accomplished with the addition of water, this is termed wet granulation. A widespread step in the formulation of many oral dosage units is wet granulation. A wet granulation step is designed to yield appropriate particle size prior to tableting. Dry granulation can lead to a rise in temperature of the particles which can cause changes to pharmaceutical ingredients. The inclusion of water into the particles during wet granulation can also change the pharmaceutical ingredients. This unfortunate side effect of these processes can be a process-induced transformation (PIT). Examples of PITs include polymorphic changes to the API brought on by heat, to hydrate form changes caused by the inclusion of water. Both of these transformations can have a dramatic impact on the dosage form. The API form in the final DP can directly impact the patient and thus it is critical to both monitor and control the form during manufacture. Therefore, the identification of PITs, an understanding of their effect on the materials during processing and the impact on the final materials is of great importance during formulation development and manufacturing.

Historically, off-line XRD QA/QC methods have been used to identify the resulting powder to either release or reject the batch. XRD, however, is both invasive and time-consuming and is susceptible to subsampling—analyzing an unrepresentative fraction of a heterogeneous mixture. Therefore, a technique that allows the mixture to be measured in situ in real-time offers potential benefits. Wikstrom et al. (2005a) compared both NIR and Raman as potential in situ monitoring tools for monitoring a wet granulation unit operation where a solvent-mediated pusedo-polymorphic hydrate transformation of theophylline, from the anhydrate (AT) to monohydrate (MT) form occurred.

The Raman spectra of samples of AT and MT show characteristic peaks that are observed at 1664 and 1707 cm^{-1} for AT and at 1686 cm^{-1} for MT. A univariate calibration model was used to determine the ratio of AT and MT in the samples. When Raman spectroscopy was used to monitor wet granulation, it was possible to follow the transformation. However, in the case of NIR, the main differences observed arose from the presence of water. Water masking and interference resulted in an inability to monitor the transformation kinetics during wet granulation by NIR.

In addition to monitoring the ratio of AT and MT following the PIT, other effects that were investigated were as follows: the effects of changing the mixing speed of the

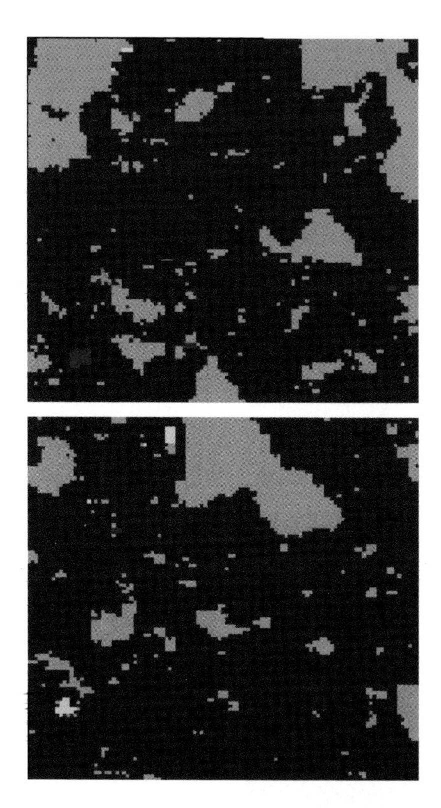

FIGURE 6.16 Euclidean distance classification results for Raman maps of mix sample 2, surfaces 2 (top), and 3 (green = I; blue = II; cyan = III; orange = diluent 1; red = diluent 2). Pixel size is 8.3 μm, and image size is approximately 920 μm × 900 μm (12,099 pixels). Reprinted with permission Society for Applied Spectroscopy.

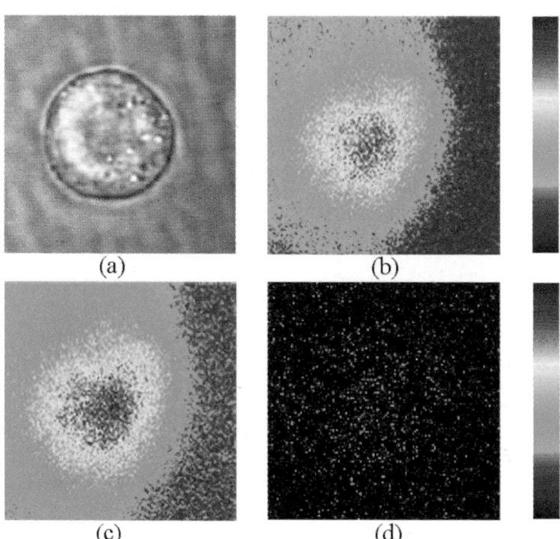

FIGURE 8.12 Example of the image record. **(a)** White light image of an MDA-435 breast cancer cell. **(b)** Raman image of the cell taken at 1000 cm^{-1} Raman band with 60× W/IR lens and exposure time of 300 seconds. **(c)** Raman image of the cell taken at 1080 cm^{-1} Raman band with 60× W/IR lens and exposure time of 300 s. **(d)** Difference of **(b)** and **(c)** before processing. The color bar indicates the relative Raman signal intensity increasing from bottom to top. Reproduced from "Direct Raman imaging techniques for study of the subcellular distribution of drugs" (*Applied Optics*, Vol. 41, No. 2) with permission of the Optical Society of America.

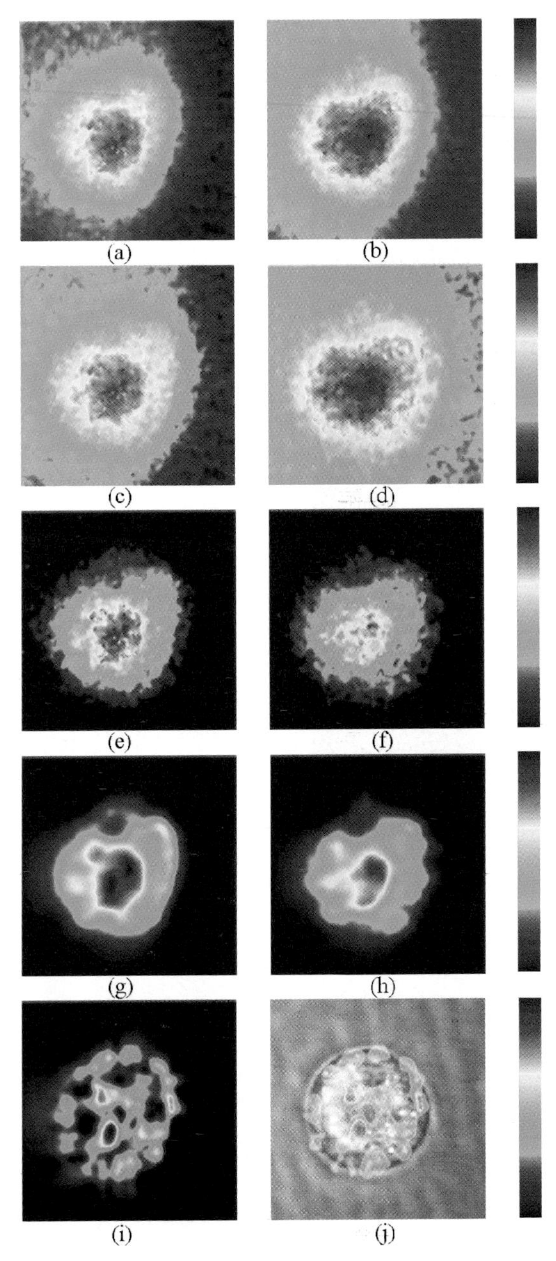

FIGURE 8.13 Postprocessing of Raman images in Fig. 8.12b and c illustrated in left and right columns, respectively. **(a and b)** Smoothed images. **(c and d)** Nonuniform illumination corrected images. **(e and f)** Constant background subtracted images. **(g and h)** Three-dimensional blur restored images. **(i)** Fluorescent signal eliminated image. **(j)** Overlay of image of **(i)** on image in Fig. 8.12a. The color bars indicate the relative Raman signal intensity increasing from bottom to top. Reproduced from "Direct Raman imaging techniques for study of the subcellular distribution of drugs" (*Applied Optics*, Vol. 41, No. 2) with permission of the Optical Society of America.

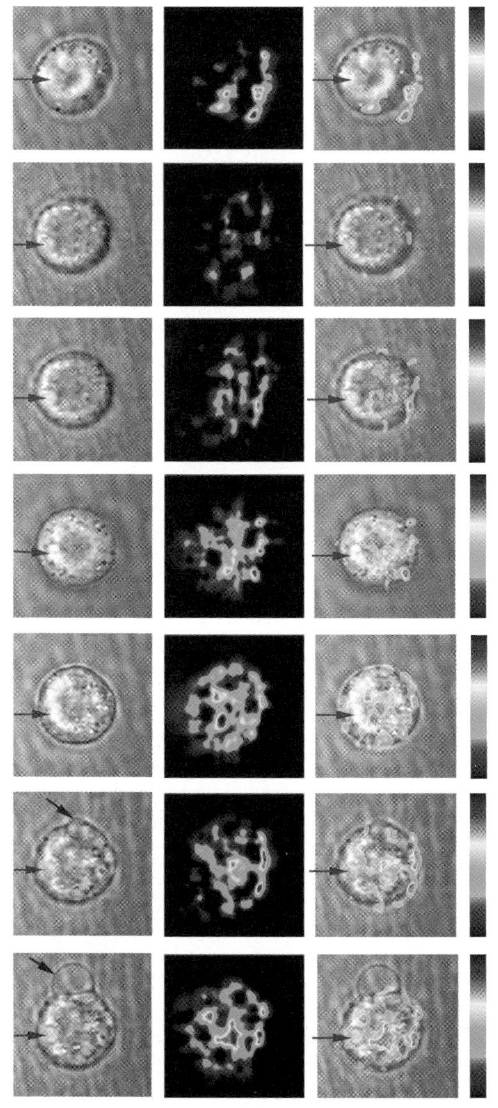

FIGURE 8.14 Images before, during, and after an MDA-435 breast cancer cell was exposed to the paclitaxel agent. The first row illustrates the images before drug treatment. The second and third rows illustrate the images 10 and 45 min during the drug treatment. The fourth to seventh rows illustrate the images 10 min, 1.75 h, 4 h, and 4.5 h after the drug treatment. The left column shows the white light images of the cell that show the cell structure; the center column shows the Raman images of the cell that show the intensity distribution in the 1000 cm^{-1} Raman band; and the right column shows the overlay of images in the left and center columns. The red arrows point to the cell nucleus region and the blue arrows point to the cell blebbing region. The color bar indicates the relative Raman signal intensity increasing from bottom to top. Reproduced from "Direct Raman imaging techniques for study of the subcellular distribution of drugs" (*Applied Optics*, Vol. 41, No. 2) with permission of the Optical Society of America.

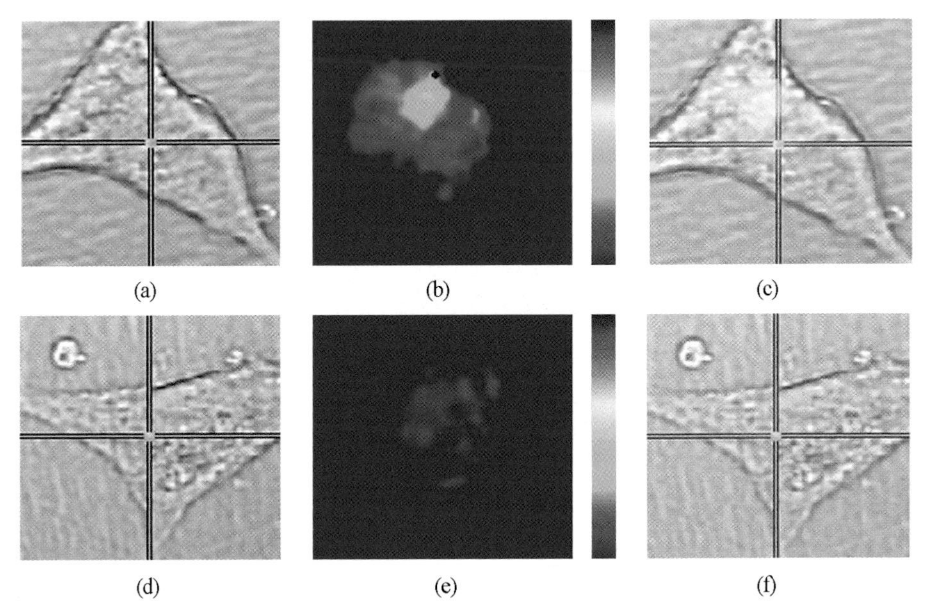

FIGURE 8.18 Comparison between Raman images of LNCaP cells treated with sulindac sulfide (top panel) and controlled cells (bottom panel). **(a)** Bright field micrograph of cell treated with sulindac sulfide. **(b)** Corresponding Raman image of **(a)**. (**c**) Overlay of images in **(a)** and **(b)**. **(d)** Bright field image of cell treated with drug solvent alone. **(e)** Corresponding Raman image of **(d)**. **(f)** Overlay of images in **(d)** and **(e)**. The color bar indicates the relative Raman signal intensity that increases from bottom to top. Reproduced from "Raman imaging microscopy—a potential cost-effective tool for drug development" (*Americal Pharmaceutical Review*, Vol. 8, No. 4) with permission of Russell Publishing, LLC.

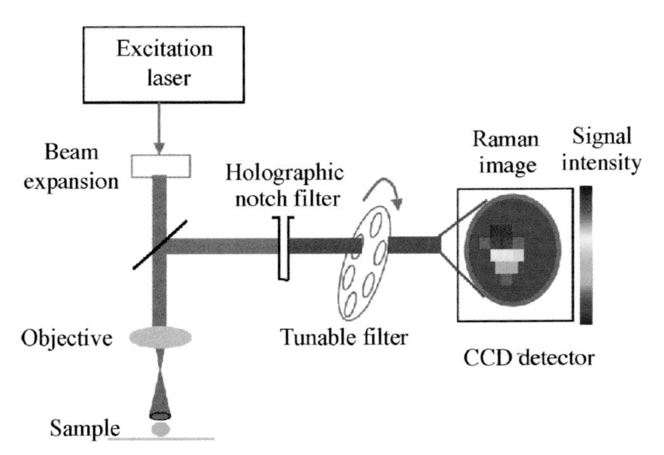

FIGURE 8.3 Schematic diagram of a CCD-based direct Raman imaging system. See text for full caption.

granulator, changing the API loading, and changing the seeding; the effects of surface properties and initial surface area of the API; and the effect of modifying the binder addition process. These experiments led to process understanding of what did not affect the granulation process. For example, no effect was observed for seeding. Also, no difference in transformation rate occurred even for significantly higher API loading, and modifying the binder addition process from adding a solid powder to adding an aqueous solution had no effect on the rate of transformation. Two factors that did influence the granulation were mixing speed and using ball-milled AT.

In the above work, *in situ* Raman spectroscopy was shown to be capable of following the solvent-mediated psuedo-polymorphic conversion of the API. NIR spectroscopy could not be used to monitor the PIT due to the large absorbance of bulk water that effectively hid spectral information related to the transformation. NIR data chiefly provided information concerning the water content during granulation. Raman was shown to be capable of generating transformation kinetics even though the timescale of the conversion was relatively short.

Raman spectroscopy provides simplicity and amenability to online and in-line sampling, but for typically sampling probes, the small sampling volume analyzed can lead, especially for solid samples, to results prone to questions regarding subsampling. A method that samples a larger area would mitigate this problem. Wikstrom et al. (2005b) evaluated such an approach by comparing a standard fiber optic probe from Kaiser Optical Systems Inc. (Kaiser) with either immersion sampling or a noncontact sampling to the P^hAT System wide-area sampling approach also from Kaiser.

The P^hAT System offers several advantages over traditional Raman probes for solids analysis. It uses a noncontact probe, so noninvasive sampling can be accomplished. This means that material does not stick to the face of the optic and contamination of the analyte is less likely with the P^hAT System than with traditional *in situ* immersion probes. Data is collected from a spot approximately 6 mm in diameter, compared to about 60 μm for the immersion probe, minimizing the potential for subsampling. The larger spot size also offers better statistical sampling of heterogeneous mixtures.

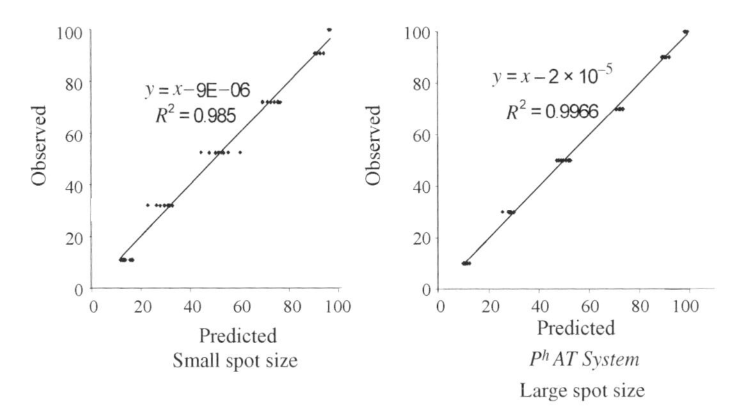

FIGURE 5.19 Calibration curves for theophylline obtained with (**a**) a standard system utilizing an immersion optic and (**b**) the P^hAT System. (Reproduced with permission. © 2006 Kaiser Optical Systems, Inc.)

Fig. 5.19 contains the calibration plots for data from theophylline obtained with the immersion optic based Raman analyzer and for the P^hAT System. The data from the P^hAT System is more repeatable, with $R^2 = 0.997$, compared to 0.985 for the immersion optic.

In addition to the P^hAT System providing more reproducible data, spectra were acquired using a 1-s exposure with four co-additions every 10 s. By comparison, the immersion optic based approach required 5-s exposures with four co-additions every 30 s. Thus, Wikstrom et al. demonstrated that the P^hAT System provided a threefold sampling time advantage. Wikstrom et al. (2005b) further demonstrated that this approach was relatively insensitive to focus over the distance of inches which facilitated initial alignment and proved to be advantageous for a bed of solid materials whose height fluctuated during the course of the operation.

During wet granulation, the granulated material is heated to drive off the water used in the mixing phase. Bryn et al. (2006) have reported on the successful use of Raman spectroscopy for the study of drying of API materials. In their study, a solvate was dried that could form two different hydrates depending on the ambient humidity. At high relative humidity (RH), the desired hydrate (hydrate II) was produced, while at low humidity, an unwanted hydrate I resulted. An online Raman probe was used to determine the appropriate conditions to maximize formation of the desired hydrate. The advantage of Raman was that it was used in a noncontact mode, and in real time could provide information on the solid form produced. Bryn et al. noted that the use of a Raman analyzer allowed production of the correct solid form at small scale as well as provided process critical information on the process of benefit to the scale-up team. Both Raman and NIR can be successfully applied to drying studies. A determination on which technology to apply for a specific drug substance or product will depend on the solvent and the API types as well as the effects of formation of the undesirable polymorph, solvate, or hydrate on the medicinal properties of the DP.

Hausman et al. (2005b) reported on the application of online Raman spectroscopy for characterizing the relationship between a drugs hydration state and the corresponding tablets physical stability. In their study, risedronate sodium was used as the API. In this study, the API was wet granulated and then fluid bed dried to moisture contents between 1 and 7%. Online Raman was used to monitoring the drying process *in situ*. It was shown that final granulation moisture had a significant effect on tablet thickness over time. During drying, dehydrated risedronate sodium can be formed; the dehydrated form may rehydrate over time, firstly causing an increase in tablet thickness and subsequently causing a loss in tablet integrity. The use of online Raman during drying enabled establishment of relationships between the fundamental hydration dynamics associated with risendronate sodium and the performance attributes of the DP tablets. It is anticipated that further publications on the opportunities for and successes of Raman spectroscopy for studying drying can be anticipated in the next few years.

Blending is a critical process in the manufacture of pharmaceutical drug products. A required result of blending is that the active pharmaceutical ingredient is homogeneously distributed. The most widely used method to assess drug content uniformity is to sample the stationary blend and then conduct some type of off-line analysis, traditionaly an HPLC method. The very process of sampling however disturbs the powder bed and can lead to nonrepresentative sampling (Mussio et al., 1997). As a known area for improvement much effort has been expended on eliminating the sampling step, and thus errors, using online in-situ PAT tools. Vergote et al. (2004) used Raman for in-line monitoring of blending of a mixture of diltiazem pellets and paraffinic wax beads (50/50 w/w).

Hausman et al. (2005a) demonstrated the use of online 785-nm excited Raman for analyzing low dose blend uniformity. They studied a 1% dose of azimilide dihydrochloride formulated with spray-dried lactose, crospovidone (disintegrant), and magnesium stearate (lubricant). The Raman-based results generated were shown to be well correlated with HPLC uniformity measured without the need of thief sampling, and much more rapidly than the by HLPC. The authors noted that the Raman approach could be extensible to automated control of blending endpoint.

LaPlant and Romero (2005) reported on the successful application of Raman to study blending, coating, lyophillization, and polymorph content in tablets. In Fig. 5.20 the experimental interface used by LaPlant and Romero (2005) for studying blending is shown. The noncontact Raman probe interfaced to the sample via an existing sapphire window. Unlike NIR analyzers used for blend monitoring, the Raman analyzer was not physically attached to the blending and thus counter balancing was not necessary. The blender was rotated at 10 Hz and Raman spectra measured using 0.1 s measurement time and a 785-nm laser. In Fig. 5.21 the Raman

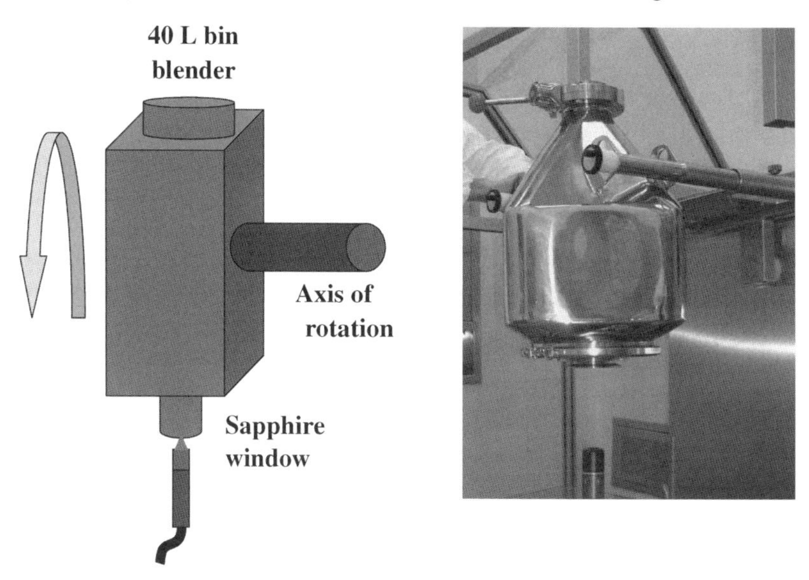

FIGURE 5.20 Diagrammatic representation and photograph of the experimental blending interface. Adapted from F. LaPlant and S. Romero, Proceedings of Midwest Pharmaceutical Process Chemistry Conference (MPPCC), Ann Arbor, MI, 2005.

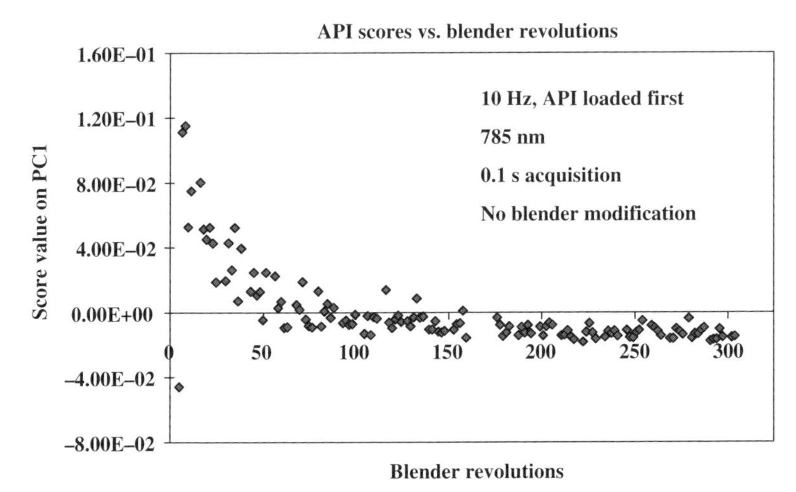

FIGURE 5.21 Results of the blending experiment. Adapted from F. LaPlant and S. Romero, Proceedings of Midwest Pharmaceutical Process Chemistry Conference (MPPCC), Ann Arbor, MI, 2005.

results are shown as a function of time. As can be seen from the figure, the API signal decays to and reaches a steady state after approximately 200 revolutions.

Jayawickrama et al. (2006) evaluated the potential for Raman spectroscopy for several drug manufacturing steps and processes. Four applications were studied:

1. Blend uniformity of granules—the team at BMS successfully interfaced a P^hAT System probe head to the feed conduit to a tablet press to assess blend uniformity. Blend uniformity can be used as an indicator of final drug product content uniformity (CU). Raman was used quantitatively to assess the concentration of API in tablets during a continuous tablet compression cycle. The results of in-line Raman were in good agreement with that of off-line HPLC. Noncontact Raman was demonstrated as a method for monitoring blend uniformity. Approximately one doseage unit was measured per analysis. The Raman method was shown to be capable of detection undesired process deviations.
2. The authors also interfaced a P^hAT System probe head to a high shear wet granulator. Raman analysis was used to monitor an API form change during high shear wet granulation. This approach was used to provide understanding on the kinetics of the transformation process.
3. Raman analysis was used to study API suspension in molten excipients—an encapsulated formulation application. Raman was shown to be capable of following the kinetics of the process, the homogeneity (or lack thereof) of API distribution, and process end point.
4. Raman analysis was used to study API solubilization in molten excipients—a second encapsulated formulation application. Raman was shown to be

capable of following solubilization. Raman was shown to be a safer approach than tradition sampling methods due to the noncontact nature of the method and the high potency of the drug material.

The work by Jayawickrama et al. (2006) further illustrated the potential of Raman spectroscopy as a viable alternative to traditional analytical methods in terms of analysis speed, nondestructive measurement, noncontact analysis, in-line, and agreement with the results of traditional methods.

Aguirre-Mendez and Romanach (2007) have studied the use of Raman for monitoring magnesium stearate in blends and final tablets. They were able to develop a method that could detect magnesium stearate down to concentration levels of 0.1% (w/w).

Tablet spray coating is a significant operation in the pharmaceutical industry with the objective product consistently coated tablets (LaPlant and Romero, (2005). The coating process has traditionally been monitored in a discontinuous fashion by removing samples and evaluating them using weight gain analysis or by a chromatographic method. Romero-Torres et al. (2005) first noted the potential for Raman to be used for measuring of tablet-to tablet coating variability. However, due to sampling limitations, inherent to small-spot Raman analysis, Romero-Torres et al. employed a complex revolving laser approach in order to obtain reproducible quantitative data.

El Hagrasy et al. (2006a,b) described a noncontact Raman fiber optic probe approach as a method for noninvasive and rapid in-line monitoring of tablet coating using a commercially available wide area, large-spot P^hAT System Raman analyzer. The team from BMS was able to successfully integrate a P^hAT System probe head to a production-scale spray coating. An example of a $\mathrm{P^hAT}$ System probe interfaced to a tablet coater is shown in Fig. 5.22.

During the coating run, the intensity of the signal arising from the titanium dioxide based coating material can be seen to increase as a function of coating time. A quantitative calibration model was developed from five batches that enabled the prediction of the amount of coating on tablets. The model developed was capable of predicting the percentage weight gain at the end of the coating run exceed the 2% (w/w) minimum set limit. For tablet coating, the effect of coating pan speed on the Raman signal was measured. For speeds of 0, 6, 9, 12, and 15 rpm it was found that no significant variability in the Raman signal was detected for pan speeds from 6 to 12 rpm. This attribute was noted by El Hagrasy et al. (2006a,b) as further evidence of the robustness of the Raman method.

The authors noted that the Raman method proved to be much faster than traditional methods including chromatographic and weighting methods, and that Raman spectroscopy proved to be a strong alternative to traditional analytical methods in terms of analysis speed, nondestructive, and noncontact nature. The approach noted provides valuable estimates of the coating weight however it does not guarantee even coating distribution. El Hagrasy et al. (2006a,b) noted that this Raman analysis could contribute to a process of consistently higher product quality.

Compton and Compton (1991) noted that Raman could be used for studying packaged consumer goods including pharmaceuticals *in situ*. The authors noted

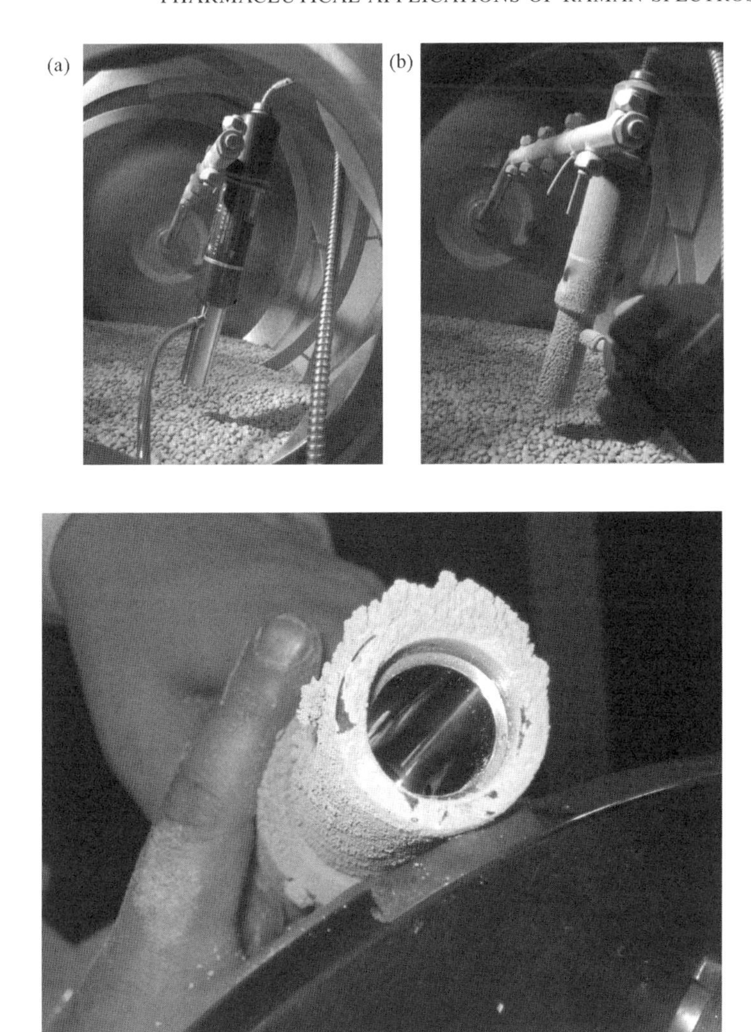

FIGURE 5.22 Picture of the P^hAT System probe head interfaced to a Bohle tablet coater: (**a**) the probe prior to coating; (**b**) the probe after the coating run was completed, and (**c**) the probe window showing the optical window after coating. (Reproduced with permission. © 2006 Kaiser Optical Systems, Inc.)

that the Raman approach could be extended to detect product tampering and product change (shelf life).

Skoulika and Georgiou (2001) developed a quantitative NIR Raman-based method for the antibacterial agent, ciprofloxacin, for several solid dosages forms from different manufacturers. Both univariate approaches using the Raman band's area and intensity indicative of ciprofloxacin and a multivariate approach using band areas were used to develop the quantitative models. The authors noted that the method developed allowed nondestructive quantitation of the sample within 30 s, which was in excellent agreement with the then United States Pharmacopoeia (US 24) and National Formulary (NF 19) method [USP (1999)], The authors noted that an extension of their work would allow for *in situ*, nondestructive measurements of the solid dosage forms through their PVC blister packs similar to the conducted by Crompton and Crompton in 1991 (Crompton and Crompton, 1991). In 2001, Taylor and Langkilde (2001) published an evaluation on the suitability of Raman spectroscopy for solid-state forms analysis for tablet quality control. The authors were able to detect a 5% polymorph impurity in a tablet containing 25% (w/w) API, with a limit of detection of 1.25% (w/w, total).

Dyrby et al. (2002) applied both NIR transmission and FT-Raman for quantitation of an API in Escitalopram tablets (5, 10, 15, and 20 mg tablets) using PLS models. The authors found that despite observing baseline separated CN band using Raman, the use of small-spot Raman gave inferior quantitative performance in comparison to transmission NIR.

LaPlant and Zhang (2005) noted that a number of techniques can be used to measure polymorph composition in pharmaceutical solids. Raman was chosen by LaPlant and Zhang because it offered high selectivity, allowing differentiation of using polymorphs and minimal sample preparation (preparation is known to change form in some drugs). It was noted that subsampling has been a limitation of Raman polymorph measurements using traditional instrumentation. Therefore, LaPlant and Zhang used large-spot Raman instrumentation. They were able to quantitate the polymorphic form in drug product tablets and chose to use multivariate analysis methods as they accounted for more possible variables in the sample. It was shown that for fluorescent samples, the effect of fluorescence on the quantitation could be minimized by optimizing the spectral range used in the model (and prediction).

Wikstrom et al. (2006) recently published on the use of Raman for the online measurement of (CU) in intact tablets. The authors concluded that in order to avoid subsampling for the Raman instrument that they used (0.5 mm spot size), the tablets should either be rotated (off-line) or for online an average along the length of the tablet be used. The authors found that the best agreement between the Raman method and the HPLC method was for off-line analysis where the tablet could be rotated and an average spectrum generated. The use of a large-spot approach would have provided for greater agreement between the online Raman and HPCL results and (S. Romero-Torres, personal communication, 2006).

Matousek and Parker (2006) conducted a study of collection geometries for large-spot Raman measurement of tablets. They compared backscatter Raman analysis of tablet versus transmission Raman analysis for the quantitation of API content in tablets. The major conclusion of this work is that transmission Raman can be accomplished even with thick compressed samples (size greater than 7 mm). The transmission approach allowed greater penetration of the sample and thus more reproducible results than could be obtained with a single-sided tablet analysis using backscatter geometry.

Ellison et al. (2004) have used Raman to quantify degradant levels in intact tablets. In this study, commercial uncoated furosemide tablets were used. A range of degradant level was achieved by storing different batches of tablets under different conditions for different periods of time. This resulted in degradant levels ranging from 0 to 1.56% of the tablet weight. The measured spectra were treated with second derivative processing to remove baseline variation and then PLS analysis for degradant quantitification. A large-spot (3 mm) fiber coupled probe was used for this study. The authors demonstrated that nondestructive quantification of a degradant of furosemide in whole tablets ranging from 0 to 1.56% of the tablet weight was shown to be possible for analyte contents greater than or equal to 0.1%.

Depending on the tablet size, a Raman method may not sample the whole tablet. As long as the subsection sample is indicative of the bulk tablet, this limitation is acceptable. Li et al. (2003) have shown for NIR analysis that analysis of a fraction of a tablet is a valid substitute for analyzing the whole tablet (as is the case with traditional wet-chemistry methods). Therefore, it is not necessary for the Raman method to sample the complete dimensions of the tablet. In the USP compendial chapter, it is recommended that if a Raman backscatter geometry is used, then analysis of both sides of the tablets and potentially several spots along the tablet are measured to verify analytical reproducibility.

The potential for Raman appears not to be limited to tablets. Niemczyk et al. (1998) used a fiber optic based Raman instrument to quantitatively determine the concentration of bucindolol in intact gel capsules. The authors stated that the method is shown to have significant potential as a nondestructive quality control method for pharmaceutical samples.

During collaborations between Kaiser and the pharmaceutical industry, the authors of this chapter used large-spot Raman to sample several different types of gel capsules with different colors including yellow and brown successfully with 785-nm excitation.

The above summary illustrates numerous examples where Raman can be used in secondary/drug product manufacturing. In some cases, other technologies may also be applied to these problems. The selection of Raman should be made based on several criteria including the chemical structure of the drug, the drug's Raman scattering cross section, the drug concentration in the tablet, the signal-to-noise ratio of the Raman system, the analysis time required, the benefits of noncontact, nondestructive sampling, the effect of the spectral signal of the solvent/excipients, and the effect of other effects such as lighting or fluorescence. In summary, the selection of a PAT tool for a specific task should be done based on an evaluation of whether the tool is appropriate for the task that it is being asked to do—this holds true whether the tool is Raman-based or not.

Prior to moving on, it should be noted that for the majority of applications described above, the measured Raman spectrum from the sample was not baseline flat from the analyzer. In general chemistry texts that introduce Raman often show a spectrum with zero background. In the industry world, this situation is rarely encountered. Often a residual undulating background attributable to excipient fluorescence can be seen. As with other techniques, it is often necessary to remove the effects of baseline in a preprocessing step such as the use of derivatives, polynomials, or singling out relative flat regions of the spectrum for the analysis. The take-home messages are firstly that appropriate data preprocessing is essential for many PAT tools including Raman and secondly the appearance of background fluorescence in the raw Raman spectrum (as long as it does not limit the SNR of the analytes of interest in the Raman spectrum to an unacceptable level) does not prevent the successful use of Raman in drug product applications.

5.9.5 Raman Opportunities Outside of PAT

Raman has been used in many non-PAT pharmaceutical applications that can be beneficial as well. Coupled with microscopy, Raman can be used to accomplish high throughput screening (HTS) for polymorphic forms (Almarsson et al., 2003; Anderton, 2004; Desrosiers, 2004; Maimald, 2006; Pommier et al., 2005; Remenar et al., 2003; Saal, 2006). Imaging and mapping can also be performed to check for counterfeiting or adulteration (Wikowski, 2005), drug characterization in DP formulations (Clark, 2002; Clarke et al., 2001; de Veij et al., 2005; Henson and Zhang, 2006; McGeorge, 2003; Sasic and Clark, 2006). Imaging may be a technique that in the future could translate into a high speed process monitoring tool. Raman has also been coupled with other techniques to aid in product understanding, such as thermal analyses (Collins et al., 2005; Kern et al., 2005), gravimetric analysis (Gift and Taylor, 2007), lyophillization (LaPlant and Torres, 2005), and TLC plates (Rao, 1993). Final, the USP (USP, 1120, 2005) and EP (European Pharmacopoeia, 2005) have recently released a compendial general chapter and a monograph on Raman spectroscopy, respectively.

For further detailed information on non-PAT applications of Raman spectroscopy, the reader is referred to other chapters in this work.

5.10 CONCLUSIONS

Raman analysis is a technique that is gaining momentum in the realm of PAT applications. Raman, when applied appropriately, can lead to better process understanding, process optimization, and process monitoring. Fiber optic coupling and the ability to do nondestructive analysis allows Raman to be used effectively for in-process monitoring. Raman provides the advantage of high spectral density leading to extensive chemical information. This means that diagnostic information concerning the molecular basis for process excursions is possible. Simple, straightforward quantitative calibrations can also be accomplished because of the potential for isolated diagnostic peaks. With the development of ATEX compliant Raman analyzers for liquid phase primary applications and large-spot analysis for solid-state analysis, the instrumental barriers to routine Raman analysis in manufacturing have been overcome. It is the authors predication that since instrumental road blocks to Raman acceptance have been removed, the value of Raman as a PAT tool in all areas, from discovery to pharmaceutical manufacturing, will become more established.

REFERENCES

Adar F, Geiger R, Noonan J, 1997. Raman spectroscopy for process/quality control. *Appl. Spectrosc. Rev.*, **32**, 45–101.

Aguirre-Mendez C, Romanach RJ, Raman A, 2007. spectroscopic method to monitor magnesium stearate in pharmaceutical blends in tablets, submitted.

Almarsson O, Hickey MB, Peterson ML, Morissette SL, Soukasene S, McNulty C, Tawa M, J MacPhee M, Remenar JF, 2003. High-throughput surveys of crystal form diversity of highly polymorphic pharmaceutical compounds. *Crys. growth Des.*, **3**, 927–933.

Anderton C, 2004. A valuable technique for polymorph screening. *Eu. Pharm. Rev.*, **9**, 2, 68–74.

Andrews JJ, Browne MA, Clarke IE, Hancewitz TM, Millichope AJ, 1998. Raman imaging of emulsion systems. *Appl. spectrosc.*, **52**, 790–796.

Andrews JJ, Hancewitz TM, 1998. Rapid analysis of Raman image data using two-Way multivariate curve resolution. *Appl. spectrosc.*, **52**, 797–807.

Arrivo S, 2003. The role of PAT in pharmaceutical research and development. *Am. Pharm. Rev.* **6**, 2, 46–53.

Barnes RJ, Dhanoa MS, Lister SJ, 1989. Standard normal variate transformation and de-trending of near infrared diffuse reflectance spectra. *Appl. spectrosc.*, **43**, 772–777.

Bauer C, Amram B, Agnely M, Charmot D, Sawatski J, Dupuy D, Huvenne JP, 2000. On-line monitoring of a latex emulsion polymerization by fiber-optic FT-Raman spectroscopy. Part 1. Calibration. *Appl. spectrosc.*, **54**, 528–535.

Beebe KR, Pell RJ, Seasholtz MB, 1998. Chemometrics: A Practical Guide, Wiley, New York, NY.

Besson JP, King PWB, Wilkins TC, McIvor MC, Everall NJ, 1996. Calcination of titanium dioxide. *European Patent Application EP* 0767222A2.

Bilhorn RB, Sweedler JV, Epperson PM, Denton MB, 1987. Spectrochemical measurements with multichannel integrating devices. *Appl. spectrosc.*, **41**, 1125–1136.

Bowie BT, Chase DB, Griffiths PR, 2000. Factors affecting the performance of bench-top Raman spectrometers. Part 1: Instrumental effect. *Appl. spectrosc.*, **54**, 164A–173A.

Bowie BT, Chase DB, Griffiths PR, 2000. Factors affecting the performance of bench-top Raman spectrometers. Part 2. Effect of sample, *Appl. spectrosc.*, **54**, 200A–207A.

Bryn S, Liang J, Bates S, Newman AW, 2006. PAT—Process understanding and control of active pharmaceutical ingredients, PAT, **3**(6), 14–19.

Chalmers J, Griffiths PR, (Eds.), *Handbook of Vibrational Spectroscopy*, Vols. 1–5. Wiley, Chichester, UK, (2002).

Chase DB, Rabolt JF, 1994. Fourier transform Raman spectroscopy: From concept to experiment. In *Fourier Transform Raman spectroscopy: From Concept to Experiment*, Chase DB, and Rabolt JF, (Eds.) Academic Press, San Diego, Chapter 1, pp. 1–48.

Clark D, 2002. The analysis of pharmaceutical substances and formulated products by vibrational spectroscopy. In *Handbook of Vibrational Spectroscopy* Vol. 5, Chalmers J, Griffiths PR, (Eds.), Wiley, Chichester, UK, pp. 3574–3589.

Clarke FC, Jamieson MJ, Clark DA, Hammond SV, Jee RD, Moffat AC, 2001. Chemical Image Fusion. The synergy of FT-NIR and Raman mapping microscopy to enable a more complete visualization of pharmaceutical formulations. *Anal. Chem.*, **73**, 2213 2220.

Claybourn M, Massey T, Highcock J, Gogna D, 1994. Analysis of processes in latex systems by FT-Raman spectroscopy. *J. Raman Spectrosc.*, **25**, 123–129.

Clegg IM, Everall NJ, King B, Melvin H, Norton C, 2001. On-line analysis using Raman spectroscopy for process control during the manufacturing of titanium dioxide. *Appl. spectrosc.*, **55**, 1138–1150.

Collins WJ, DuBois C, Cambron RT, Redman-Furey NL, Bigalow Kern AS, 2005. Development and evaluation of a TG/DTA/Raman System. *J. ASTM Int.*, **2**(9), 1–10.

Compton DAC, Compton SV, 1991. Examination of packaged consumer goods by using FT-Raman spectrometry. *Appl. spectrosc.*, **45**, 1587–1589.

Cooper JB, Flecher PE, Vess TM, Welch WT, 1995. Remote fiber-optic raman analysis of xylene isomers in mock petroleum fuels using a low-cost dispersive instrument and partial least squares regression analysis. *Appl. spectrosc.*, **49**, 586–592.

Cooper JB, 1999. Process control applications for Raman spectroscopy in the petroleum industry. In *Analytical Applications of Raman Spectroscopy*, Pelletier MJ, (Ed.) Blackwell Science, Chapter 5, and references therein.

Dell Orco P, Diederich A, Rydzak JW, 2002. Designing for crystalline form and crystalline physical properties: The role of *in situ* and *ex situ* monitoring techniques. *Am. Pharm. Rev.*, **5**(2), 46–54.

Desrosiers PJ, 2004. The Potential of preform. *Mod. Drug Discov.*, **7**, 1, 40–43.

de Veij M, Vandenabeele P, Moens L, 2005. The rise of Raman. *Eu. Pharm. Rev.*, **3**, 86–89.

Dyrby M, Engelsen SB, Norgaard L, Bruhn M, Lundsberg-Nielsen L, 2002. Chemometric quantitation of the active substance (containing CN) in a pharmaceutical tablet using near-Infrared (NIR) transmission and NIR FT-Raman Spectra. *Appl. spectrosc.*, **56**, 579–585.

El Hagrasy A, Chang S-Y, Desai D, Kiang S, 2006. Application of Raman spectroscopy for quantitative in-line monitoring of tablet coating. *Am. Pharm. Rev.*, **9**(1), 40–45.

El Hagrasy A, Chang S-Y, Desai D, Kiang S, 2006. Raman spectroscopy for the determination of coating uniformity of tablets: Assessment of product quality and coating pan mixing efficiency during scale-Up. *J. Pharm. Innov.*, (Sept./Oct.) 37–42.

Ellison C, Lyon R, Jefferson E, Faustino P, Strachan D, Kemper M, Lewis I, Hussain A, 2004. Raman spectroscopy using a large spot laser for degradant quantification in an intact tablet. In AAPS *Annual Meeting and Exposition*, November 7–11, Baltimore, MD, Poster T3050.

European Pharamacopiea, Monograph 2248—Raman Spectrometry Council of Europe, Strasbourg, (2005).

Everall NJ, Owen H, Slater J, 1995. Performance analysis of an integrated process Raman analyzer using a multiplexed transmission grating, CCD detection, and confocal fiber-optic sampling. *Appl. Spectrosc.*, **49**, 610–615.

Everall NJ, 1998. Measurement of orientation and crystallinity in unixial drawn poly(ethylene terephthalate) using polarized confocal Raman microscopy. *Appl. spectrosc.*, **52**, 1498–1504.

Everall NJ, King B, 1999. Raman spectroscopy for polymer characterization in an industrial environment. *Macromol. Symp.*, **141** 103–116.

Everall NJ, Clegg I, King B, 2002. Process measurements by Raman spectroscopy. In *Handbook of Vibrational Spectroscopy* Vol. 4, Chalmers J, Griffiths PR, (Eds.), Wiley Chichester, UK, pp. 2770–2801.

Farquharson S, Simpson SF, 1992. Application of Raman spectroscopy to industrial processes, SPIE, Bellingham, WA, SPIE 1681, 276–290.

Feld MS, Fisher DO, Freeman JJ, Gervasio GJ, Hochwalt MA, Laskowski LF, Thomsen EE, 1993. Process for preparing phosphoru trichloride. *US Patent* 5,260,026, November 9.

Feld MS, Fisher DO, Freeman JJ, Gervasio GJ, Hochwalt MA, Laskowski LF, Thomsen EE, 1994. Process for preparing phosphorus trichloride. *US Patent* 5,310,529, May 10.

Folestad S, Johansson J, 2003. Opening the PAT Toolbox. *Eu. Pharm. Rev.*, **8**(4), 36–42.

Frank CJ, 1999. Review of pharmaceutical applications of Raman spectroscopy. In *Analytical Raman spectroscopy*, Pelletier MJ, (ed.), Blackwell Science, Oxford, UK, 224–275.

Freeman JJ, Fisher DO, Gervasio GJ. 1993. FT-Raman on-line analysis of PCl_3 reactor material *Appl. Spectrosc.*, **47**, 1115–1122.

Garrisson AA, Moore CF, Roberts MJ, Hall PD, 1992. Distillation process control using Fourier transform spectroscopy. *Process Control Qual.*, **3**, 57–63.

Garrison AA, Martin MZ, 1993. Fourier transform Raman spectroscopy—Application to process control. In *Proceedings of 9th International Conference on Fourier Transform Spectroscopy*, Bertie JE and Wieser H, (Eds.), SPIE, Bellingham, WA, SPIE 2089, pp. 210–211.

Geladi P, MacDougall D, Martens H, 1985. Linearization and scatter correction for near-infrared reflectance spectra of meat. *Appl. Spectrosc.* **39**, 491–500.

Gervasio GJ, Pelletier MJ, 1997. On-line Raman analysis of PCl_3 reactor material. At-*Process*, **3**, 7–11.

Gift AD, Taylor LS, 2007. Hyphenation of Raman spectroscopy with gravimetric analysis to interrogate water–solid interactions in pharmaceutical systems. *J. Pharm. Biomed. Anal.*, **43**, 14–23.

Gilbert AS, Hobbs KW, Reeves AH, Hobson PP, 1994. Automated headspace analysis for quality assurance of pharmaceutical vials by laser Raman spectroscopy in optical measurements and sensors for the process industries, Gorecki C, Preater RWT, (Eds.); *SPIE Process*. Vol. 2248, SPIE, Bellingham, WA, pp. 391–398.

Hancewicz TM, Petty CJ, 1995. Quantitative Analysis of vitamin A using fourier transform Raman spectroscopy. *Spectrochim. Acta.*, 51A, 2193–2198.

Hausman DS, Cambron RT, Sakr A, 2005a. Application of Raman spectroscopy for on-line monitoring of low dose blend uniformity. *Int. J. Pharm.*, **298**, 80–90.

Hausman DS, Cambron RT, Sakr A, 2005b. Application of on-line Raman spectroscopy for characterizing relationships between drug hydrate state and tablet physical stability. *Int. J. Pharm.*, **299**, 19–33.

Helland IS, Naes T, Isaksson T, 1995. Related versions of the multiplicative scatter correction method for preprocessing spectroscopic data. *Chemom. Intell. Lab. Syst.* **29**, 233–241.

Hendra PJ, Jones C, Warnes G, (Eds.): *Fourier Transform Raman Spectroscopy: Instrumentation and chemical applications*. Ellis Horwood, Chichester (1991).

Henson MJ, Zhang L, 2006. Drug characterization in low dosage pharmaceutical tablets using Raman microscopic mapping. *App. Spectrosc.*, **60**, 1247–1255.

Heys JR, Powell ME, Pivonka DE, 2004. Real time Monitoring of tritium gas reactions using Raman spectroscopy. *J. Labelled Compd. and Radiopharm.*, **47**, 983–995.

Hirschfeld T, Chase DB, 1986. FT-Raman spectroscopy: Development and justification. *Appl. spectrosc.*, **40**, 133–137.

Hu Y, Liang JK, Myerson AS, Taylor LS, 2005. Crystallization monitoring using Raman spectroscopy—Simultaneous measurement of desupersaturation profile and polymorphic form in flufenamic acid systems. *Ind. Eng. Chem. Res.*, **44**, 1233–1240.

Hu Y, Wikstrom H, Byrn SR, Taylor LS, 2006. Analysis of the effect of particle size on polymorphic quantitation by Raman spectroscopy. *Appl. Spectrosc.*, **60**, 978–984.

Jayawickrama D, El Hagrasy A, Chang S-Y, 2006. Raman applications in drug manufacturing processes. *Am. Pharm. Rev.*, **9**(7), 10–17.

Jestel N, 2005. Raman spectroscopy. In *Process Analytical Technology.* Katherine Bakeev A, (Ed), Blackwell Science, pp. 133–169, Chapter 5.

Kaiser Optical Systems, Monitoring grignard production in real-time. Applications Note 302 V1.0 (2001).

Kaiser Optical Systems, Diels–Alder reactions with real-time. Analytics, Applications Note 314 V2.0 (2006a).

Kaiser Optical Systems, Kinetics of a catalytic hydrogenation reaction, Applications Note 300 V2.0 (2006b).

Kern AS, Collins WJ, Cambron RT, Redman-Furey NL, Use of a TG/DTA/Raman system to monitor dehydration and phase conversions. *J. ASTM Int.*, **2**(7), 1–10.

Kramer R, 1998. *Chemometric Techniques for Quantitative Analysis*, Marcel Dekker, New York, NY.

LaPlant F, Zhang X, 2005. Quantitation of polymorphs in drug product by Raman spectroscopy. *Am. Pharm. Rev*, **8**(5), 88–95.

LaPlant F, Romero S, 2005. *Proceedings of the Midwest Pharmaceutical Process Chemistry Conference* (*MPPCC*), Ann Arbor, MI.

Lewis IR, Griffiths PR, 1996. Raman spectrometry with fiber-optic sampling, *Appl. spectrosc.* **50**, 12A–30A.

Lewis IR, 2000. Process Raman spectroscopy. In Handbook of Raman Spectroscopy: From Research Laboratory to Process Line, Lewis IR, Edwards HGM, (eds), Marcel Dekker, New York, NY, pp. 919–974.

Lewis IR, Edwards HGM, Eds. Handbook of Raman Spectroscopy: From Research Laboratory to Process Line, Marcel Dekker, New York, NY.

Lewis IR, Rosenblum SS, 2002. General introduction to fiber-optics. In *Handbook of Vibrational Spectroscopy*, Chalmers J, Griffiths PR, (eds.), Wiley, Chichester, UK, Vol. 2, 1533–1540.

Lewis IR and Lewis ML, Fiber-optics probes for Raman spectrometry. In *Handbook of Vibrational Spectroscopy* Vol. 2, Chalmers J, Griffiths PR, (eds.), Wiley, Chichester, UK, pp. 1587–1597.

Lewis IR, 2005. Mechanistic study of a microwave-assisted organic reaction. *Spectrosc. Eur.*, **17**(4), 34–36.

Li TL, Donner AD, Choi CY, Frunzi GP, Morris KR, 2003. A statistical support for using spectroscopic methods to validate the content uniformity of solid doseage forms. *J. Pharm. Sci.*, **92**, 1526–1530.

Lipp ED, Grosse RL, 1998. On-line monitoring of chlorosilane streams by Raman spectroscopy. *Appl. Spectrosc.*, **52**, 42–46.

Long RL, Marrow D, 2003. On-line measurement and control of polymer properties by Raman spectroscopy. *International Patent* WO 03/042646 A3, May 22.

Long RL, Impelman RW, Chang S-Y, Audrey TJ, Yanh DA, Marrow D, 2005. On-line measurement and control of polymer properties by Raman spectroscopy. *International Patent* WO 2003/049663 A3, June 2.

Maiwald M, 2006. Examining the range of methods available to derive numerous polymorphic forms. *Am. Pharm. Rev.*, **9**, 4, 95–99.

Marteau P, Hotier G, Zanier-Szydlowski N, Aoufie A, Cansell F, 1994. Advanced control of C8 aromatic separation process with real time multipoint on-line Raman spectroscopy. *Process Control Qual.*, **6**, 133–140.

Marteau P, Zanier-Szydlowski N, Aoufie A, Hotier G, Cansell F, Remote F, 1995. Raman spectroscopy for process control. *Vib. Spectrosc.*, **9**, 101–109.

Matousek P, Parker AW, 2006. Bulk Raman analysis of pharmaceutical tablets. *Appl. Spectrosc.*, **60**, 1353–1357.

McCreery R, 2000. *Raman Spectroscopy for Chemical Analysis*. Wiley-Interscience, Wiley, Chichester, New York, NY.

McGeorge G, 2003. Combining Raman spectroscopy and microscopy to support pharmaceutical development. *Am. Pharm. Rev.*, **6**(3), 94–99.

Muzzio FJ, Robinson P, Wightman C, Brone D, 1997. Sampling practices in powder blending, *Int. J. Pharm.*, **155**, 153–178.

Niemczyk TM, Delgado-Lopez MM, Allen FS, 1998. Quantitative determination of bucindolol concentration in intact gel capsules using Raman spectroscopy. *Anal. Chem.*, **70**, 2762–2765.

Owen H, 2002. Volume phase holographic optical elements. In *Handbook of Vibrational Spectroscopy*, Vol. 1, Chalmer J and Griffiths PR, (eds.), Wiley, Chichester, UK pp. 482–489.

Parker SF, 1994. A review of the theory of Fourier-transform Raman spectroscopy. *Spectrochim. Acta*, 50A, 1841–56.

Pelletier MJ (ed.), *Analytical Raman Spectroscopy*, Blackwell Science, Oxford, UK (1999).

Pelletier MJ, 2003. Quantitative analysis using Raman spectrometry. *Appl. spectrosc.*, **57**, 20A–39A.

Pivonka DE, Empfield JR, 2004. real time *in situ* Raman analysis of microwave-assisted organic reactions. *Appl. Spectrosc.*, **58**, 41–46.

Pommier CJ, Rosso V, Song A, 2005. Challenges in data analysis of Raman spectra from crystallization in 96 wellplates. *Am Pharm. Rev*, **8**(2), 19–24.

Powell LP, Campion A, 1986. 1986. Rapid headspace analysis in sealed drug vials by multichannel Raman spectroscopy. *Anal. Chem.*, **58**, 2350–2352.

Raman CV, Krishnan KS, A New type of secondary radiation. *Nature* **121**, 501 (31 March, 1928), cabled to Nature on 16 February.

Rau A, 1993. Basic experiments in thin-layer chromatography-Fourier transform Raman spectrometry. *J. Raman Spectrosc.*, **24**, 251–254.

Ray C, Wethman R, Wasylyk J, 2005. Effectively using PAT in a process development environment to expedite processing in a pilot plant facility. *PAT*, **2**(2), 17–20.

Remenar JF, MacPhee JM, Larson BK, Tyagi VA, Ho JH, McIlroy DA, Hickey MB, Shaw PB, Almarsson O, 2003. Salt selection and simultaneous polymorphic assessment via high-throughput crystallization: The case of sertraline. *Org. Proc. Res. Dev.*, **7**, 990–996.

Roberts MJ, Garrisson AA, Kercel SW, Muly EC, 1991. Raman spectroscopy for on-line, real time, multi-point industrial chemical analysis. *Process Control Qual.*, **1**, 281–291.

Rodriguez-Hornedo N, Nehm SJ, Seefeldt KF, Pagan-Torres Y, Falkiewicz CJ, 2006. Reaction crystallization of pharmaceutical molecular complexes. *Mol. Pharm.*, **3**, 362–367.

Romero-Torres S, Perez-Ramos JD, Morris KR, Grant ER, 2005. Raman spectroscopic measurement of tablet-to-tablet coating variability. *J. Pharm. Biomed. Anal.*, **38**, 270–274.

Saal C, 2006. Quantification of polymorphic forms by Raman spectroscopy. *Am. Pharm. Rev.*, **9**(4), 76–81.

Sasic S, Clark DA, 2006. Defining a strategy for chemical imaging of industrial pharmaceutical samples on raman line-mapping abd gloal illumination instruments. *App. Spectrosc.*, **60**, 494–502.

Savitzky A, Golay M, 1964. Smoothing and differentiation of data by simplified least squares procedures. *Anal. Chem.*, **36**, 1627–1639.

Sawatski J, Charmot D, Amram B, Macron C, Huvenne J-P, Agnely M, Simon A, Lehner C, Asua J, Leiza J, Armitage P, 1998. NIR FT-Raman spectroscopy: On-line control of an emulsion polymerization for the laboratory to the production plant. In *Proceedings of the XVIth International Conference on Raman Spectroscopy*, Heynes A, (Ed.), Wiley, Chichester, UK, pp. 728–729.

Seasholtz MB, Archibald DD, Lorber A, Kowalski BR, 1989. Quantitative analysis of liquid fuel mixtures with the use of fourier transform near-IR Raman spectroscopy. *Appl. Spectrosc.* **43**, 1067–1072.

Shaver J, 2000. Chemometrics for Raman spectroscopy. In *Handbook of Raman Spectroscopy: From Research Laboratory to Process Line*, Lewis IR, and Edwards, HGM (ed). Marcel Dekker, New York, NY, pp. 275–306.

Skoulika SG, Georgiou CA, 2001. Rapid quantitative determination of ciprofloxacin in pharmaceuticals by use of solid-state FT-Raman spectroscopy. *Appl. Spectrosc.*, **55**, 1259–1265.

Slater JB, Tedesco JM, Fairchild RC, Ian R. Lewis, 2000. Raman spectrometry and its adaptation to the industrial environment. In *Handbook of Raman Spectroscopy: From Research Laboratory to Process Line*, Lewis IR and Edward, HGM (ed). Marcel Dekker, New York, NY, pp. 41–144.

Starbuck C, Spartalis A, Wai L, Wang J, Fernandez P, Lindemann CM, Zhou GX, Ge Z, 2002. Process optimization of a complex pharmaceutical polymorphic system via *in situ* Raman spectroscopy. *Crys. Growth Des.*, **2**(6), 515–522.

Strobl GR, Hagedon W, 1978. Raman spectroscopic method for determining the crystallinity of polyethylene. *J. Polym. Sci. Polym. Phys. Edn.*, **16**, 1181–93.

Svensson O, Josefson M, Langkilde FW, 2000. The synthesis of metoprolol monitored using Raman spectroscopy and chemometrics. *Eur. J. Pharm. Sci.*, **11**, 141–155.

Swierenga H, De Weijer AP, Buydens LMC, 1999. Robust calibration model for on-line and off-line prediction of PET yarn shrinkage by Raman spectroscopy. *J. Chemom.*, **13**, 237–249.

Taylor LS, Langkilde FW, 2001. Evaluation of solid-state forms present in tablets by Raman spectroscopy. *J. Pharm. Sci.*, **89** (10), 1342–1353.

Tumuluri VS, Kemper MS, Sheri A, Choi S-R, Lewis IR, Avery MA, Avery BA, 2006. The use of Raman spectroscopy to characterize hydrogenation reactions. *Org. Process Res. Dev.*, **10**(5) 927–933.

Tumuluri VS, Kemper MS, Lewis IR, Prodduturi S, Avery BA, Repka MA, in preparation.

Turrell G, Corset J, 1996. *Raman Microscopy—Developments and Applications*, Academic Press, London, UK.

US FDA, Draft Guidance for Industry PAT—A framework for innovative pharmaceutical development, manufacturing, and quality assurance, August 2002

US FDA, Guidance for Industry PAT—A framework for innovative pharmaceutical development, manufacturing, and quality assurance, September 2004.

The United States Pharmacopeia, The National Formilary, The United States Pharmacopeial Convention, Inc., Rockville, MA, US 24, 420 (1999).

The United States Pharmacopeia, USP General Chapter 1120, Raman Spectrometry, The United States Pharmacopeial Convention, Inc., Rockville, MA, US 29, (2005).

Vergote G, De Beer T, Vervaet C, Remon J, Baeyens W, Diericx N, Verpoot F, 2004. In-line monitoring of a pharmaceutical blending process using FT-Raman spectroscopy. *Eur. J. Pharm. Sci.*, **21**, 479–485.

Wang F, Wachter JA, Antosz FJ, Berglund KA, 2000. An investigation of solvent-mediated polmorphic transformation of progesterone using *in situ* Raman spectroscopy. *Org. Process Res. Dev.*, **4**, 391–395.

Wang Y, McCreery RL, 1989. Evaluation of a diode laser/charge coupled device spectrometer for near-infrared Raman spectroscopy. *Anal. Chem.*, **62**, 2647–2651.

Warman M, Hammond SV, 2005. Advantages of applications specific measurement systems. In *Proceedings of the Pittsburgh Conference*, Chicago, pp. 1680–1682.

Wethman R, Ray C, Wasylyk John, 2005. Development and implementation of an in-line quantitative Raman method for in process pharmaceutical monitoring. *Am. Pharm. Rev.*, **8**, 6, 57–63.

Wikstrom H, Marsac PJ, Taylor LS, 2005a. In-line monitoring of hydrate formation during wet granulation using Raman spectroscopy. *J. Pharm Sci.*, **94**, 209–219.

Wikstrom H, Lewis IR, Taylor LS, 2005b. Comparison of sampling techniques for in-line monitoring using Raman spectroscopy. *Appl. Spectrosc.*, **59**, 934–941.

Wikstrom H, Romero-Torres S, Wongweragiat S, Williams JAS, Grant ER, Taylor LS, 2006. On-line content uniformity determination of tablets using low-resolution raman spectroscopy. *Appl. Spectrosc.*, **60**, 672–681.

Williams AC, 2000. Some pharmaceutical applications of Raman spectroscopy. In *Handbook of Raman Spectroscopy: From Research Laboratory to Process Line*, Lewis IR and Edwards HGM, (eds.), Marcel Dekker, New York, NY, pp. 575–592.

Williams KPJ, Everall NJ, 1995. Use of micro Raman spectroscopy for the quantitative determination of polyethylene density using partial least squares calibration. *J. Raman Spectrosc.*, **26**, 427–433.

Williams P, Norris K, 1984. Optimization of mathematical treatments of raw near-infrared signal in the measurement of protein in hard red spring wheat. I. Influence of particle size. *Cereal Chem.*, **61**, 158–165.

Williamson JM, Bolling RJ, McCreery RL, 1989. Near-infrared Raman spectroscopy with a 783 nm diode laser and CCD array detector. *Appl. spectrosc.*, **43**, 372–375.

Wiss J, Zilian A, 2003. Online spectroscopic investigations (FTIR/Raman) of industrial reactions: Synthesis of tributyltin azide and hydrogenation of chloronitrobenzene. *Org. Proc. Res. Dev.*, **7**, 1059–1066.

Witkowski MR, 2005. The use of Raman spectroscopy in the detection of counterfeit and adulteration of pharmaceutical products. *Am. Pharm. Rev.*, **8**(1), 56–62.

Zhou G, Wang J, Ge Z, Sun Y, 2002. Ensuring robust polymorph isolation using in-situ Raman spectroscopy. *Am. Pharm. Rev.*, **5**(4), 74–80.

6

RAMAN CHEMICAL IMAGING OF SOLID DOSAGE FORMULATIONS

SLOBODAN ŠAŠIĆ
Pfizer Global Research and Development, Ramsgate Road, Sandwich CT13 9NJ, UK

Chemical imaging combines spectroscopic techniques with optical microscopes and digital imaging technologies to produce images in which the chemical distribution in a sample is revealed. In other words, the chemical identity of the components is identified in addition to their spatial distribution. Without using chemical imaging, a sample may appear as a uniform material: internal composition cannot be assessed by visual microscopy. Although the concept of chemical imaging is relevant to a variety of different analytical methods, only the vibrational spectroscopic techniques, Raman, IR, or NIR (near-infrared) will be considered in this chapter. The specific purpose of this chapter is to describe the Raman imaging techniques and data analysis tools that are typically used and also to detail various applications in the pharmaceutical industry.

In the context of dealing with pharmaceutical samples, use of chemical (Raman) imaging methods helps to transform the overwhelmingly white, featureless appearance of tablets or powders into detailed images where the constituent materials are indicated by colored pixels. Obviously, chemical imaging can be carried out with other sorts of solid materials as well, such as minerals, semiconductors, and foods. In all these application areas, imaging techniques are becoming increasingly popular because of more widely recognized benefits of learning from images and the continuous progress in the capabilities of the instrumentation. In the pharmaceutical industry, chemical distinction of the constituents and details of their spatial distribution (including adjacencies of components) may help to describe morphology

Pharmaceutical Applications of Raman Spectroscopy, Edited by Slobodan Šašić

(architecture) of tablets, homogeneity of blends, identity and distribution of impurities, and so on. The wealth of information provided may be used for various purposes in product development and during manufacturing. Normally, chemical images are obtained with minimal sample preparation and certainly without the use of any contrast agent (e.g., a dye). Also, although the acquisition time for an image may be significant (from hours to days), the actual procedures are automated so that labor cost is minimized. While in most cases the collection of experimental data is straightforward, the analysis of the data obtained can be much more complicated.

6.1 METHODS FOR CHEMICAL IMAGING

Conceptually, there are two ways to conduct a Raman imaging experiment. In the first approach, which is much more popular, Raman spectra are collected from a number of positions on the sample and each spectrum is tagged with spatial coordinates. The spectra can be obtained by exciting the sample by either point or line illumination and therefore this method is called point or line mapping. The chemical images are obtained after preprocessing the raw spectra and the method is called "Raman mapping", because images are not obtained directly.

The newer approach is dubbed wide-field or global illumination imaging (or global imaging) and refers to the collection of a series of images of the surface of the sample at a limited number of preselected wavenumbers. The major difference with regard to the previous approach is the simultaneous illumination and collection of the back-scattered Raman signal at discrete wavenumbers from an area, and not from a spot on the surface. This is a true imaging technique for the simple reason that chemical images thus produced are a direct result of the experiment.

A scheme describing these two concepts is given in Fig. 6.1 with a more detailed description being given below.

Particularly informative sources on this subject can be found in the literature by Treado and Nelson (2001), and McCreery (2000).

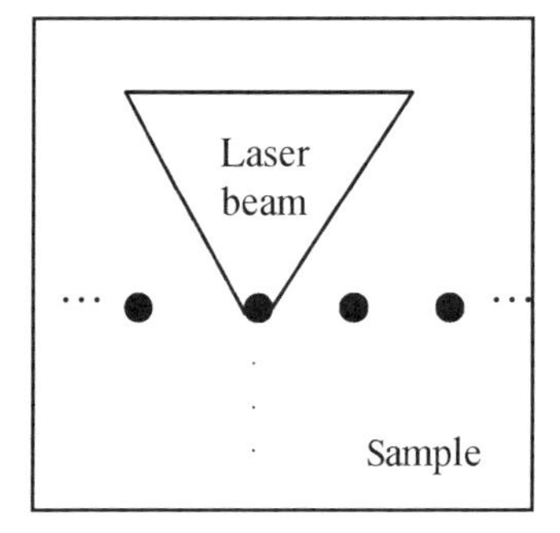

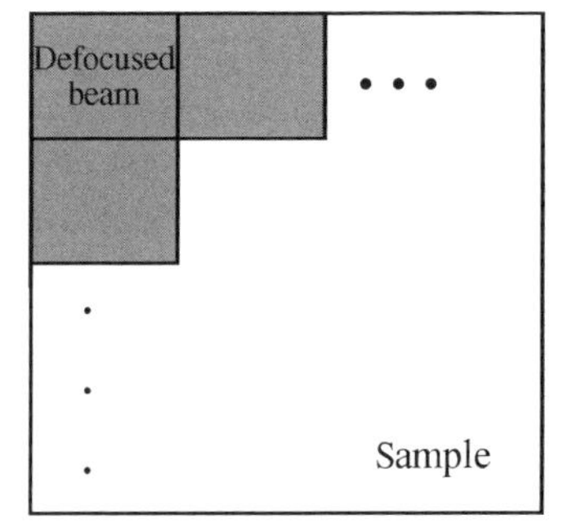

FIGURE 6.1 Schematic representation of principles of point mapping and global imaging.

6.1.1 Point and Line Mapping

The idea to use a Raman microscope to collect a large number of spectra from the surface of the sample via automated shifts of the microscope stage is a straightforward extension of the first Raman microprobe instruments developed in the 1970s (Delhaye and Dhamelincourt, 1975). Such instruments can be technically considered as combinations of a Raman spectrometer and a microscope. One possible layout of such instruments is given in Fig. 6.2. In the configuration shown in the Fig. 6.2, the laser beam is projected onto the sample via a holographic notch filter and a mirror. The back-reflected light, elastically back-scattered laser light, and inelastically scattered Raman scatter are introduced into the spectrometer after passing back through the holographic notch filter, which transmits Raman-shifted radiation but reflects the laser light so that only the Raman signal is collected on the charge coupled device (CCD) camera.

The result of an experiment is typically a very large number of spectra that is directly proportional to the area imaged and the desired spatial resolution (i.e., set by the preset shift of the microscope stage). For example, imaging of a $1 \times 1\,\text{mm}^2$ area in 10 μm steps (spatial resolution) would require 10,000 spectra. Such numbers lead to experiments that require very long acquisition times because, as already mentioned at several places in this book, Raman scattering is a very inefficient process and therefore long acquisition times have to be employed (no less than several seconds per stage position) for each spectrum.

The number of spectra collected is obviously equal to the number of pixels in a chemical image, which in turn corresponds to the quality of the image. Thus, obtaining high quality images normally requires long measurement time either because of the time needed for acquisition of spectra of acceptable quality (that the images are built from) or because of a large number of positions on the surface that are being imaged.

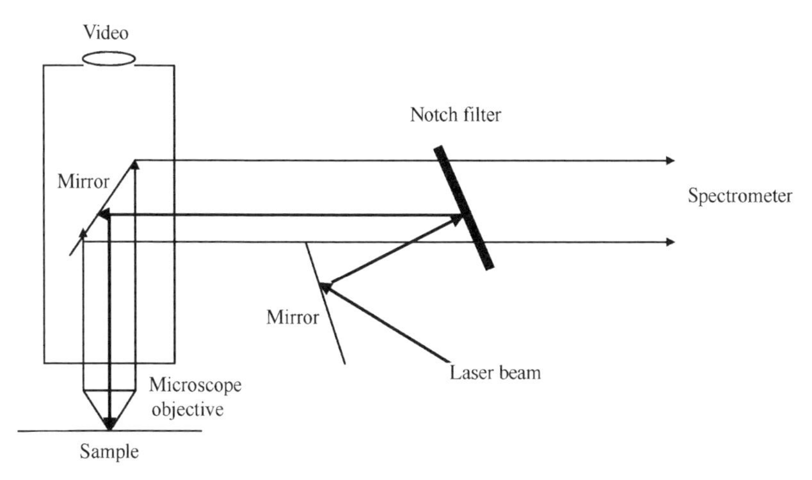

FIGURE 6.2 Scheme of the layout of a point/line Raman mapping system.

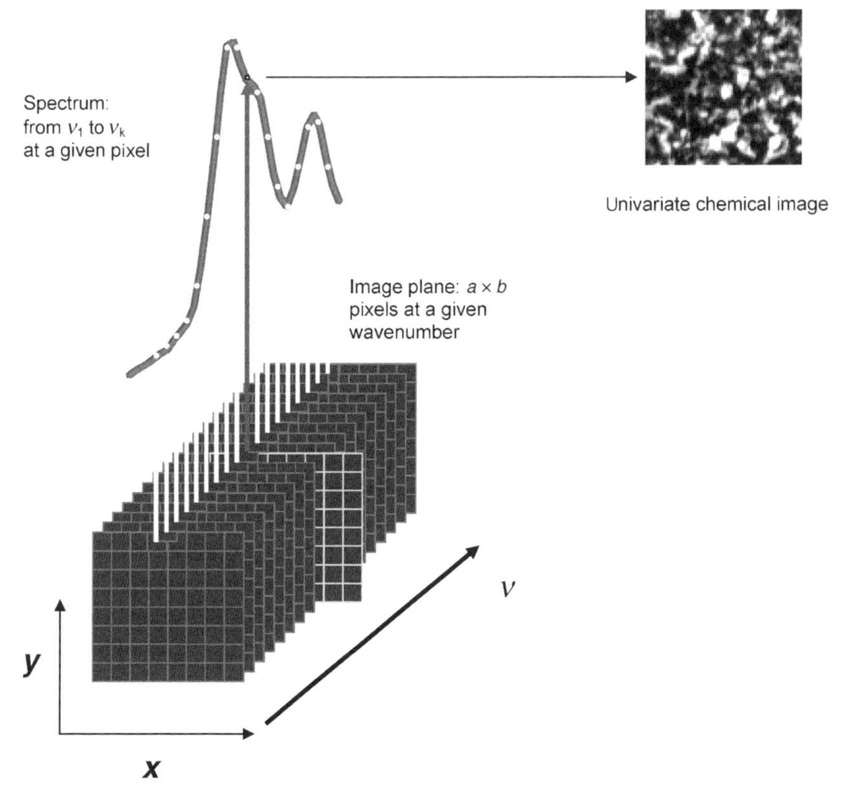

FIGURE 6.3 Schematic representation of the so-called hyperspectral data cube. There are three dimensions (or four if the intensity is counted), two spatial X and Y and the wavenumbers. The maps are obtained from the spectra that are acquired from different points on the sample. The images can be obtained at every wavenumber. A slice through those images produces the spectra that are associated with the same point (pixel) on the grid.

The spectra employed are ordered in a structure that is called a hyperspectral cube, shown schematically in Fig. 6.3. It consists of three dimensions (or rather four): two spatial dimensions, X and Y, the wavenumber axis Z, and the measured intensities are the elements that fill the array. These arrays are usually reshaped and spectra ordered so that the data analysis methods applicable to matrices of data can be applied (Andrew and Hancewizc, 1998). More details on this aspect are disclosed below. The 3D structures can be left intact and 3D data analysis approaches applied, although this method of analysis is less popular than those that operate on the matrices of data.

Line mapping systems use cylindrical optics or scanning mechanisms to distribute a laser beam so that it is parallel to the direction of the slit of a dispersive spectrometer. This configuration leads to spectra being collected on the CCD detector all along the laser line. In other words, a single scan theoretically produces as many spectra as there are pixels along the laser line instead of only one as in the systems with point excitation. The dimension on the sample that is perpendicular to the laser line is then scanned mechanically by shifting the stage, as in the point illumination systems described above. The use of line illumination increases the fidelity of the chemical image because of the much better spatial resolution achieved along the laser line (which is a function of microscope magnification and pixel size), while the spatial resolution in the direction perpendicular to the laser line is obviously still a function of the preset stage shift. The number of spectra collected significantly increases in comparison with the point systems, but their intensity is much weaker because of the elongation of the excitation beam that leads to lower power density per unit illumination area.

The Raman mapping spectra, obtained with either line or point illumination, are usually of very good quality and this is due to the use of dispersive spectrometers. These data sets can be considered as collections of a very large number of micro-Raman spectra that are expected to be of good quality if obtained from solid materials since the structural density of such samples is high and back-scattering is intense. Also the focusing power of the microscope objectives used contributes to this intensity by producing a large power density on the surface of the sample. Unfortunately, for samples that fluoresce, the large power density can significantly increase the fluorescence background that may entirely mask the Raman signal. Similarly, high optical power density (e.g., particularly when large magnification objectives are used) may ignite samples that absorb significant amounts of the incident laser light.

While the spectroscopic component of these systems is undoubtedly of superior quality to those produced on the global illumination platforms, the spatial (or imaging) aspect is not such a strong point. The major problem with the spatial aspect is that large parts of the area on the sample are passed over (i.e., coverage is incomplete). Therefore, the area imaged is approximated by a number of points/pixels/spectra and this does not necessarily describe the characteristics of the area in the best way. The shift of the stage (typically no less than 5 or 10 μm, depending on the size of the area) and the time constraints (the need to image during a reasonable amount of time limits the number of points that can be taken) define the quality of the image. However, this problem can be reasonably addressed if some information about the sample being imaged is available. For example, if the size of particles that form the sample is known, the experimental parameters can be set with a high degree of confidence.

6.1.2 Global Illumination Instruments

Global imaging systems are based on an entirely different principle. The excitation beam is defocused so that a much larger area of the surface of the sample is covered. The Raman-scattered light is then collected via special filter that transmits light only through a very narrow range of wavenumbers. In this way the complete Raman image (at specific wavenumbers) of the irradiated surface is obtained on the CCD detector.

The absence of dispersive optical elements raises the reasonable question of whether full spectra can be obtained by these instruments. This is possible since

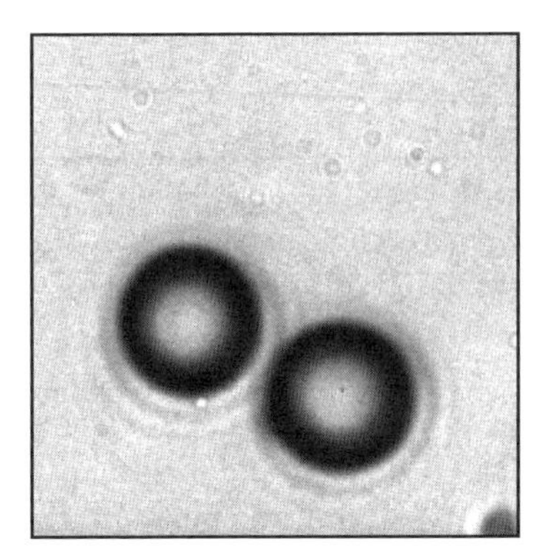
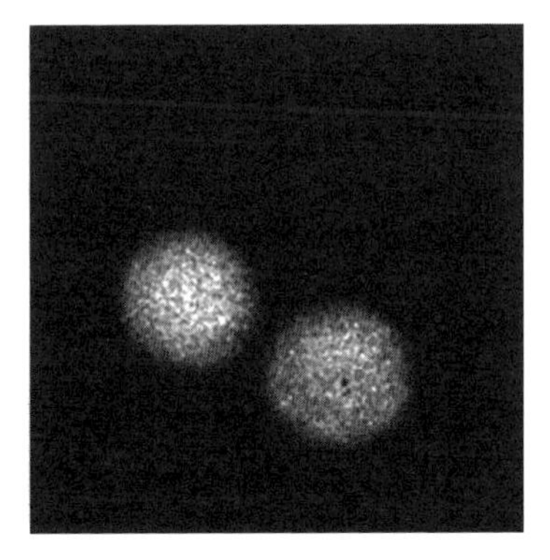

FIGURE 6.4 Brightfield and univariate Raman image of 2 μm diameter polystyrene microspheres separated by 200 nm obtained by a global illumination system. Reprinted with permission Wiley.

the filters can be sequentially tuned to a number of wavenumbers so that, if appropriate, full spectra can be collected. However, the philosophy behind these instruments accentuates the spatial aspect, not the spectral component of the experiment. The spatial resolution of the global illumination instruments is theoretically quite superior to that of the point or line mapping instruments, because the image of the full area is obtained on the CCD; that is, the complete surface is scanned. In practice, this means that the spatial resolution is identical in both dimensions. A powerful demonstration of the ability of these instruments to distinguish between very close particles whilst being able to chemically characterize them is given in Fig. 6.4 [in which polystyrene microspheres are imaged (Treado and Nelson, 2001)].

Several classes of the wavenumber-selective filters that are indispensable for global imaging instruments are described in detail by Treado and Nelson (2001). The most important class appears to be liquid crystal tunable filters (LCTF), which are based on birefringent interference geometry. The key features of these filters are their spectral resolution and transmittance. The first factor refers to their ability to reject signal outside the selected wavenumber range across which imaging takes place. Typically, the spectral resolution can be $<10\ cm^{-1}$, a value that is comparable with that obtained with a single-stage dispersive monochromator. This factor is important for distinguishing between bands that are overlapped. The second factor determines the amount of light passing through to the detector and thus it determines the strength of the signal on the CCD detector.

The general characteristics of these instruments are as follows:

(i) The images are obtained with a very large number of pixels employed but this does not affect the length of experiments as directly as in the case with the point or line mapping (where the proportion is direct). The number of pixels employed is, however, linked to the spatial resolution and Raman intensity and one can change the spatial resolution in order to achieve a stronger signal via binning (averaging or adding) of the pixels. Obviously,

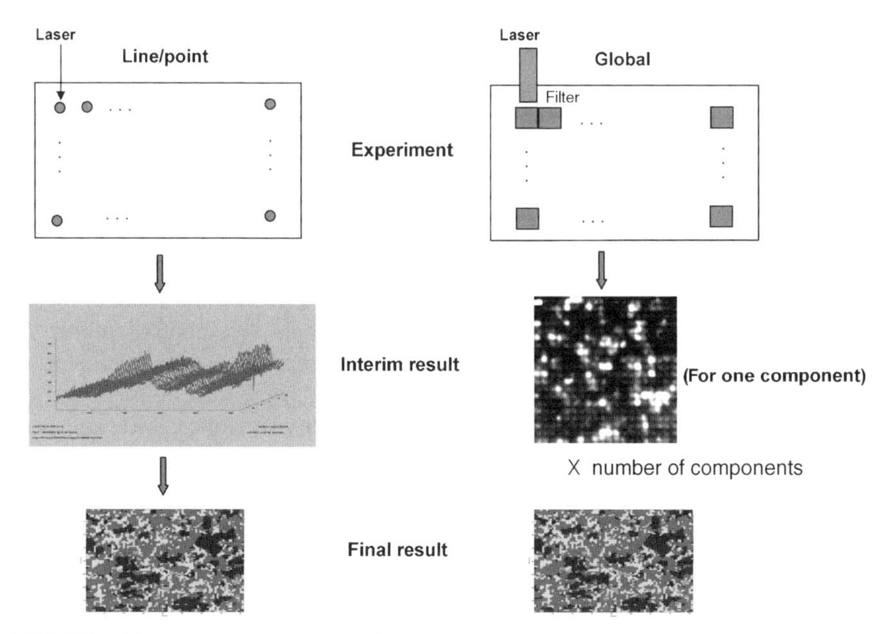

FIGURE 6.5 From experiment to image: The sketch showing the experiment and its immediate result and how these are converted into the most desirable outcome (composite image). There is one image per component on the global illumination platform, thus multiplication with the number of components. The images on the mapping system are obtained by chemometrics.

the larger number of pixels employed leads to superior visual quality of the resultant chemical images.

(ii) The direct result of the experiment is an image so there is relatively little need for data preprocessing (which normally includes only basic operations such as baseline correction).

(iii) Conducting experiments is quite simple.

(iv) In most cases, the chemical composition of the sample needs to be known to carry out optimal experiments.

Samples of unknown composition can still be analyzed but in such cases mapping is a more suitable choice (especially for samples that are compacted solids).

A cartoon showing the differences between the two approaches is given in Fig. 6.5. The performances of the two imaging systems are compared in detail in this chapter.

6.2 DATA ANALYSIS

There is relatively little need for much further data analysis for the chemical images obtained on a global illumination system, because such images consist of a fairly simple and easily understandable data structure in which there is a straightforward relationship between the wavenumbers employed and the corresponding images. However, as can be inferred from Figs. 6.3 and 6.5, the complexity and abundance of data (i.e., spectra) in hyperspectral data cubes offer various options for employing advanced data analysis tools, some of which will be described here in detail.

An ideal situation in chemical imaging is to find that the pure component spectra of all the constituents of the sample can be detected and there is at least one nonoverlapped, uniquely assignable, wavenumber for each component. In such cases, it is easy to produce chemical images through the simplest, univariate approach whereby chemical images are represented as spatially dependent intensity variation of such nonoverlapped wavenumbers, one per each component. This approach is based on knowledge of the chemical identity and thus the Raman spectra of the components of a formulation, a situation that is frequently met in product development. If the pure component spectra are not known, it is much more difficult, although still possible, to produce chemical images that are based on the least overlapped features of the underlying spectra.

However, such a situation is rarely met in practice. For many formulations, some components are present at low concentration and it is unlikely that their spectral signatures will be readily recognized, and spectral overlap is frequently a serious problem. Determination of how many components can be detected and the extent of spectral variation across a very large array of spectra (normally measured in thousands) demands application of complex data analysis tools based on linear algebra.

6.2.1 Data Preprocessing

Data preprocessing is an important preliminary step in the analysis of image data. The goal of data preprocessing is to reduce interference effects arising, for example, from variability in particle size, morphological difference such as surface roughness, undesirable optical effects, detector artifacts, and differences in Raman scattering coefficients. A particularly important example of preprocessing is equalization of Raman responses. Components of the formulation may have significantly different Raman scattering coefficients so that some of them may disproportionately contribute to the overall signal leading to bias. Normalization is a simple way of bringing all Raman responses onto the same scale.

For all the mapping data from Pfizer presented in this review, two preprocessing data steps were invariably employed. The mapping spectra were first baseline variability corrected with the polynomials of various degrees so that baseline effects, which were quite significant in some cases, were eliminated. Every spectrum was then divided by its mean (a process known as mean-normalization), which eliminated the effect of variations in Raman scattering efficiency. This data were then analyzed with principal component analysis (PCA) or self-modeling curve resolution (SMCR) algorithms. Multiplicative scatter correction (MSC), a method very popular in the analysis of NIR spectra, was not used.

6.2.2 Principal Component Analysis PCA

PCA has traditionally been the core technique in multivariate image analysis and its application in imaging data analysis has been discussed in detail by Geladi and Grahn (1996). The monograph by Malinowski (1991) and the book by Vandeginste et al. (1998) are recommended as very useful sources of information on PCA. More information on PCA is given here, because this method appears to be quite important in the analysis of spectra from mapping experiments and in the subsequent production of images.

In general, PCA decomposes a data matrix **D** into matrices of scores **T** and loadings **P**, and residual **E** as follows

$$\mathbf{D} = \mathbf{T}\mathbf{P}^{\mathrm{T}} + \mathbf{E} \tag{6.1}$$

where **T** and **P** each consists of N orthogonal column vectors associated with N principal components (PCs). In spectroscopy, the matrix of experimental spectra **D** is meant to comprise the data obeying Beer's law so that **D** can be also represented as

$$\mathbf{D} = \mathbf{C}\mathbf{A}^{\mathrm{T}} \tag{6.2}$$

where **C** $(n \times c)$ is matrix of concentration of components (with n being the number of samples and c the number of components), while **A** $(w \times c)$ is matrix of pure component spectra (with w being the number of wavenumbers).

Obviously, the score matrix **T** can be correlated with the concentration matrix **C**, while the loadings **P** can be correlated with the pure component spectra in **A**. The principal role of PCA is to distinguish between the signal and noise in **D**. The denoising through PCA can be achieved by reconstruction of **D** by Equation 6.2 after eliminating PCs that are not related to chemical components (removing corresponding columns in **T** and **P**). These columns are ideally characterized by noisy features, as opposed to the signal-related columns that are linear combinations of the pure component spectra. In this way, any matrix of experimental data can be considerably denoised, which is particularly important in Raman spectroscopy because of the inherently weak signal-to-noise ratio in Raman spectra. In addition, the regions of spectral and concentration significance are highlighted. The procedure described above has frequently been used in our laboratory for determining how many species can be detected based on their spectra, and to produce chemical maps.

Caution is advised, however, with interpretations of results and conclusions based on PCA data. PCs are extracted by mathematically preserving the maximal amount of variance in the mapping spectra and therefore a PC cannot be unambiguously assigned to a chemical species. Occasionally, if a loading is similar to one of the pure component spectra the corresponding score image approximates to the spatial distribution of that component. Our experience, however, indicates that in several cases a surprisingly high level of similarity is achieved and thus so much emphasis is put on PCA in this chapter.

PCA can be used for denoising of the mapping data, interactive exploratory image analysis, feature extraction for cluster analysis, and so on. It is usually applied after the mapping data are reordered (or unfolded) so that the 3D data structure (two spatial and one wavenumber axes) is temporarily suspended. This is achieved by the controlled appendance of the two spatial dimensions so that a 2D data structure is obtained $(x \times y, \times v)$ that is PC analyzed and then reordered (folded) back into the original 3D layout. In this way, each of the first few $x \times y$ matrices in the 3D array is meant to be an image of a given component (again, provided there is enough similarity among the loadings and pure component spectra).

6.2.3 Self-Modeling Curve Resolution

Another family of data analysis tools that is of interest is known as SMCR (Tauler and Kowalski 1995; Jiang et al., 2004). SMCR techniques search for the variables/wavenumbers or samples/spectra that are most unique in a data set. These are then used for alternating least square (ALS) regression, the second part of most SMCR techniques, to approximate the pure component spectra and corresponding concentration profiles (chemical images). Only two SMCR algorithms are mentioned here: orthogonal projection approach (OPA) (Cuesta Sanches et al. 1996) and Simplisma (Windig and Guilment, 1991).

Both these methods are based on the principle that the spectra can be represented as vectors in an n-dimensional space (where n is the number of wavenumbers) and that therefore dissimilarity/uniqueness of the spectra is equivalent to the angles among the vectors (small angle $\rightarrow$ high similarity; high angle $\rightarrow$ very dissimilar spectra). The algorithm proceeds with a series of determinants being calculated in which the spectrum that is geometrically most distant from the mean spectrum of the complete data matrix is identified. (In principle, the same can be done with the wavenumbers, or variables, a process in which the wavenumber with the most unique profile is sought.) In the next step, the spectrum most orthogonal to the one found above is extracted, and then in the third step the spectrum most orthogonal to the plane spanned by the first two spectra found is identified, and so on. The calculations stop when the angle between the spectra and spaces spanned becomes quite small. This means that the remaining spectra can be shown as linear combinations of the previously extracted spectra. The spectra selected in this way represent the boundaries inside which all other spectra can be found. Ideally, if the experimental spectra contain the pure component spectra of all the components of the tablet, these spectra would be serially extracted and all other spectra in the set would be obtained as linear combinations of these pure component spectra. However, this is a very rare occurrence in practice due to spectral mixing. The Raman spectra generally appear to be thoroughly overlapped with each other. Unique Raman signals of minor components are rarely detected and exist as only minor, hardly perceivable features on top of the strong spectra and in most cases they cannot be identified with SMCR.

OPA and Simplisma have been used in this chapter mostly to determine the most dissimilar (unique) spectra in the analyzed mapping data and to assess the prospects for univariate imaging based on these spectra. In some cases, the following step,

ALS regression, is also used. The role of ALS is to try to refine the results of SMCR and to eliminate the effect of overlapping by extracting the pure component spectra from the sets of unique, though still overlapped spectra.

6.3 EXPERIMENTAL

6.3.1 Sample Preparation

It is probably not an exaggeration to claim that sample preparation is in most cases a trivial task. Practically all the potential problems with sample preparation when imaging tablets are solved by using a trimming instrument that flattens the surface of the tablet. Trimming normally ensures that the entire surface that is being imaged is at the same focal plane. The imaging area and stage shifts are key parameters of mapping/imaging experiments that should be specified when the size distribution of the components of the tablet is taken into account. Under normal circumstances (i.e., with no agglomerates expected), it should suffice to image across a few square millimeters on the surface.

Imaging of blends is even simpler as they are poured onto a microscope slide and compressed slightly with the cover. For beads or inhalation materials some other techniques are used. Beads are halved under a microscope using a scalpel, while the inhalation materials are spread by simply dispersing them onto a microscope slide.

6.3.2 Instruments

Most of the details concerning the instruments used for obtaining results presented here are described above. Majority of the data shown here were obtained with the point/line mapping systems unless said otherwise. It is worth mentioning that in most cases the spectra were obtained by exciting the sample with laser lines above 700 nm. In general, the green laser lines, 500–570 nm, are rarely used due to the risk of exciting fluorescence and burning the sample. However, there are no strict rules and each case should be considered separately. Although the use of longer wavelength for illumination significantly reduces the risks of fluorescence, the S/N ratio is much better when using the shorter wavelengths and this can considerably shorten total acquisition time. Our experience with defocused irradiation with the 532 nm line is that in most cases samples do not burn while the level of fluorescence can be tackled by manipulating the stage settle time and spatial binning.

6.3.3 Software

Various software routines, both commercial and custom written, were used for analyzing the data. For univariate images, the software that normally accompanies the instrument was sufficient because producing univariate images without analyzing details of the spectra is a relatively trivial task. However, for more complex approaches such as SMCR, more often than not custom written routines were employed. SMCR routines are quite popular in academic research (and are constantly improving) and some versions of SMCR algorithms can be found in commercially available software as well.

The platform for the custom written routines has almost invariably been Matlab (MathWorks, Natick, MA), and it is very likely that this programming language is the most popular tool for exploring chemical images. Some of the routines, such as Simplisma, have been described in sufficient detail in the literature so that their incorporation in Matlab is rather trivial (Windig and Guilment, 1997).

The two most specialized chemical imaging programs, to the best of my knowledge, are ChemXpert (ChemImage Corp., Pittsburgh) and Isys (Spectral Dimensions, now part of Malvern Instruments) that is Matlab-based. All the operations on the images from this laboratory were carried out by using ChemAnalyze (now replaced by ChemXpert) with some support from Matlab. Isys appears to be popular with the users of NIR imaging instruments. Both products handle the mapping/imaging data irrespective of how these were obtained (subject to solving formatting issues).

6.4 APPLICATIONS

6.4.1 Tablets

The following example illustrates how all the components in a typical tablet can be imaged through the mapping technique. A stepwise (and quite generic) approach to analyzing the mapping data is presented here which starts with simple univariate images, addresses its issues and limitations, proceeds with the chemometrics techniques described above, and finishes with prospects for composite imaging in which all the components are shown. These operations are equally applicable to any other imaging/mapping problem. A more thorough description can be found in the references Šašić and Clark, (2006), Šašić, (2007b), and Clark and Šašić, (2006).

The pure component spectra of the components of the tablets are normally known in advance so that selecting the univariate (nonoverlapped) points should

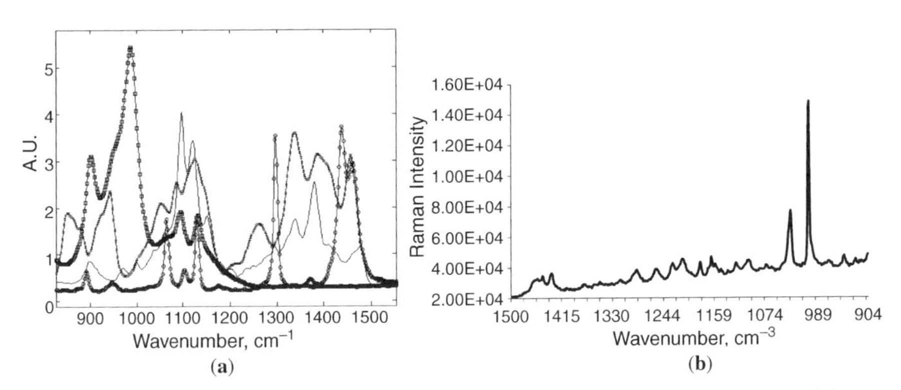

FIGURE 6.6 (**a**) The pure component spectra of Avicel (–), DCP (□), Explotab (x), and Mg stearate (○). (**b**) The spectrum of pure API. This figure exemplifies that overlapping is substantial and that thus imaging via wavenumbers that are unique for the components present (one wavenumber for each component) is very difficult. Reprinted with permission Society for Applied Spectroscopy.

FIGURE 6.7 The univariate Raman chemical images of API, Avicel, and DCP. Reprinted permission Society for Applied Spectroscopy.

be quite straightforward. Figure 6.6 shows the pure component spectra of the components of the tablet analyzed with the three univariate points selected. No univariate points are selected for the two components (Explotab (sodium starch glycolate) and magnesium stearate) accounting for 3 and 1% (w/w), respectively. Caution is always advised with univariate imaging of minor components because their Raman signals may be overwhelmed by the signal of major components (or they may simply be not detected) so that such univariate images can, in fact, be misleading. An illustrative example of this problem will be mentioned later on. The univariate images of the three major components are shown in Fig. 6.7.

The prospects for univariate imaging can best be assessed by scanning the variability of the spectra across the entire set of mapping spectra. This is carried out by a SMCR method that selects the most dissimilar (i.e., unique) spectra. Once those spectra have been found, the reliability of the images shown in Fig. 6.7 can be better judged, and also a definitive answer can be given to the question of whether the minor components have distinguishable Raman responses. SMCR is indispensable for this task because one has to use an automated data analysis method to compare $\sim$ 6000 spectra that are collected during this particular experiment ($2 \times 2\,\text{mm}^2$ area with 25 μm step). Figure 6.8 shows the most unique spectra found by OPA.

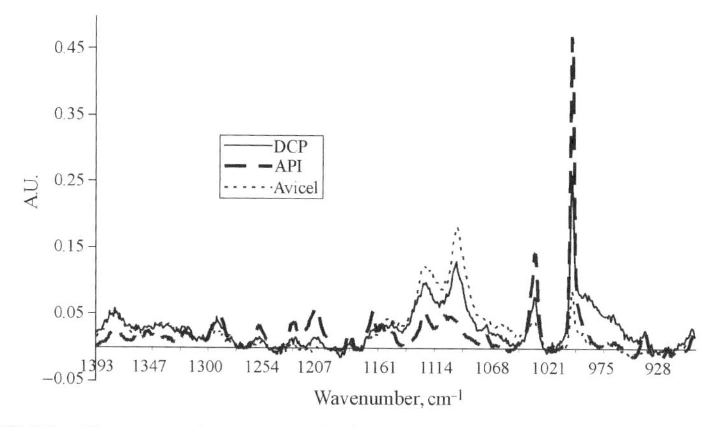

FIGURE 6.8 The most unique spectra in the entire set of CIRCO 6000 Raman mapping spectra. The legend assists to recognize to which component are these spectra related. For example, the 'API' spectrum is taken at the spatial position rich with API and hence the Raman spectrum at that position as very strong API peaks. Note that the signal of API and Avicel can be detected in each of these three spectra, while the strongest indication of DCP is just a shoulder to the API peak at 1000 cm^{-1}.

The comparison with the pure component spectra in Fig. 6.6 shows that API and Avicel peaks dominate in all the spectra in the mapping data set (API > 20%, w/w). Despite being present in the same quantity as the API, dicalcium phosphate (DCP) has a much weaker Raman response and is thus indicated via a barely visible shoulder at 980 cm^{-1} to the major API peak at 1003 cm^{-1}. No indications of the peaks of Explotab or Mg stearate can be seen. Consequently, there are no grounds to believe that these two components can be imaged univariately while the univariate image of DCP becomes doubtful given the rather weak Raman response of the major DCP peak (relative to the API peak) at which the univariate image is obtained, and also the vicinity with the much stronger API peak.

In order to obtain more information on the distribution of minor components, PCA was applied. The PC loadings and accompanying score images are shown in Fig. 6.9. The loadings from Fig. 6.9 reveal that the spectral indications of all the components can be recognized. The first loading comprises the anticorrelated bands of API and Avicel, meaning that the bands of API are positive, while those from Avicel are negative. The third loading very clearly features the major band of DCP at 980 cm^{-1}. Some spectral indications of Explotab are recognized in loading #5. It is rather difficult to recognize any bands due to Explotab in any loading as clearly as in the case of the previous three component system. This is because

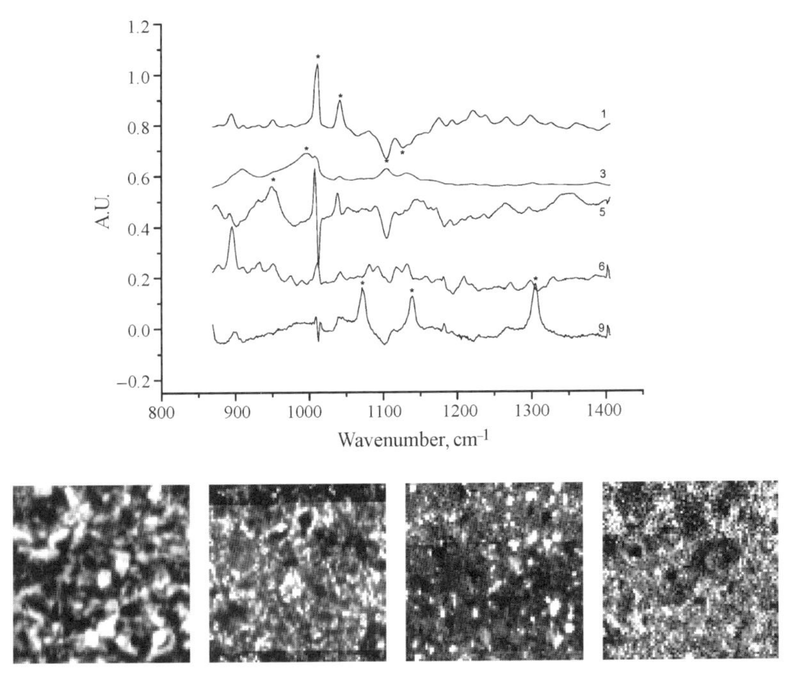

FIGURE 6.9 The PC loadings from the Raman mapping spectra and the PC score images ($2 \times 2\,mm^2$). The first score image from the left features both API and Avicel (white and black pixels, respectively), followed by DCP (PC2), Explotab (PC5), and Mg stearate (PC 9). Reprinted with permission Society for Applied Spectroscopy.

Explotab is present at a much lower concentration and its spectrum is significantly overlapped with that of Avicel. Nevertheless, the peak at 940 cm^{-1} and the broad spectral features in the long wavenumber range hint that this loading may be associated with Explotab. However, the assignment of the corresponding score image would be rather tentative on the basis of this evidence. Finally, and most strikingly, loading #9 features three strong bands all assignable to Mg stearate. It is very surprising that a loading at such a high position is so free of noise and has obvious bands present. In addition, the use of the criteria for retaining the signal-related components (via analysis of the eigenvalues, for example) would indicate that only the first two or three PCs should be retained, certainly not the first 9. In conclusion, the analysis of the PC loadings shows that it is quite likely that all the components of the tablet can be imaged.

The PC #1 score image analyzed is that due to API and Avicel. The white pixels in this image refer to the API, the black ones refer to Avicel which is a consequence of the anticorrelated signal of these two components in PC loading #1. Through a similar assessment, the DCP image is also easily obtained. As for Explotab, the appearance of PC #5 score image is quite interesting, because it features white egg-like shapes that correspond to the shapes of the Explotab particles. Thus, despite not particularly convincing spectroscopic evidence of Explotab in the fifth loading, the appearance of the score image seems to support the view that this loading can indeed be linked with Explotab. Finally, the only information that can be obtained from the Mg stearate score image is that it is finely spread across the tablet with no agglomerates present. The similarity between the univariate and multivariate images (Figs. 6.7 and 6.9) is obvious for all the components that could have been imaged by both methods. This is an important lesson given that the univariate data for DCP were rather doubtful, as already commented on above.

When it can be shown that all the components can be imaged separately, a question arises regarding the prospects of producing a composite image that will feature all the components. This is not feasible by simply combining the images from Fig. 6.9, because there is far too much overlap between the pixels belonging to various components. For example, all the pixels available can be (and in fact are) assigned to either the API or Avicel on the basis of the strong and ubiquitous Raman signal of these two components so that there is no room for any other component. If there were pixels at which no response from either API or Avicel were observed, one would have been able to assign some pixels to other components. However, the spectral mixing is so strong that every experimental spectrum obtained is dominated by signals from API or Avicel in various proportions. The only way to produce a composite image is to artificially raise the thresholds for API and Avicel and thereby free pixels for the minor components (or those with relatively weak Raman signal such as DCP).

de Juan et al. (2004) used a more advanced chemometrics approach for analyzing the mapping data by carrying out chemometric scanning of the spectra from the surface of the sample. They created and employed the method termed fixed-size image window evolving factor analysis (FSIW-EFA) in which the full data set is divided into a large number of local data matrices (this is equivalent to carefully

TABLE 6.1 Percent Cumulative Image Variance Explained by PCA Using Different Bilinear Model Sizes (de Juan et al., 2004).

Model size	Percent API				
	0%	20%	40%	60%	80%
1	96.3	83.4	58.3	51.7	89.6
2	96.9	95.3	93.1	80.4	94.4
3	97.1	95.9	96.8	97.9	98.6
4	97.2	96.2	97.1	98.2	99
5	97.3	96.4	97.4	98.4	99.1

analyzing small areas on a tablet) that are then carefully inspected based on the number of components present. The full (rather complex) sketch of how the data are reorganized is given in the literature (de Juan et al. 2004). This way of analysis increases the sensitivity because the weak signal from minor components can be detected more readily than when global methods are applied (e.g., PCA on the full data set) in which weak signals from those components are subdued by much stronger signals from other components.

FSIW-EFA has been applied to the two-component tablet and the chemometric results are summarized in Table 6.1 and Fig. 6.10. The results in Table 6.1 reveal that three components are identified, one of which is declared an impurity and its spatial distribution is given in Fig. 6.10. This conclusion could not have been reached by visual inspection of the thousands of spectra in the data set and it is also questionable whether the global PCA would be able to detect the Raman signal and position of the impurity. Some other interesting spectroscopic, imaging, and chemometrics points are also described in this article (de Juan et al., 2004).

The two examples described above show how Raman mapping can be effectively used to determine the distributions of the components on the surface of the tablets. It is very important to mention that all the components (even if present at 1%) were recognized and relatively reliably imaged despite the vast differences in concentration and Raman scattering efficiency. Significant amount of computation (chemometrics) was required to extract this data and this could not be achieved in any other way. The analysis presented is actually a very persuasive example of the usefulness of the techniques such as PCA or SMCR. In addition, an interesting debate can, and should, ensue from these studies as the results obtained are not fully in line with the theoretical (mathematical) expectations.

6.4.1.1 Comparison with NIR Mapping Given that for various reasons (some of which are described in previous chapters) the NIR technology seems to be the most popular spectroscopic approach for industrial needs, it is worth comparing how the NIR mapping performs on the same sample as (Šašić, 2007b).

The major advantage of using a NIR mapping instrument is speed (in addition to the cost): the acquisition times for corresponding experiments (the same area and spatial resolution) took only 15–20% of the time needed for Raman mapping. However, in terms of the information content, the Raman data look more promising.

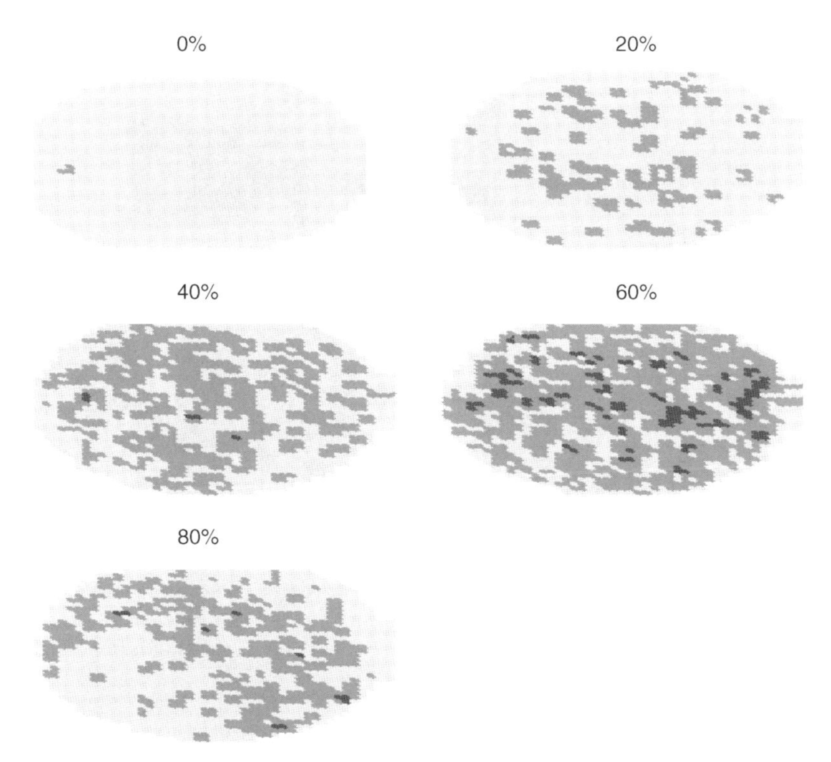

FIGURE 6.10 Local rank maps of the images from five tablets. The darkest points show places where Raman responses from three components were detected (most mixed spectra). The percentage refers to the active compound. Reprinted with persmission, Elsevier.

Figure 6.11 summarizes results of the NIR mapping of the tablet. The most unique spectra detected and the corresponding images are shown. Two cases are shown in which the imaged tablets were manufactured in different ways. In the first case, the most unique spectra obtained with OPA only feature the signal from API and Avicel. No indications of DCP or minor components were obtained. Consequently, only the images of API and Avicel can be produced in the univariate manner.

PCA does not add more information because of the far too heavy overlap between the pure component spectra which diminishes the prospects for associating loadings to the pure component spectra. Thus, the final conclusion for this tablet would be that only the two major components can be imaged.

For the second tablet, API and Avicel are again easily recognizable, but this time an indication of DCP is obtained as well. DCP actually has no NIR signal, but it can be recognized through the appearance of the background signal and this is illustrated in Fig. 6.11. In addition to the two spectra featuring the bands of API and Avicel in different proportion, one more unique signal is detected that resembles the background and is thus assigned to DCP. The corresponding images prove that this

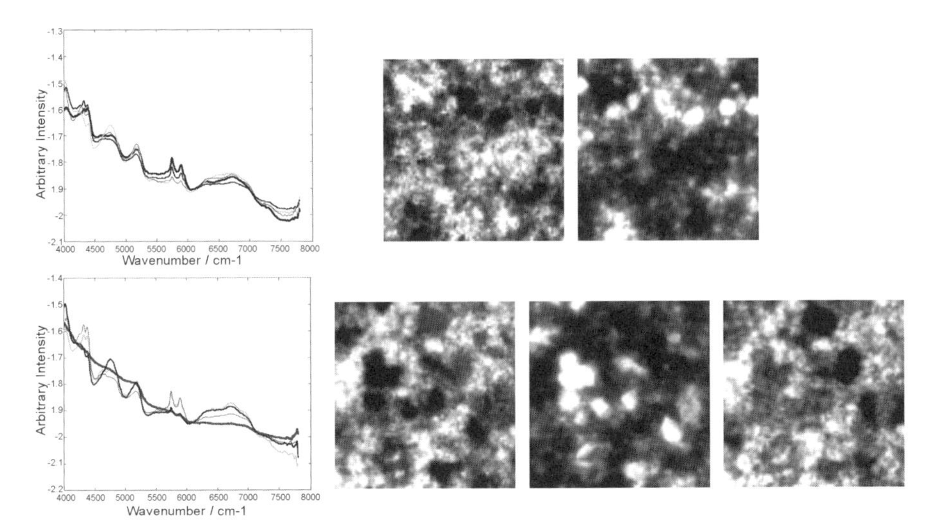

FIGURE 6.11 The most unique NIR mapping spectra from two tablets found by OPA. (–) The first unique spectrum, (–) the second, (—) the third, and (—) for the fourth. Shown also are the corresponding univariate images ($2 \times 2\,mm^2$) of Avicel, API, and DCP (left to right). DCP is found only in one tablet, seen as compact areas with black pixels because at those positions there is no signal but only background (see text for explanation). Reprinted with permission Society for Applied Spectroscopy.

assignment is correct, because the image associated with the background spectra appears structured (showing the domains with the component that does not absorb). Since this tablet was manufactured with DCP being added to the granules and not to the blend, it is more likely to appear in separated domains which is exactly what is seen in Fig. 11. PCA results fully support these conclusions by showing two components, one of which corresponds to the anticorrelated bands of API and Avicel, while another one is associated with the background.

In short, it appears that the NIR technique is less effective for imaging all the components of the tablet because it does not seem to be sensitive enough to detect the minor components (such as Mg stearate). Similarly, DCP, which is a very important excipient, is only recognized under some specific conditions—it seems that the penetration depth of the NIR signal and heavy overlap of the pure component spectra prevent detection of the background spectra (i.e., DCP) for finely mixed material. However, the speed of acquisition and cost are unquestionably significant advantages of the NIR mapping in comparison with Raman mapping. Given that there is absolutely nothing unusual that would distinguish these tablets analyzed here from any other tablet with similar API loading, the results obtained can actually be considered generic and applicable in a much wider context.

6.4.1.2 Comparison Between the Point/Line and Global Illumination The same area on a placebo tablet is imaged on the systems with line and global illumination, respectively (Šašić and Clark, 2006). The placebo formulation mimicked the composition of the tablet analyzed above containing predominantly Avicel and

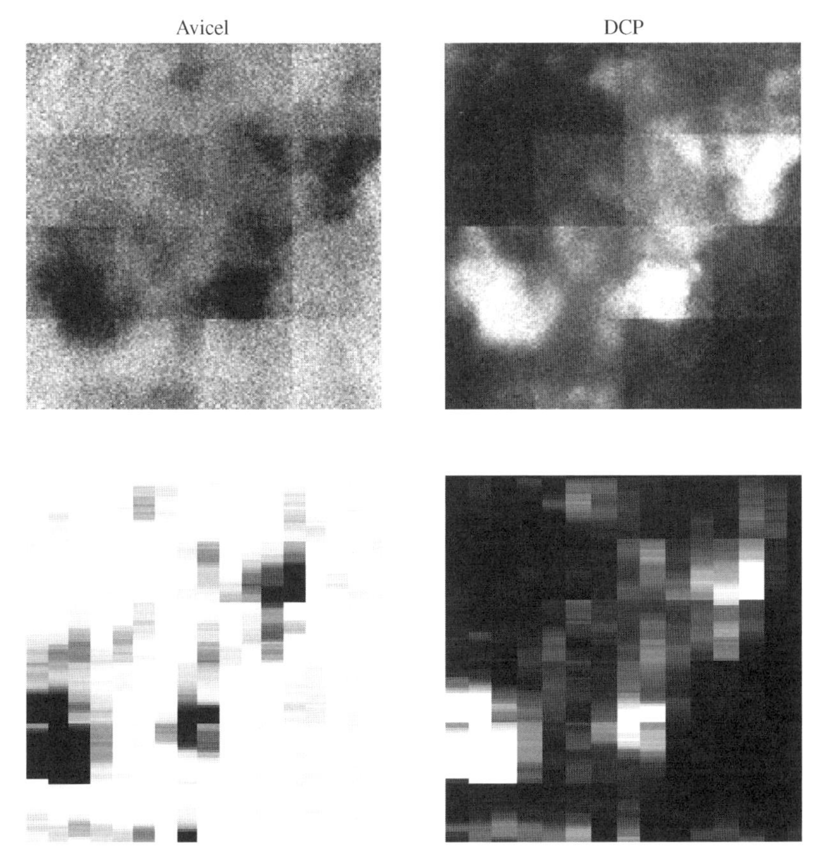

FIGURE 6.12 Comparison of chemical images of Avicel and DCP obtained on the Raman global illumination (top) and line mapping systems (bottom). The size of all images is $150 \times 150\,\mu m^2$. Reprinted with permission Society for Applied Spectroscopy.

DCP together with Explotab and Mg stearate as minor components. The data analysis approach in the examples described above was applied to these data as well: (that is, univariate images were first produced, followed by the OPA selection of the spectra and PCA to produce the score images.

The univariate images of the two major components are very similar on both systems (Fig. 6.12). The images from the global illumination are of much better visual quality because of a much higher number of pixels being employed. The spatial resolution is also much better (theoretically it can be made incomparably better provided the complete area on the sample is imaged, literally). However, this system does not appear to be able to detect minor components. Figure 6.13 shows the OPA results, which reveal that the spectral differences are far more persuasive on the line illumination system. The DCP peak is clearly visible as well as some of the peaks of Mg stearate, which points to good prospects of obtaining the images of minor components via multivariate analysis. On the contrary, the spectral

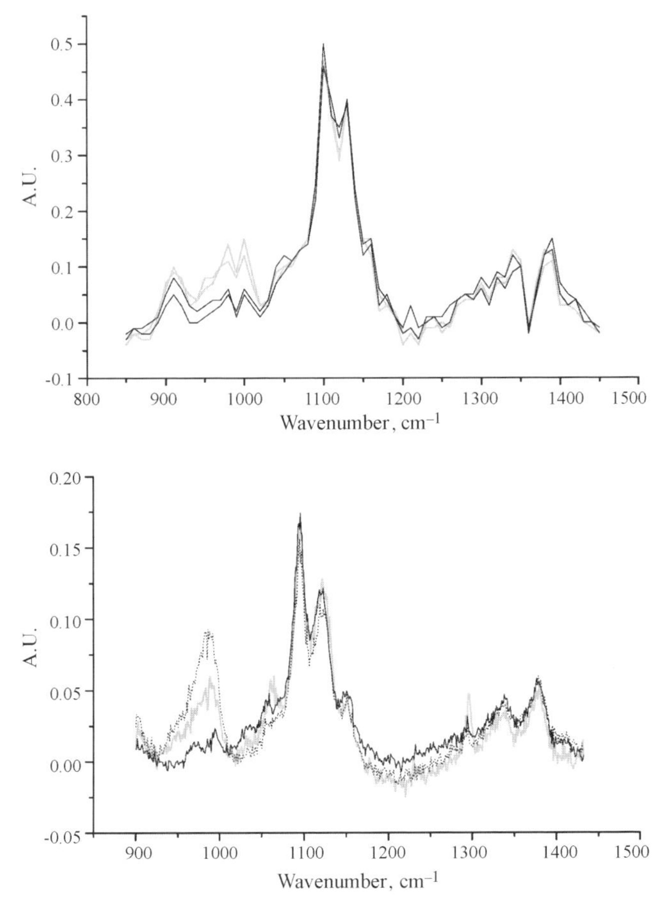

FIGURE 6.13 Pairs of the most unique spectra found by OPA in the global illumination spectral set (×20 objective) and the most unique spectra for the line mapping Raman system. The third identified spectrum (gray) contains some features of Mg stearate at 1065 and 1295 cm^{-1}. Reprinted with permission Society for Applied Spectroscopy.

differences on the global illumination system are much less distinct with the DCP peak being only a small shoulder to the major Avicel peak and with no indications of the minor components.

As expected, the PCA results for the minor components quite closely resemble those shown in Fig. 6.9, while reliable images could not be obtained on the global illumination system. In many respects, this conclusion is analogous to that obtained from the comparison between the NIR and Raman mapping. This is not surprising given that the penetration depth for the global illumination system is substantial, while the sensitivity and spectral resolution are far worse than that can be achieved on the instruments equipped with standard optical elements such as the grating (sensitivity is strongly dependent on the LCTF elements). The acquisition times on global illumination instruments are generally much shorter because only eight or nine wavenumbers would suffice to produce visually excellent images; but similar to the NIR mapping, global illumination also seems to be burdened with limited ability to handle minor spectral responses. The purpose of showing the full spectra obtained by the global illumination platform was simply to demonstrate the abilities of that platform with respect to the line illumination, but in practice it is not recommended to collect full spectra because this eliminates one of the strongest points of that approach, that is, fast imaging.

6.4.1.3 Imaging API at Low Concentration As mentioned in Section 1.6, Raman signals of API tend to be relatively strong and in many cases not too overlapped with the bands of typical excipients so that one can hope to obtain a fairly acceptable Raman signal even from minor amounts of API in formulations. Such prospects are even better if a multivariate data analysis tool is used to enhance weak API signals (as illustrated in previous section dealing with imaging Mg stearate and Explotab). This section summarizes two very recent publications in which very small amounts of API in tablet formulations, $<1\%$ (w/w), were successfully mapped.

A recovery problem in the tablets of Xanax and Alprazolam has been addressed by Šašić (2007a). The goal of the Raman mapping exercise was to determine the spatial distribution of the API in the tablets with good and bad recovery, the API content being 0.4% and 0.8% (w/w) for two types of tablets investigated. The analysis of the pure component spectra revealed that Alprazolam (the name of the API) has a strong and practically nonoverlapped band at 687 cm^{-1} so that one begins with producing a univariate image at that wavenumber. The image obtained is shown in Fig. 6.14 together with the spectra obtained by randomly clicking on the white pixels in the image. As one can see, the image appears rather indistinct with uniform distribution of the API across the imaged area (2×2 mm^2 in 25 μm steps). While this result is actually good, because it reveals that there are no agglomerates of the API, the spectra shown indicate poor reliability of that image. The signal of the API at the univariate wavenumber is quite weak and the image produced does not convincingly represent the API because the S/N ratio is too small. Thus, it would be wrong to consider the image in Fig. 6.14 as the spatial distribution of the API. This case illustrates the danger of unconditionally relying on previously selected wavenumbers without checking the strength of the Raman signal, that is spectra from which the image is to be obtained. The only way to produce a more reliable image is to denoise the spectra, and this is accomplished with PCA.

Quite surprisingly, the fourth loading of the PCA of the Raman spectra from the 0.4% formulation is very comparable with the spectrum of Alprazolam, thus facilitating much more reliable multivariate imaging of Alprazolam (Fig. 6.15). This is conceptually similar to the case shown in Fig. 6.9. The image produced is now based on strong evidence and indeed represents unique experimental evidence for the distribution of the API in a commercial formulation. This image can further be processed, thresholded, and binarized leading finally to the comprehensive quantitative information about the distribution of the API (Šašić, 2007a).

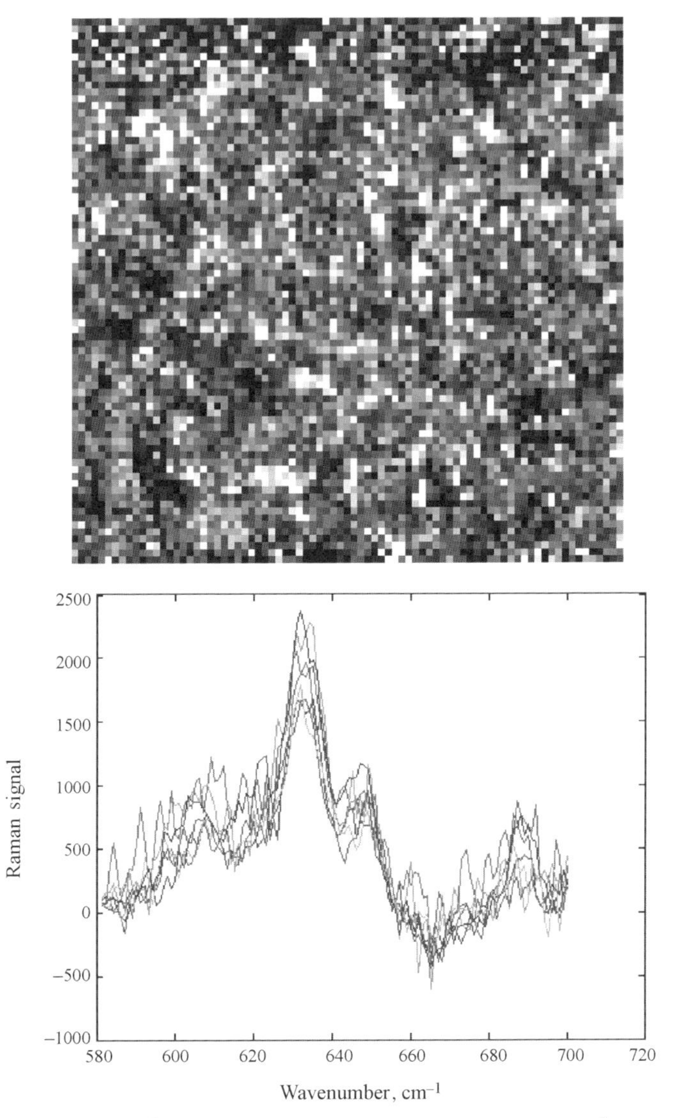

FIGURE 6.14 $2 \times 2\,\text{mm}^2$ univariate image of Alprazolam at 687 cm^{-1} and the Raman spectra from randomly selected white pixels. Reprinted with permission Springer.

Henson and Zhang (2006) used Raman mapping as a sensitive method for assessing partial transformation of a low concentration API during tablet manufacture (or stability testing). They comprehensively analyzed tablets with the API present at 0.5% (w/w): the API was known to have three different solid-state forms. In order to classify the spectra of these three forms and produce reliable chemical images, which is unquestionably a very demanding challenge, they first constructed a spectral library that contained the spectra of the three forms and the two spectra of

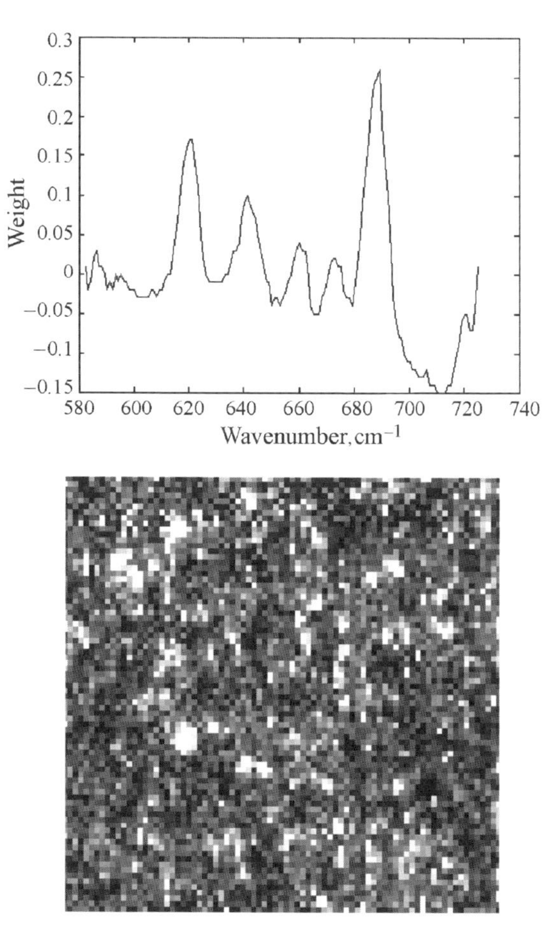

FIGURE 6.15 The fourth PC loading obtained by PCA of the Raman mapping spectra from a Xanax tablet that is very similar to the spectrum of Alprazolam. The PC4 score image that, correspondingly, features Alprazolam is shown too. The size of the image is $2 \times 2\,mm^2$, which is about 50% of the image-able area on the tablet. Reprinted with permission Springer.

the most abundant excipients. Partial least square (PLS) regression was then applied as a classification tool with arbitrary thresholds being applied in the first instance to determine class membership. This step was replaced by using another classifying algorithm, Euclidean distance (ED), to achieve more objective assessment of the relative content of the API forms.

The method was first tested on the tablets containing only one of the three API forms. In practically all the tablets analyzed, it was found that using the PLS approach supported by ED classification the form of the API present in the tablet was exactly recognized. After this encouraging result, the tablets containing all three forms were analyzed. The forms were present in the ratio 1:1:8 (which corresponded to 10% conversion of the desired API form) with the total API concentration unchanged at 0.5% (w/w).

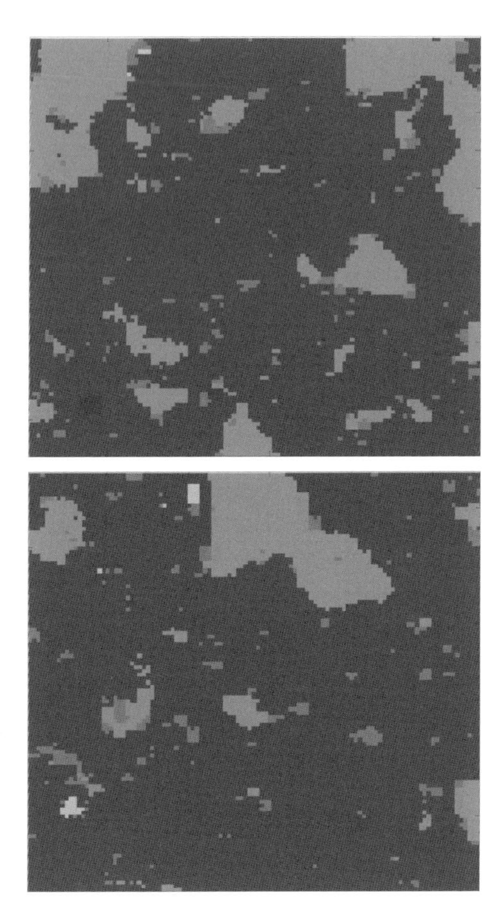

FIGURE 6.16 Euclidean distance classification results for Raman maps of mix sample 2, surfaces 2 (top), and 3 (green = I; blue = II; cyan = III; orange = diluent 1; red = diluent 2). Pixel size is 8.3 μm, and image size is approximately 920 μm × 900 μm (12,099 pixels). Reprinted with permission Society for Applied Spectroscopy. See color plates.

Figure 6.16 shows one of the seven images that were produced with spatial distributions of all five components shown. These images were quantified and it was found that the average number of pixels associated with the three components was 6:6:88, which is in a good agreement with the set ratio of 1:1:8. This study convincingly demonstrated that polymorphic impurities as low as 0.05% can be reproducibly detected and mapped which is indeed an excellent result.

6.4.1.4 Confocal Raman Mapping The detection and quantification of amorphous materials have become quite a significant theme in formulation development. Precise characterization of such materials cannot always be achieved by bulk measurements particularly for inhalation formulations in which the amorphous component can alter adhesion or cohesion, thus modifying the performance of the product.

Confocal Raman imaging can be used for both axial and lateral characterization of inhalation formulations, as nicely demonstrated in the study by Ward et al. (2005). They analyzed a model compound, sorbitol, with two distinct forms: the naturally occurring crystalline form and the amorphous form. The amorphous form is produced by a novel localized heating method in which micron sized amorphous domains are obtained. The amorphous domains on the surface of crystalline sorbitol disks were then analyzed by atomic force microscope (AFM) and confocal Raman imaging. The purpose of Raman imaging was to provide information about the size of the thermally modified domain within the matrix of the crystalline phase: a measurement that could not be achieved by AFM.

The spectral differences between the crystalline and amorphous forms of sorbitol allowed relatively simple creation of chemical images of the amorphous component. These differences, typical for the Raman spectra of crystalline and amorphous forms of many materials, were identified in the intensity of the bands and the full width at half-maximum (FWHM) in the 850–900 cm^{-1} spectral region. The mapping spectra were collected with a $\times 100$ objective at various positions inside the sorbitol disk and the chemical images (Fig. 6.17) of the amorphous form were produced on the basis of both intensity differences and FWHM (of the peak at 878 cm^{-1}). Figure 6.17 reveals that the signal assigned to the amorphous content is strongest on the surface of the sample. As the confocal plane (that defines the depth of measurement) is lowered into the sample, the contours of the circular domain diminish gradually, disappearing finally at about 20 μm below the surface. The reconstructed spectra from both the crystalline and amorphous domains confirmed that the applied thermal treatment did not induce degradation of the sorbitol.

Breitenbach et al. (1999) employed Raman confocal mapping to determine the physical state of the active, physicochemical stability of the formulation (solid solution of ibuprofen in polyvinylpyrrolidone, PVP), and the content and homogeneity of the distribution of the API. The physical state of the API in the polymer matrix was also compared with the solutions of the API in a solvent.

The confocal mapping spectra showed that the ibuprofen form in the solid solution obtained by the hot-melt extrusion technique was equivalent to that in the solution. The spectrum of the crystalline powder of ibuprofen was found to be different from the spectra from both solid and liquid solutions. The confocal acquisitions were performed at a depth of 10 μm across the $45 \times 25\ \mu m^2$ area. The chemical images of ibuprofen, created by comparing the most suitable bands of ibuprofen and PVP, demonstrated that the API was homogeneously distributed inside the polymer matrix.

6.4.2 Imaging of Spatially Resolved Materials with Global Illumination Platform

Our experience with the global illumination imaging platform suggests that such systems are particularly suitable for imaging spatially isolated objects. While the problems with the depth of penetration/spectral mixing and sensitivity to low-concentration components are exacerbated with the compacted nature of materials being imaged, spatially resolved samples can be quickly and efficiently imaged

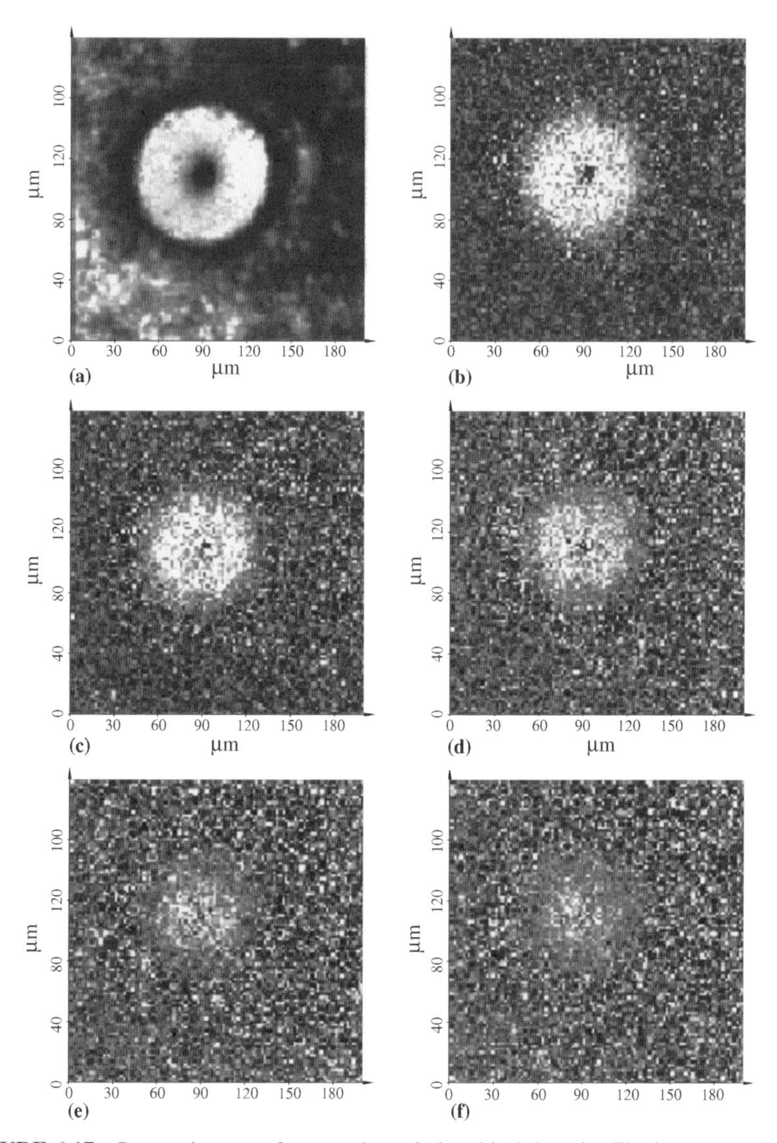

FIGURE 6.17 Raman images of a quench-cooled sorbitol domain. The images on the left are based on peak intensity, the ones on the right on FWHM. The z-values (depth of measurement) are 0 µm for (**a**) and (**b**), −5 µm for (**c**), −10 µm for (**d**), −15 µm for (**e**) and −20 µm for (**f**). Reprinted with permission Springer.

as the following example illustrates. The example selected is an inhalation formulation, but conceptually the learning is transferable to any material that can be easily separated into individual particles (such as nasal sprays).

A two-component inhalation blend was analyzed (lactose and API), a sample was dispersed onto a microscope slide, and then both components were imaged. Four

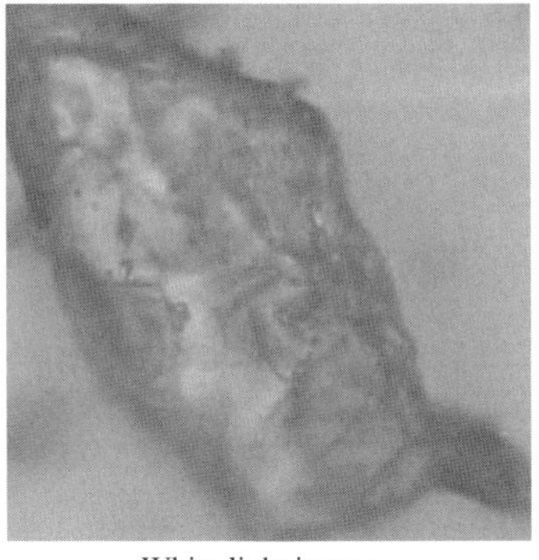

White light image

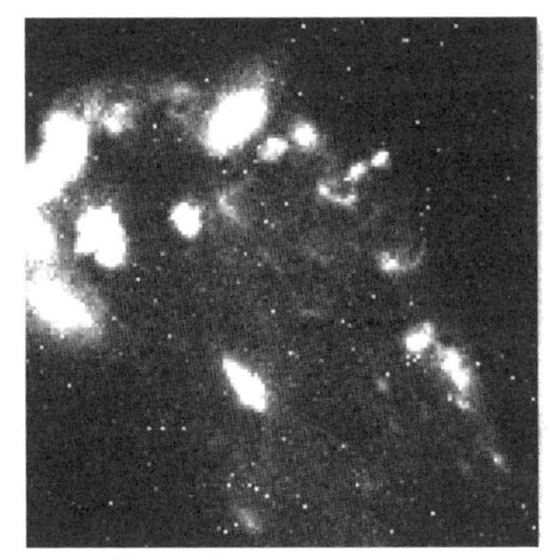

Raman image

FIGURE 6.18 White light and Raman images of the API on the surface of a lactose particle. The size of the images is $100 \times 100\,\mu m^2$.

wavenumbers were used for both API and lactose with various combinations of acquisition times per wavenumber and pixel binning. Higher binning leads to better S/N ratio and thus shorter acquisition times. Figure 6.18 shows the image of the API on a single particle of lactose obtained with a ×20 objective with the field of view approximately covering $100 \times 100\,\mu m^2$ so that the microscope stage was not moved during this experiment (Šašić, 2005). On average, images like this can be obtained in a few minutes. The API can be clearly visualized on the surface of the particle. The Raman signal of the API is quite strong so that there is no doubt that the white pixels indeed represent the API. Interestingly, although the API accounts for less than 5% of the formulation, its Raman signal is stronger than that of lactose and thus the API image in Fig. 6.18 is of better quality than that of lactose (not shown).

Depending on the goal of the experiment, a number of particles can be automatically imaged and their identity confirmed. Such an experiment cannot be effectively done using other chemical imaging platforms simply because it is impossible to illuminate all the particles with point or line illumination. Compacted materials, such as tablets, are probed discontinuously but this is not considered to be a significant issue, because dramatic morphological changes are not supposed to occur in the material between successive stage shifts. Covering all the particles of a spatially resolved material would imply "infinitely" small stage shifts and an extremely high number of spectra to collect, that is, a very long acquisition time, which would obviously be unacceptable. In the global illumination approach, all the particles on the slide can be imaged and the user sets the spatial resolution by taking into account several parameters such as the strength of the Raman signal, size of the particles being imaged, and so on. Similarly, the visual quality of the chemical image in Fig. 6.18 can be accomplished only with the true imaging global illumination technique.

6.4.3 Mapping of Beads

Beads are another kind of spatially resolved material and an example of their analysis is described here. Because of their compacted nature, the layered beads can

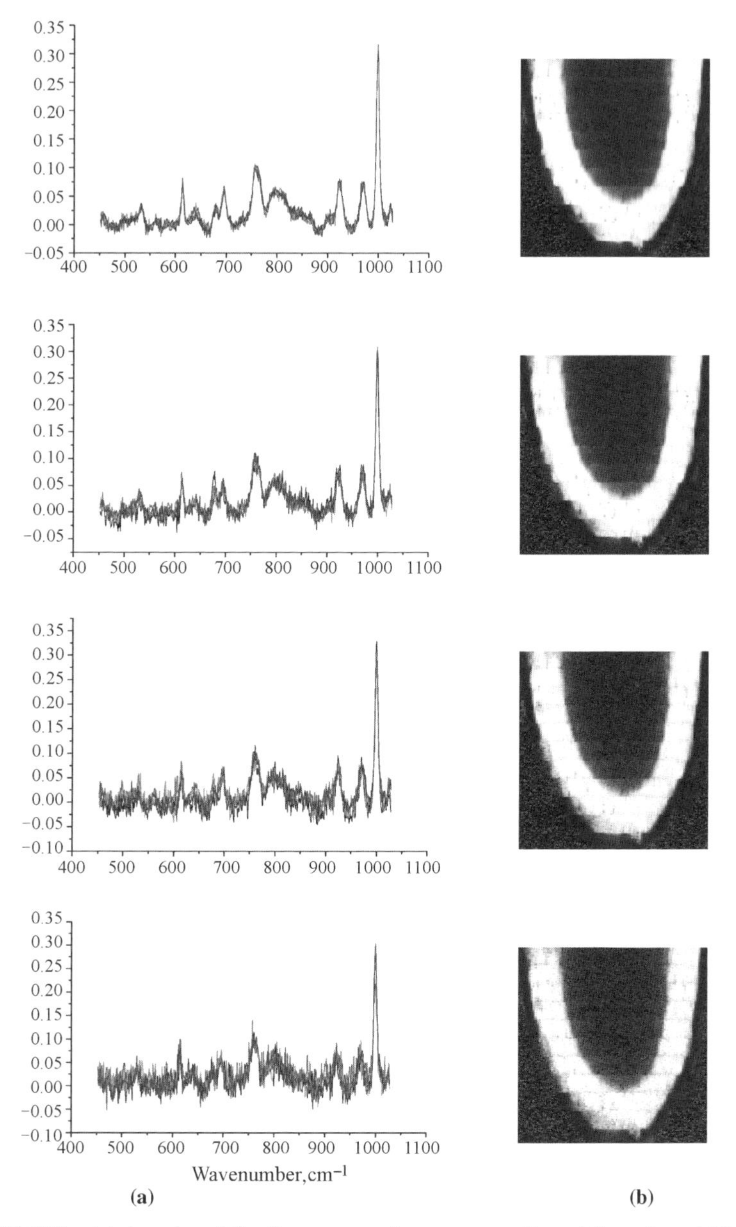

FIGURE 6.19 (**a**) A series of the Raman mapping spectra collected from the middle layer of the bead (API) for 30, 10, 5, and 3 s (top to bottom). (**b**) The corresponding univariate chemical images of API obtained after noise reduction with PCA. The differences among these images after binarization are minimal. Reprinted with permission Royal Society of Chemistry.

also be considered to be a material for which the global illumination system is the platform of choice for fast imaging. However, a line or point mapping system can also be applied as a fast and reliable imaging method.

The key element for redefining the inherently slow mapping techniques into fast methods is in the recognition that relatively simple data analysis procedures can significantly improve the quality of the Raman spectra obtained from the beads. The structure of the beads greatly facilitates such applications—the Raman spectra obtained from the layers are largely nonoverlapped thus being ideal (but still real) targets for applications of noise-reducing routines based on linear algebra. The approach that is briefly described here is not suitable for mapping of tablets, because they have incomparably more complex data structures.

Šašić et al. (2004, 2005) exemplified how Raman mapping of three-layered beads can be conducted in less than 1 h with the quality of the images being nearly equivalent to that obtained after about 10 h without using advanced data analysis. The idea presented is quite simple. The spectra obtained with short acquisition times are of poor S/N ratio and thus are not likely to provide the basis for high quality images. However, if the noise in such spectra is mathematically reduced, then the images based on them may be as good as the images from the spectra obtained with much higher S/N ratio (at much longer acquisition times). Figure 6.19 shows a series of spectra obtained by sequentially reducing acquisition times from 30 to 3 s. The quality of the spectra significantly deteriorates with shortening acquisition times and it seems unlikely that the images produced from the 3-s spectra would be comparable with those obtained from the 30 s spectra. However, if PCA is applied to the 3 s spectra, a large amount of noise is relatively easily eliminated because the Raman signal from each of the three well-defined layers is easily distinguished from noise. With only the first three PCs retained, the mapping data are reconstructed and univariate images of the three layers are produced. These images are then compared with the images obtained from the spectra acquired at 30 s (Fig. 6.19), and the visual difference is almost imperceptible. In order to achieve a more precise comparison, the pixels were counted and their positions were compared, and it was found that the PCA-based method allowed for a 10-fold reduction in acquisition time with only a ~3% loss in the quality of the images. These results can, theoretically, be improved even further by using other mathematical tools that employ the pure component spectra (which can be trivially obtained). This extra information certainly improves prospects for fast mapping. As a result, by Raman mapping one can quickly determine the structure of the beads and the chemical identity of the layers.

REFERENCES

Andrew JJ, Hancewizc TM, 1998. Rapid analysis of Raman image data using two-way multivariate curve resolution. *Appl. Spectrosc.*, **52**, 797.

Breitenbach J, Schroff W, Neumann J, 1999. Confocal Raman spectroscopy: Analytical approach to solid dispersion and mapping of drugs. *Pharm. Res.*, **16**, 1109.

Clark DA, Šašić S, 2006. Chemical Images: An introduction to technical approaches and issues. *Cytometry Part A*, 69A 815.

Cuesta Sanches F, Toft J, van den Bogaert B, Massart DL, 1996. Orthogonal projection approach applied to peak purity assessment. *Anal. Chem.*, **68**, 79.

de Juan A, Tauler R, Dyson R, Marcolli C, Rault M, Maeder M, 2004. Spectroscopic imaging and chemometrics: A powerful combination for global and local sample analysis. *Trends Anal. Chem.*, **23**, 70–79.

Delhaye M, Dhamelincourt PJ, 1975. Raman microprobe and microscope with laser excitation. *J. Raman Spectrosc.*, **3**, 33.

Geladi P, Grahn H, 1996. *Multivariate Image Analysis*. Wiley, New York.

Henson M, Zhang L, 2006. Drug characterization in low dosage pharmaceutical tablets using raman microscopic mapping. *Appl. Spectrosc.*, **60**, 1247.

Jiang J, Liang Y, Ozaki Y, 2004. Principles and methodologies in self-modeling curve resolution. *Chemometr. Intell. Lab. Sys.*, **71**, 1–12.

Malinowski ER, 1991. *Factor Analysis in Chemistry*, Wiley, New York.

McCreery RL, 2000. *Raman Spectroscopy for Chemical Analysis*, Wiley, New York.

Šašić S, 2005. Chemical imaging of pharmaceutical samples by a Global illumination Raman imaging instrument. Abstract, XXXII FACSS, Quebec City, Canada.

Šašić S, 2007a. Raman mapping of low-content API pharmaceutical formulations. I. Mapping of Alprazolam in Alprazolam/Xanax tablet. *Pharm. Res.*, **24**, 58.

Šašić S, 2007b. An in-depth analysis of Raman and near-infrared chemical images of common pharmaceutical tablets. *Appl. Spectrosc.*, **61**, 239.

Šašić S, Clark DA, 2006. Defining a strategy for chemical imaging of industrial pharmaceutical samples on Raman line-mapping and global illumination instruments. *Appl. Spectrosc.* **60**, 494.

Šašić S, Clark DA, Mitchell JC, Snowden MJ, 2004. Univariate versus multivariate Raman imaging—A simulation with an example from pharmaceutical practice. *Analyst*, **129**, 1001.

Šašić S, Clark DA, Mitchell JC, Snowden MJ, 2005. Raman line mapping as a fast method for analyzing pharmaceutical bead formulations. *Analyst*, **130**, 1530.

Tauler R, Kowalski B, 1995. Selectivity, local rank, three-way data analysis and ambiguity in multivariate curve resolution. *J Chemometrics*, **9**, 31–58.

Treado PJ, Nelson MP, 2001. *Raman Imaging* in Chalmers JM and Griffiths PR, (eds) *Handbook of Vibrational Spectroscopy*, Wiley, New York.

Ward S, Perkins M, Zhang J, Roberts CJ, Madden CE, Luk, SY, Patel N, Ebbens SJ, 2005. Identifying and mapping surface amorphous domains. *Pharm. Res.* **22**, 1195.

Windig W, 1997. Spectral data files for self-modeling curve resolution with examples using the Simplisma approach. *Chemometrics Intell. Lab. Sys.*, **36**, 3–16.

Windig W, Guilment J, 1991. Interactive self-modeling mixture analysis. *Anal. Chem.* **63**, 1425–1432.

7

IN VIVO RAMAN CONFOCAL MICROSPECTROSCOPY OF SKIN

ANDRE VAN DER POL, WILLIAM M. RIGGS and PETER J. CASPERS
River Diagnostics BV, Dr. Molewaterplein 50, Ee 1979, 3015 GE, Rotterdam, Netherlands

7.1 INTRODUCTION

The ability to measure the chemical composition of living biological tissues nondestructively is a valuable tool for pharmaceutical research. Raman spectroscopy has qualities that make it unusually attractive for such measurements, especially its ability to measure tissue noninvasively using confocal optics. However, the chemical complexity of biological tissues makes such measurements challenging. Because of this, Raman spectroscopy has only recently begun to make significant contributions in this area.

Since skin is the most readily accessible biological tissue, it has been the first to be studied extensively using Raman spectroscopy for *in vivo* measurements on human subjects. This chapter will focus on applications of Raman spectroscopy to measurement of the composition of skin, including compounds that have been applied to skin and the effects of such compounds on skin composition.

The ability to obtain *in vivo* data on skin noninvasively, with microscopic resolution, allows new information to be obtained on many important and previously difficult-to-address questions, including those relevant to treatment of skin with active pharmaceutical ingredients (APIs; Caspers et al., 2003). Various areas of skin research increasingly depend on such detailed knowledge of the molecular composition of skin and the spatial distribution of skin constituents.

On a microscopic scale, the skin is highly heterogeneous. Its molecular composition and structure vary tremendously, depending on the depth from the skin

Pharmaceutical Applications of Raman Spectroscopy, Edited by Slobodan Šašić

surface and the location on the body. In the stratum corneum especially, concentration gradients of components (e.g., water gradients, pH gradients, diffusion kinetics) often play a role in biochemical or skin physiological processes. The composition of the skin is also affected by skin disorders, environmental factors such as sun exposure, seasonal variation, and cosmetic or medical treatments. In addition, skin treatments may also bring about changes in dimension, such as increased stratum corneum thickness due to swelling. These spatial complexities of the skin yield best to Raman tissue analysis with a confocal approach, where spatial resolution consistent, with the size of many features of interest ($\approx$5 μm deep and $\approx$1 μm long), can be achieved. This is not meant to suggest that confocality is a requirement for measurements on all biological tissues or even for all *in vivo* measurements. Other approaches have, in fact, been used for a number of applications of Raman to skin (some are described in this chapter) and also for nonskin applications of Raman to biological tissue (see, e.g., Hanlon et. al., 2000).

For understanding topics such as the transdermal delivery and action of APIs, noninvasive methods are particularly welcome. This is partly because they cause less discomfort to the patient or volunteer subject, since the skin is not damaged, and also because noninvasive methods enable investigation of the skin in its natural state without affecting its integrity, morphology, or molecular composition. Noninvasive measurements can be performed repeatedly on the same skin area in vivo and can thus be used to monitor time-dependent changes in the skin brought about by the application of APIs.

Numerous chemical changes in the skin that can be studied via *in vivo* Raman spectroscopy are of interest in relation to APIs. These include the rate of penetration of topically applied materials, residence time of applied materials in the skin, metabolic or chemical changes that an applied material may undergo once in the skin, biochemical effects that may be induced by the presence of an active compound in the skin, and the appearance in the skin of materials delivered to the body orally, by topical application, or by other means. Interest in these phenomena is not limited to pharmaceuticals. Some companies in the cosmetics industry have been the leaders in adopting Raman microspectroscopy for these applications. Some of the examples to be described in this chapter originated in the areas of skin science of greatest interest to the cosmetics industry. They nevertheless illustrate the possibilities generally offered by Raman that are applicable in pharmaceutical research. Application of confocal Raman microscopy in pharmaceutical research and development has begun, and promises to become widespread.

7.1.1 Major Methods Used Prior to *In vivo* Raman Spectroscopy

Prior to the advent of *in vivo* Raman spectroscopy, a number of other methods have been used to monitor possible paths of APIs or other substances as they penetrate the skin, or to detect and follow reactions of interest in skin tissues. Some of the major methods, including Franz cells, ^{14}C labeling, biopsies, tape stripping, near infrared (NIR), and infrared (IR) spectroscopy, have been summarized in this chapter for reference to help frame and highlight the novel utility that the Raman technique is now bringing to the pharmaceutical arena.

Franz cells, both static and flow, are standard tools for measuring the flux of a material through a membrane, which can be excised skin. The material of interest is placed on the membrane and the amount of the material appearing in a reservoir on the other side is measured by chemical analysis. Probably, the major disadvantage of this method is that it is an *in vitro* method that requires excised human or animal skin. The advantages, of course, are that the equipment is relatively inexpensive, accessible to most laboratories with a need for such information, and the method is standard and well known.

^{14}C labeling is a sensitive and accurate means to measure penetration through the skin *in vivo*. The compound to be tested is first synthesized with the ^{14}C label incorporated. Then it can be applied and its penetration into the body monitored by a method, such as analysis of urine, using standard radiation detection methods. A disadvantage of this method is the need to expose the subject to a radioactively labeled compound during the time required for elimination. Because of this difficulty, and the added cost of synthesis of the labeled compound, the method is generally used only for high priority applications.

Skin biopsies and tape stripping have been extensively developed and are widely utilized. Both allow for direct chemical analysis of materials in the skin. They are, of course, inherently destructive, in the sense that the same material cannot be measured, treated, and then measured again. They are also invasive, albeit only minimally, especially in the case of stratum corneum tape stripping. The possibility of using Raman microspectroscopy as an alternative, providing an "optical biopsy," has clear appeal.

NIR and IR analysis potentially offer spectroscopic "fingerprint" information closely similar to that available from Raman spectroscopy, and therefore have been used with some success for skin science applications. NIR reflectance spectroscopy has been used for assessing the water content of the stratum corneum, for example (Woo et al., 2001; Attas et al., 2002). IR spectroscopy has been used in the attenuated total reference (ATR) mode fairly extensively for *in vivo* analysis of skin as well. Many such studies have been reported in the fields of cosmetics, drug delivery, penetration enhancement, and sunscreens. One of the earliest examples is the work by Tanojo et al. (1997) on penetration of the enhancer oleic acid. However, these methods have significant disadvantages compared to Raman spectroscopy in these applications. NIR reflectance spectroscopy does not provide the wealth of specific molecular identification information inherent to IR and Raman spectroscopy, and it is not sufficiently spatially resolved to offer composition profiles as a function of depth.

IR spectroscopy using ATR sampling devices, while surface sensitive, is neither well depth-resolved nor able to sample micrometer size areas. An alternative IR spectroscopic technique was presented recently (Notingher and Imhof, 2004) that allows for the high molecular specificity of IR spectroscopy as well as some degree of depth resolution. This technique, thermal emission decay-Fourier transform infrared spectroscopy (TED-FTIR) applies pulsed thermal excitation in combination with gated FTIR emission detection. The gating time allows to some extent for depth selection. The authors successfully demonstrated penetration of propylene glycol (PG) into the stratum corneum using this technique.

7.1.2 *In vivo* Raman Methodology

Developing high spatial resolution microspectroscopic Raman measurements on tissue to be fast enough to use for *in vivo* clinical studies is a significant technical challenge, requiring instrumentation especially optimized for the application (Caspers et al., 2002). Even the excellent general purpose Raman instruments available in most well-equipped analytical laboratories are not capable of conducting practically useful measurements on skin. Recently, however, Raman instrumentation has been developed and made commercially available that is capable of practical use in these demanding applications. Figure 7.1 shows a photo of the first commercially available Raman skin analyzer. It employs advanced technologies that make this difficult application useful for practical analysis.

Key technical advances in the past decade in optical technology, especially in lasers and detectors, have helped enable the development of Raman instrumentation capable of meeting the demands of this application area. The most important considerations for an optimized Raman skin analyzer are selection of lasers, choice of optical materials, detector quality, optomechanical stability, and, for practical utility, a software that is easy to use and can effectively handle the large volumes of data that are generated by *in vivo* panel studies. Laser safety considerations also call

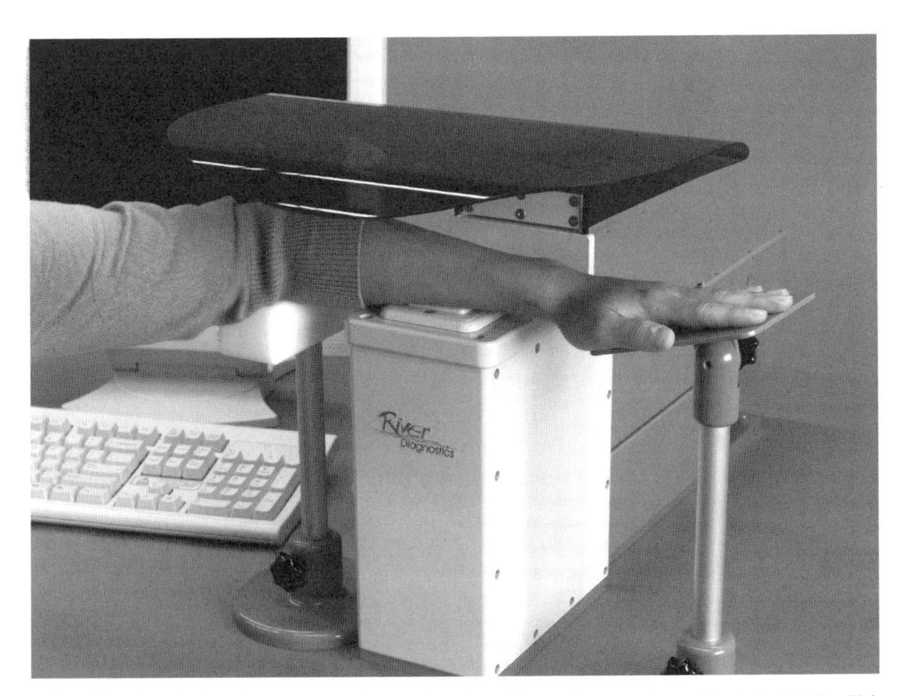

FIGURE 7.1 The instrument pictured, the River Diagnostics Model 3510 Skin Composition Analyzer, is the first Raman instrument optimized for *in vivo* analysis of skin.

for stringent technical requirements that must be met, thereby strongly influencing the overall systems engineering of a Raman skin analyzer.

A capable Raman skin analyzer can be thought of as being composed of four components, each of which must meet critical requirements: (a) a laser light source and associated light conditioning optics, (b) a near-infrared (NIR) optimized microscopic measurement stage, (c) the Raman spectrometer, and (d) specialized operating and data analysis software. The key issues for each of these components are discussed below. In addition, laser safety considerations impose certain limiting conditions on the technical solutions that can be implemented. These will be briefly discussed as well.

7.1.2.1 Laser Excitation Source The laser(s) used must emit light at wavelengths at which no photo(bio)chemical reactions are brought about and minimal fluorescence is stimulated in the skin. This places a lower limit on the usable laser wavelength at approximately 660 nm. Furthermore, no significant tissue heating (mainly due to rovibrational absorption by water molecules and electronic excitation of the chromophores melanin and hemoglobin) can be allowed, putting an upper limit on the laser excitation wavelength of about 850 nm. Thus the choice of laser wavelengths is restricted to a 'biological window' in the near infrared, approximately in the range from 660 to 850 nm. The lasers must be single mode emitters to enable diffraction limited laser focusing into a single spot, necessary to achieve the highest spatial resolution. They should be stable in power output and wavelength, and exhibit a narrow emission line to allow for achievement of high spectral resolution. The laser light must be coupled into the light conditioning optics needed for removal of unwanted laser diode background radiation or satellite emissions. Single mode optical fiber, optimized to the specific laser wavelength used, is normally utilized for this. The lasers must have moderately high output power available (There is a safety limitation on how high the power is permitted to be. This is discussed in Section 7.1.2.5.) All optical materials in the laser light path must exhibit minimal fluorescence or other background contributions. The laser light, after optical filtering and collimating, must be coupled to an appropriate microscope.

7.1.2.2 Microscope Measurement Stage The microscope must have uncompromised confocal optics, since the Raman signal is collected back through the microscope objective. The optical train must be capable of a very efficient transmission of this signal back to the spectrometer entrance aperture. The microscope must provide spatial resolution well below the thickness of the stratum corneum if useful information about the distribution of materials in the skin layer is to be obtained. This requires a microscope objective optimized for NIR wavelengths. Although theoretical microscope resolution is on the order of the wavelength of the light ($\approx$1 μm), the practical limit is on the order of 3–5 μm primarily because of refractive index variations due to the inhomogeneity of the skin. The objective focal position must be movable in the axial direction (z-axis) under precise control so that measurements can be made at successive depths in the skin to generate composition profiles as a function of depth.

The microscope system must be physically designed and oriented to be convenient and comfortable for human subjects. Commercial Raman skin analyzers presently available use an inverted microscope system with a platform on which to place the area to be measured (most commonly the volar aspect of the forearm) to satisfy this criterion.

7.1.2.3 Raman Spectrometer The spectrometer used for *in vivo* measurements on skin must have very high laser wavelength rejection, high transmission at optical interfaces (low reflection and scattering losses) to preserve as many of the information-bearing photons as possible, and a high performance (sensitive) detector. These considerations are, of course, desirable in any Raman spectrometer, but they become critical for this application area because measurements on skin always involve measuring a complex sample where a change in the abundance of a minor constituent is often the information sought. Further, high quality spectra must be routinely obtainable in a short time to obtain meaningful information from measurements made on a time scale compatible with the endurance (and patience) of volunteer panelists. Therefore, maximizing the signal-to-noise ratio (S/N) by employment of an optimized spectrometer design is of great importance.

For broad utility in a clinical research environment, stability and repeatability are further important considerations in order to allow data to be rigorously compared over time. This means that the spectrometer design must incorporate features to assure very high mechanical and optical stability.

In clinical environments where more than one spectrometer is in use, it is further required that results obtained on one skin analyzer are directly and reliably comparable with results obtained on another skin analyzer. This places very high demands on the accuracy and reproducibility of instrument calibration and correction for instrument response effects.

7.1.2.4 Software Data acquisition software for *in vivo* Raman measurements must have specialized features to handle the often large number of measurements in typical panel studies and to satisfy requirements that are not normally encountered in other types of Raman analysis. For example, the software must enable the operator to quickly select locations of interest on the skin surface. Also, since depth information (usually changes in composition as a function of depth) is important, the software must incorporate a reliable and accurate means of locating the skin surface for reference. Further, the software must be structured to minimize data acquisition "dead time" between sequential spectrum acquisitions to maximize throughput when thousands of spectra are typically acquired in a day.

The data processing software must also incorporate special features. Due to the many experimental variables in a typical study designed to extract information on the skin *in vivo*, the number of spectra to handle can become very large. In conventional spectroscopic processing software, spectra can be manipulated and analyzed typically one-by-one or batch-by-batch. For the larger numbers of spectra typical for *in vivo* studies, the time simply to read in each single spectrum and export the result after analysis can become prohibitive. Even for processing spectra batchwise, the time to sort, select and read in the spectra to define the batches for analysis can become a bottleneck. Therefore, the software must feature ways to enter the experimental design, and use this to select and process the spectra accordingly.

The spectral analysis routines that extract the desired composition of the skin must be mathematically stable and unambiguous. These routines must also be easy to use for scientists who may not be spectroscopy experts. Given the large amounts of spectra there should be automatic outlier detection available in the analysis. Finally, the extracted composition information must be accessible in a common format (e.g., ASCII), to enable import into other (statistical) analysis software programs.

Caspers et al. (2003) developed an algorithm suitable for calculation of the main composition of the stratum corneum from a Raman spectrum. In this method, the experimental spectra are fitted to a built-in library of spectra of pure stratum corneum compounds. The fitting algorithm is a nonrestricted multiple least squares fit. The fit coefficients (the weighting factors for the contribution of each library spectrum) are calculated from the matrix product:

$$C = \mathrm{inv}(\mathbf{x}^{\mathrm{T}} \cdot \mathbf{x}) \cdot \mathbf{x}^{\mathrm{T}} \cdot \boldsymbol{y}$$

where C is the fit coefficient, $\mathbf{x}$ is the matrix with library spectra (arranged columnwise); $\boldsymbol{y}$ is a vector containing the experimental spectrum, $\mathbf{x}^{\mathrm{T}}$ is the transpose of $\mathbf{x}$, and inv is the inverse. This equation has only one unique solution (thereby minimizing the ambiguity of the analysis), and from a computational point of view this one-step calculation is extremely fast. For example, on a common personal computer the calculation of the composition of hundreds of spectra takes only a few seconds. These characteristics make this approach the favored one for analysis of *in vivo* Raman spectra of skin.

7.1.2.5 Laser Safety Considerations Laser safety is a necessary consideration for experiments in which panelists skin will be exposed to laser light. In practice, laser exposure of the skin during measurement is required to be at a power level below the maximum permissible exposure (MPE) limit, a formally defined level of exposure below which no reddening or other visible skin irritation or damage will occur. The MPE is determined by international standards bodies and specified in relevant standards (IEC, 60825-1; ANSI Z136.1). The laser wavelength and the overall geometric configuration of the analyzer sampling arrangement are important determinants of the MPE. Since instrument designs differ, the MPE must be determined for each individual instrument design. To provide a "feel" for MPE magnitude, the configuration of the instrument shown in Fig. 7.1 results in an MPE limit of 30 mW for 785 nm laser excitation and 20 mW for 671 nm excitation. These values are not general for the wavelengths cited, but must be determined for any instrument design intended for *in vivo* skin analysis.

More laser power on the sample yields more signal and correspondingly faster analyses. However, an instrument capable of delivering laser power to the skin at the MPE limit can gain further in sensitivity and speed only by gaining efficiency in the spectrometer—for example, by conserving every Raman photon to the extent practically feasible. This means that efficient optical design is critical for *in vivo* measurements on skin.

Eye safety is also an obviously important consideration. There must not be any significant risk of eye damage from exposure to the laser beam while the measurement window is not covered by the skin to be measured. Practically speaking, laser exposure of the eye is not a difficult risk to manage in a properly designed instrument since the laser beam diverges at a high angle when emerging from the microscope objective, but the risk must nevertheless be properly managed. With the instrument referred to above, incidental direct observation of the beam is not an eye hazard at the power levels used provided no optical instruments are used to observe the beam.

When these elements—an appropriate laser light source, microscopic measurement stage, NIR-optimized Raman spectrometer, specialized software, and *in vivo* laser safety provisions—are combined in a Raman instrument, valuable information hitherto unavailable to researchers becomes accessible.

7.2 APPLICATIONS

In this section we review a number of applications of confocal Raman microspectroscopy that illustrate the utility of the method for qualitative and quantitative measurement of compounds in skin. Included are applications to date of the technique to problems in pharmaceutical research, and others that illustrate capabilities that are clearly transferable to and potentially useful in pharmaceutical science.

7.2.1 Effects of Topical Moisturizers

Considerable initial interest in *in vivo* Raman spectroscopy on skin has originated in the cosmetics industry, where the region that is of greatest relevance is most often the stratum corneum. Since the active ingredients employed are, in at least some cases, utilized to influence the chemistry of the stratum corneum, they are also of pharmaceutical interest by analogy. The results of some of these studies are suggestive of productive application areas in the pharmaceutical arena. The study described in this section is one of these.

Sieg et al. (2006) compared the effects of three different moisturizers *in vivo*. They performed a blinded, randomized 3-week study with 14 volunteers. Three cosmetic moisturizers (A, B, and C), one of which contained niacinamide (A), and an untreated control site (U) were tested in randomized order on the volar forearms of the panelists. Measurements were carried out in a temperature-controlled room (21°C) and the panelists were given 10 min to acclimatize before measurements. After baseline readings, 2 μg/cm^2 of product was applied over a 25 cm^2 area.

Panelists returned 24 h later for a day 1 reading. From this point on, panelists applied the product twice daily for a period of 2 weeks. On measurement days no product was applied in the morning. After 2 weeks, the treatment was stopped and after an additional 1 week with no further treatment the final reading was taken.

The Raman spectra were used to construct concentration profiles of water and natural moisturizing factor (NMF), the naturally occurring mixture of amino acids, lipids, and sweat constituents that help to stabilize and maintain the moisture content of the stratum corneum. Figure 7.2a shows the part of the spectrum from which water concentration is calculated. Water concentration profiles were used to determine stratum corneum thickness at each measurement location and time (Fig. 7.2b). Stratum corneum thicknesses of different body sites, as determined by Raman spectroscopy, were compared to optical coherence tomography (OCT) measurements on the same sites, confirming the thickness determination.

Detailed profiles of the water distribution across the stratum corneum were obtained, showing changes in thickness and hydration upon treatment. From these profiles the area under the curve was calculated, representing the total stratum corneum hydration. Figure 7.3 shows the hydration results for the different moisturizing products. Total stratum corneum hydration is seen to increase significantly after 2 weeks of treatment with the niacinamide-containing product, and to persist into the 3rd week. The authors conclude that "The approach... allows a hitherto unavailable quantitative depth comparison of different treatments... This has given tremendously valuable insight into the mode of action of moisturizing formulas."

7.2.2 Drug Uptake into the Skin

Dennis et al. (2004) demonstrated an effective application of Raman spectroscopy for studying the uptake of anesthetics in the stratum corneum. Their method is indirect, with Raman measurements made *in vitro*. However, the experimental design allowed *in vivo* information to be inferred. They applied a gel containing tetracaine and TiO_2 (as an internal standard), using a patch on the forearm of a volunteer. At several times between 0 and 60 min after application the gel was removed and homogenized, and Raman spectra of the gel were recorded. TiO_2 is known to absorb in the skin to a very limited extent, if at all. Therefore, its concentration in the gel was assumed to remain constant, and the TiO_2 Raman signal was used as an internal signal intensity reference. The tetracaine Raman signal in the spectrum of the gel sample was observed to decrease as a function of application time. This decrease was used to calculate the quantity and rate of tetracaine uptake in the stratum corneum. Penetration of the anesthetic into the skin was independently confirmed by standardized measurements of the reduction of pain response in the treated area. This indirect measurement method provides a relatively simple and objective drug analysis tool that can provide hitherto difficult or impossible to obtain *in vivo* information on the skin. The major disadvantages of this method are the necessarily indirect inference of the amount of material taken up by the skin, and lack of spatially resolved information.

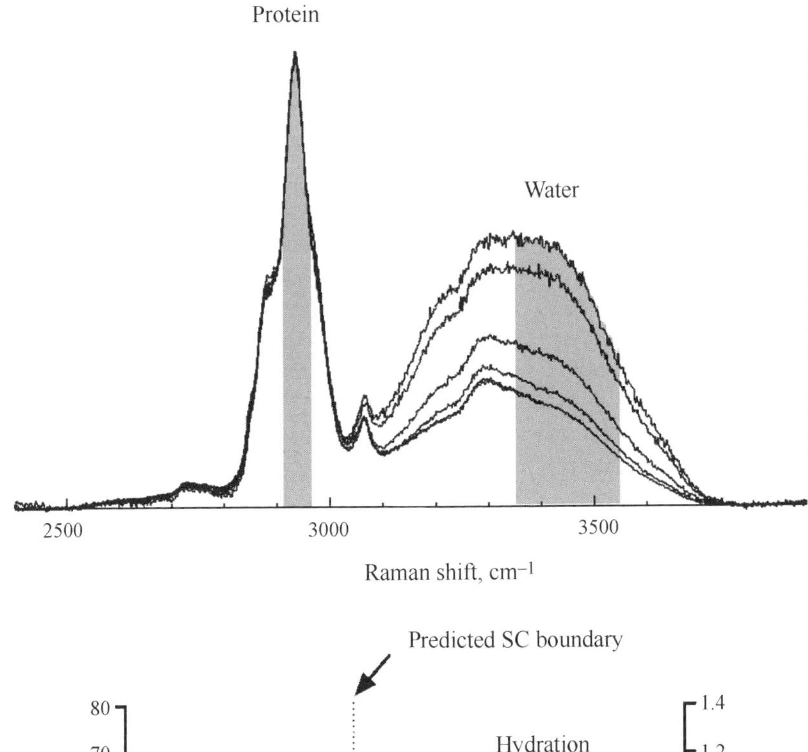

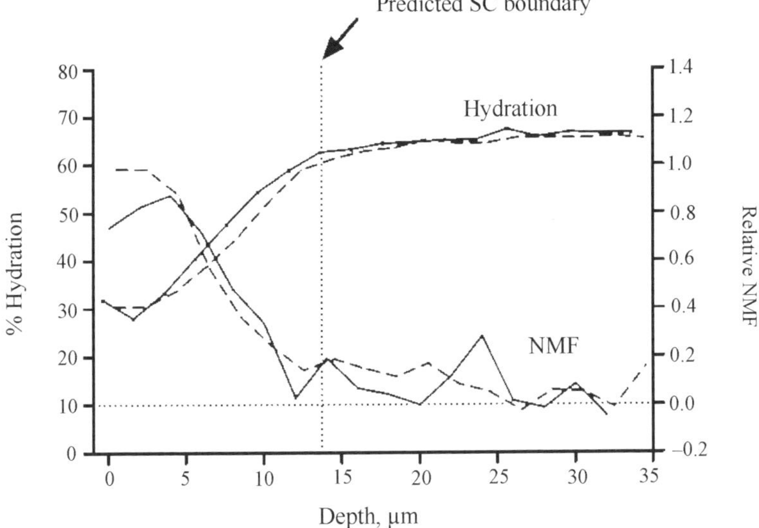

FIGURE 7.2 (**a**) Typical (baseline corrected) Raman spectrum containing peaks due to mainly keratin and water. Indicated are the regions of the peaks that are integrated for calculation of the water concentration (mass%). (**b**) Determination of the stratum corneum (SC) boundary with the viable epidermis by simultaneous skin hydration and NMF measurements. The thickness was confirmed by OCT.

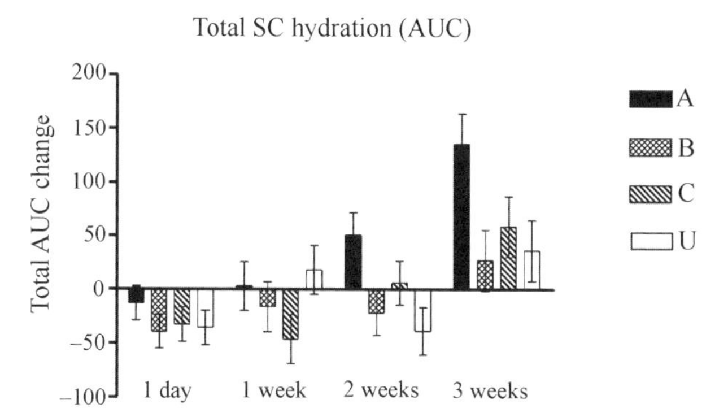

FIGURE 7.3 Change in the total stratum corneum hydration as a function of treatment.

7.2.3 Monitoring Transdermal Drug Delivery

There are three major approaches to enhancing the generally slow rate of penetration of drugs across the stratum corneum barrier. These are (1) encapsulation of the drugs in lipid vesicles, (2) the use of formulations incorporating penetration enhancers, and (3) conversion of the drug, usually by derivatization, to a chemical form that penetrates readily.

7.2.3.1 Encapsulation Encapsulation of drugs in lipid vesicles as a strategy to improve transport rates has been studied by Mendelsohn and his group using *in vitro* confocal Raman microscopy in conjunction with IR microscopy to understand the permeation of different types of lipid vesicles into pig skin (Xiao et al., 2005a,b). The exogenous substances, various deuterated phospholipids, were applied on the surface of excised pigskin. Following application and a waiting period to allow penetration to occur, maps were made of cross-sections through the skin using IR microscopy. These maps were used to visualize the distribution of components. Confocal Raman microscopy was then used to generate depth profiles, measured from the skin surface down to 50–100 μm below the surface. In addition, lateral profiles were recorded at a fixed depth of about 10 μm below the surface. The use of deuterated products in these experiments produced Raman bands due to CD stretching vibrations in the 2000–2200 cm^{-1} region where other Raman bands are sparse. This enabled a clear differentiation between the deuterated exogenous components and intrinsic skin constituents, allowing the penetration process to be mapped.

Although these experiments were performed *in vitro*, they highlight the advantages of vibrational spectroscopy for monitoring both exogenous and endogenous components in skin permeation investigations. Extension of confocal Raman spectroscopy to *in vivo* application expands the possibilities in this field even further.

7.2.3.2 Penetration Enhancers The use of penetration enhancers is well documented in the literature (Williams and Barry, 2004), and has begun to be studied by

Raman spectroscopy. Caspers et al. (2002) first demonstrated the use of *in vivo* confocal Raman microspectroscopy for *in situ* monitoring of the penetration and persistence in the stratum corneum of topically applied dimethyl sulfoxide (DMSO), a well-known penetration enhancer.

DMSO distribution in the stratum corneum of palmar skin was monitored at various depths, exposure times, and recovery times. Most of the dose permeated through the stratum corneum within 20 min. Surprisingly however, some DMSO remained within the tissue for hours and a small fraction was still detected after 3 days. Small shifts in DMSO peak positions were observed upon absorption in the stratum corneum. This can be taken as evidence for interactions with water and possibly other polar tissue moieties. Likewise, changes in band positions and shapes in the amide I region, mainly due to keratin, were recorded from locations at the skin surface. However, with the relatively low doses applied, firm conclusions regarding conformational changes of keratin due to DMSO interaction could not be drawn.

Song et al. (2005) have used confocal Raman microscopy on cadaver skin in a study of the activity and toxicity of iminosulfuranes, a group of novel chemical penetration enhancers. Samples of human cadaver skin were treated with one of the iminosulfurane analogs for 1 h. The distribution of the enhancer was then monitored at various depths using confocal Raman measurements. The results showed penetration of the enhancer into the stratum corneum to a depth of $\approx 20\ \mu m$, with a maximum at 5–10 μm from the skin surface. They also showed that the enhancer remained primarily in the stratum corneum without significant penetration into the viable epidermis.

Recent work (Van der Pol et al., 2006) describes quantitative *in vivo* confocal Raman measurements of an API applied on human skin showing the effect of ethanol, a common penetration enhancer. The work demonstrates enhancement of penetration of caffeine into the skin, showing differences in the extent of penetration as a function of time and of the delivery medium, water in one case, and a water/ethanol solution in the other. Solutions of 1.8 mass% of caffeine in water and in ethanol/water (1:2) were applied to the volar aspect of the lower forearm for 1 or 2 h. Four volunteers participated in the study; five concentration depth profiles per volunteer were recorded across the stratum corneum. Data were taken at 2 μm depth intervals. Figure 7.4 shows a typical Raman spectrum of untreated skin and skin treated with the aqueous caffeine solution. Raman signals due to the caffeine molecule can be observed readily.

Quantitative measurements of caffeine content were made for each sampled depth. The caffeine concentration is expressed in mmol of caffeine per gram of keratin. Quantification was based on fitting of the *in vivo* Raman spectra using a linear superposition of model spectra of caffeine and the main skin components and on the relative Raman intensity response due to caffeine and keratin.

The uptake of caffeine from the ethanol–water solution was compared for different volunteers at two exposure times. Figure 7.5 shows the quantitative concentration depth profiles (each an average of the five taken) for three volunteers after 1 h of exposure, and for one volunteer after a 2 h application of the caffeine solution. The data show clearly the increased caffeine concentration at the longer exposure time.

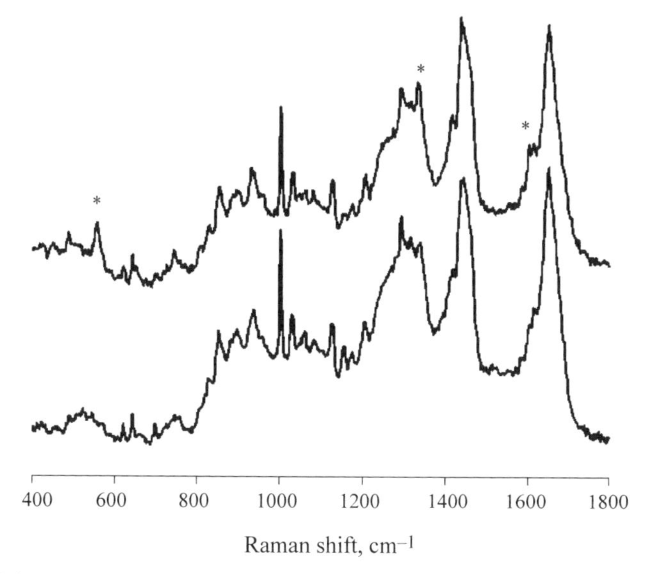

FIGURE 7.4 Typical Raman spectra (10 s acquisition time) of the stratum corneum for untreated skin (lower spectrum) and skin treated with caffeine in water solution (upper spectrum).

The stratum corneum thickness typically lies in the range 12–25 μm on the volar aspect of the forearm. In this experiment a depth range of about 40 μm was measured, ensuring coverage of the whole stratum corneum. For the three volunteers, the content of caffeine in the stratum corneum was found to be similar after a 1 h application. For example, at a depth of about 20 μm the caffeine content was around 0.1 mmol/g, corresponding to about 15 mg/g of keratin. The 2-h application time results in caffeine content at least two times higher throughout the stratum corneum.

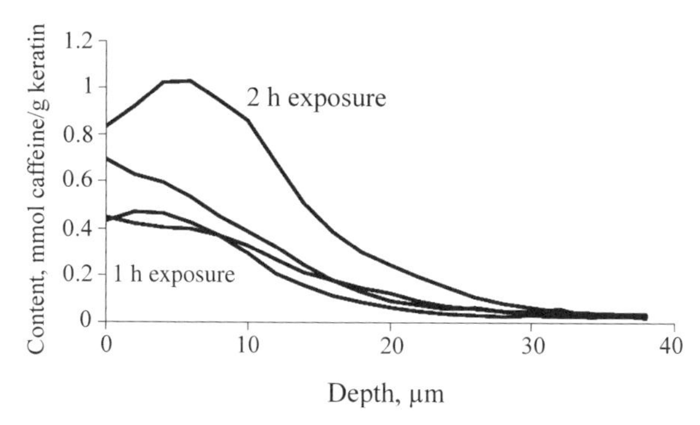

FIGURE 7.5 Concentration depth profiles (average of five repeats) of caffeine from ethanol–water (1:2) solution, for one volunteer at 2 h exposure (top curve) and for three volunteers at 1 h exposure (clustered curves).

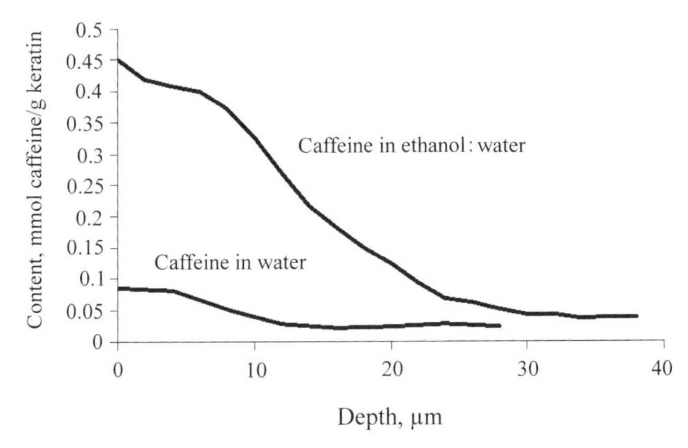

FIGURE 7.6 Comparison of concentration depth profiles of caffeine dissolved in ethanol–water (1:2) or water alone, volunteer 2. One hour application.

A second experiment was performed to compare the caffeine uptake from two solvents, water and ethanol–water. The averaged concentration depth profiles are shown in Fig. 7.6

The caffeine content after application from the ethanol–water solution is much higher than for the water solution. In the water-only case, a concentration of about 0.03 mmol/g of keratin (about 5 mg/g) is observed, whereas about five times that much was found in the ethanol–water case, illustrating the expected penetration enhancing effect of ethanol. Thus, as expected, increasing time of exposure results in greater absorption of the API into the skin, as does the use of ethanol for its penetration enhancing properties.

7.2.3.3 Derivatization Versus Use of an API Penetration Enhancer Confocal Raman microspectroscopy has recently been used to conduct proof of principal experiments that demonstrated dynamic *in vivo* monitoring of a topically applied API by confocal Raman spectroscopy (Van der Pol and Caspers, 2006). The effectiveness of using a penetration enhancer was compared with application of the active compound after derivatization with a lipophylic moiety.

The component investigated is denoted here as compound "X". (The material is proprietary; identification of the component at this time would raise intellectual property issues.) Compound X itself does not absorb into the stratum corneum in its pure form. This experiment was conducted to investigate whether it would be absorbed into the stratum corneum in two alternative formulations (Fig. 7.7).

Both experimental formulations were very simple. The first utilized the penetration-enhancing qualities of ethanol by using it as a solvent for the API, component X. For the second formulation, component X was modified through chemical derivatization with a lipophylic carrier molecule to enhance penetration into the skin. Both formulations were applied to the skin using a standard patch chamber. Raman depth profiles were recorded from two volunteers after application times of 0.5 and 4.5 h.

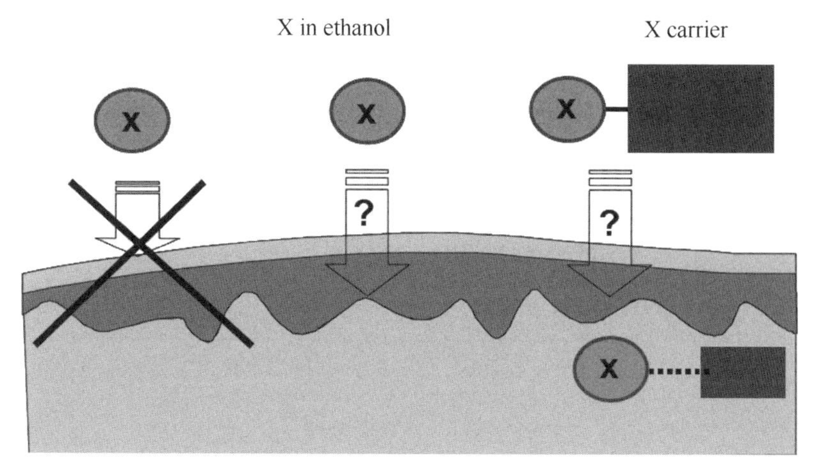

FIGURE 7.7 Assessment of uptake in the SC of active ingredient X, in different forms.

The concentration profiles shown in Fig. 7.8 are typical of results obtained. Profiles were measured at a number of different locations in the test site on the volar aspect of the forearm to average out biological variability. The example profiles are average profiles obtained from one of the volunteers. Each of the profiles shown was obtained by averaging nine profiles measured at different locations within each 3 cm × 3 cm test site. The results indicate no uptake of compound X in the pure form, as expected (Raman signal was only observed at the surface of the stratum corneum), while clear diffusion-like penetration profiles were obtained for X in the other two formulations. Furthermore, it was found that the observed concentration of component X was substantially lower after 4.5 h, indicating depletion at the surface, presumably by processes such as further transport into the viable epidermis, lateral diffusion, or metabolism.

The data available to date in this area are very sparse, as implied by the preliminary nature of the work described above. Nevertheless, these results suggest substantial potential value.

7.2.4 Direct Monitoring of Retinol in the Stratum Corneum

Retinol (Vitamin A), well known in the cosmetics industry for its antiwrinkle properties, improves the elasticity of the skin by increasing the synthesis of collagen. The biological role of retinol and the products of its metabolism have been reviewed by Perlmann (2002), who also discusses its biomedical effects. Retinoid deprivation during pregnancy (or exposure to excesses of retinoids) is thought to be correlated to birth defects such as spina bifida and cleft palate. Niederreither et al. (1999) have proposed treatment of pregnant women with all-*trans* retinoic acid to promote embryonic tissue differentiation. Thus, the retinoid class of substances find applications in both cosmetic and medical fields.

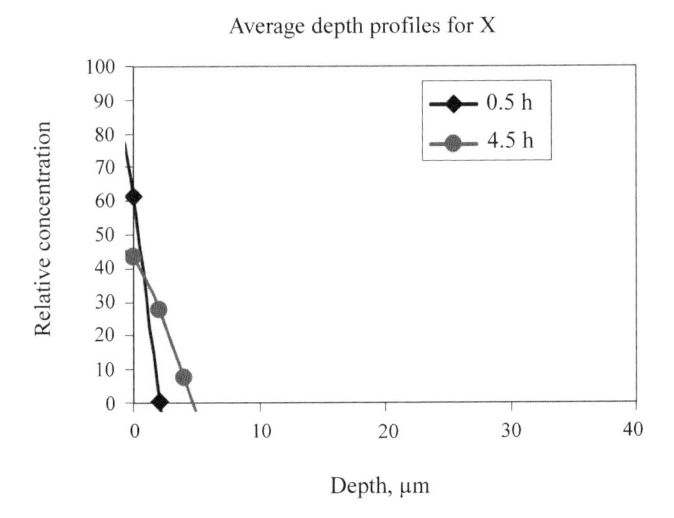

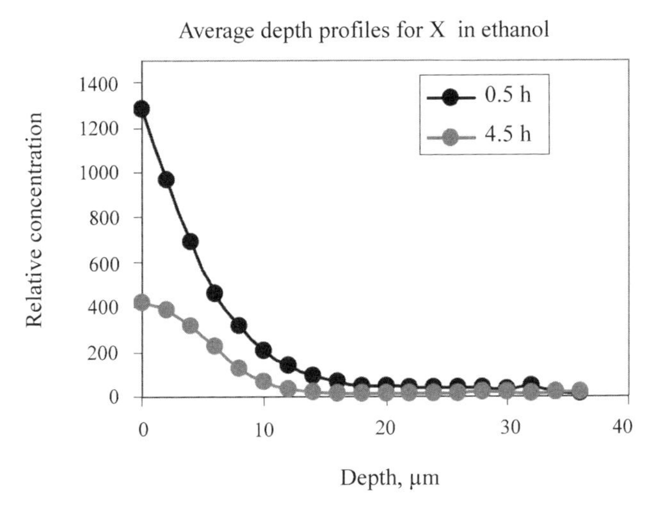

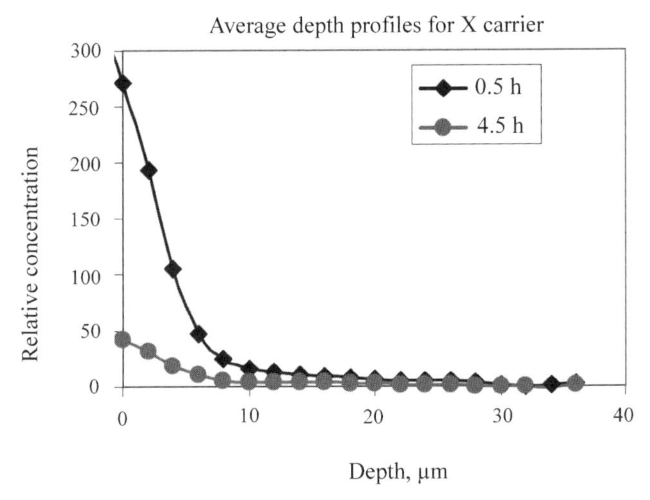

FIGURE 7.8 Relative abundance of component X (with respect to keratin) in the stratum corneum of a volunteer as a function of depth and for different application times and formulations.

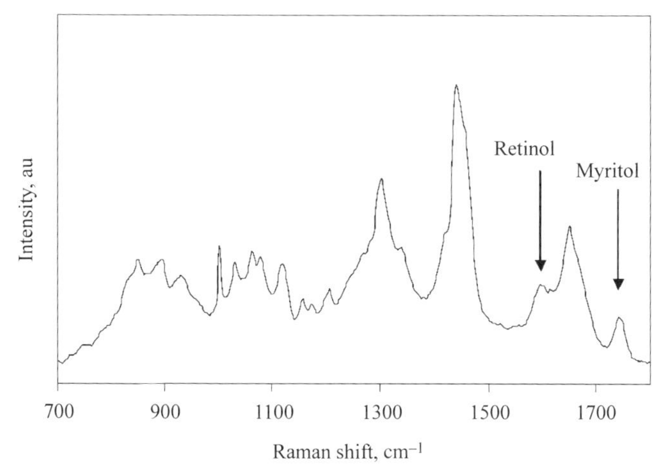

FIGURE 7.9 *In vivo* Raman spectrum of the surface of the stratum corneum of human skin (volar aspect of the forearm), treated with a solution of retinol (0.3 wt%, 2 h) in myritol oil. Spectrum acquisition time: 30 s.

Retinol is unstable under the influence of oxygen, water, and UV irradiation. Therefore, it is important in topical applications to ensure either very rapid transport through the stratum corneum, or controlled (and stabilized) release. Raman spectroscopy is an excellent tool to make the spatially resolved *in vivo* chemical composition measurements required to monitor experiments for this purpose, since retinol produces a very strong Raman signal. Figure 7.9 shows an example *in vivo* Raman spectrum of the surface of the stratum corneum showing signals due to myritol, a nonskin penetrating oil commonly used in skin creams, and retinol, both applied from a topical preparation.

Topical delivery of retinol was studied *in vitro* by Failloux et al. (2004) using Raman microspectroscopy in conjunction with a Franz cell apparatus to measure penetration and diffusion rates. They compared penetration and diffusion of retinol into the epidermis when retinol was introduced in either an oil-in-water (oiw) emulsion or in encapsulated form.

First, a sample of mouse skin was used in a Franz cell experiment. Retinol penetration was assessed by measurement of the time dependent retinol concentration in the Franz cell reservoir using high pressure liquid chromatography (HPLC). Retinol storage was assessed similarly by measurement of release of retinol from a mouse skin (loaded with retinol). Then the same two formulations containing retinol were applied to biopsy skin from a human abdomen. These biopsy samples were then studied with Raman microspectroscopy mapping of the retinol signals. Lateral maps of retinol abundance at the surface and 20 μm below the surface were measured. Both Franz cell and Raman experiments showed that retinol releases more slowly from its formulation in the encapsulated form. Consistent with this, the Franz cell experiments showed that the residence time of retinol in the skin is increased for an encapsulated formulation. The Raman data showed that retinol in its encapsulated form absorbs faster into the skin than retinol in an oil emulsion formulation.

In discussion, the authors emphasized that very fast uptake in or through the stratum corneum is desirable to protect retinol from potential breakdown from UV irradiation. Penetration through only the first few micrometers of skin thickness is sufficient for this due to the very shallow penetration depth of UV radiation in lipid materials. Direct results to assess the speed of uptake through the stratum corneum were not available from this experimental approach.

New depth resolved *in vivo* data on the delivery of retinol as an active ingredient for promotion of skin health have recently appeared (Pudney et al., 2007). In this work, confocal Raman microspectroscopy was used to study retinol delivery *in vivo*. Initially, the method was tested with retinol in propylene glycol(PG)/ethanol, a highly effective model delivery system identified from earlier *ex vivo* experiments. The retinol was successfully measured penetrating into the skin of the volar forearm, first into the stratum corneum and then into the viable epidermis. This was monitored over a 12-h period after which retinol could no longer be detected. Penetration of the retinol was observed to be highly correlated with the depth of penetration of the PG. Figure 7.10 illustrates a typical result. Shown are spectra recorded from the surface of the stratum corneum down to 10 μm depth, in steps of 2 μm. Peaks due to PG and *trans*-retinol are observed clearly.

This experiment was followed by an investigation of delivery from myritol. Since myritol is nonskin-penetrating, this system was applied both with and without

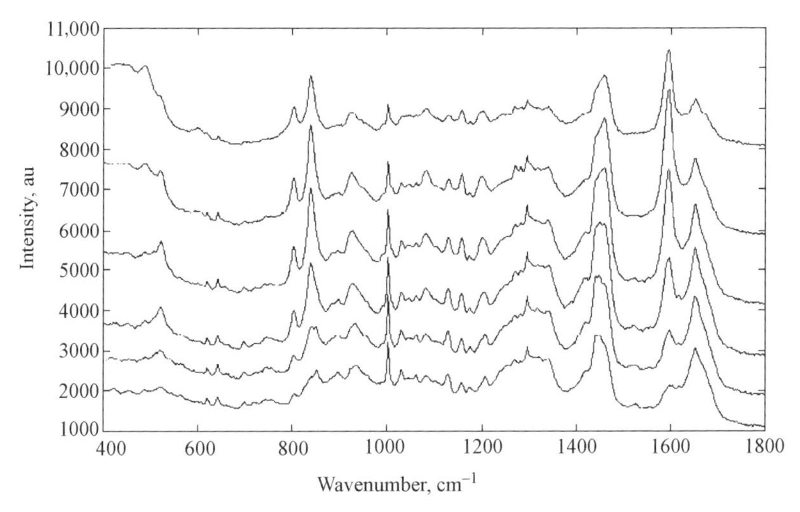

FIGURE 7.10 Spectra of the skin treated with *trans*-retinol in PG/ethanol, measured every 2 μm from the surface (top spectrum) to 10 μm depth, showing the prominent *trans*-retinol band at 1594 cm^{-1} and bands due to PG at 803 and 840 cm^{-1}. Spectra vertically offset for clarity.

the use of penetration enhancers. Two penetration enhancers believed to act by different mechanisms were chosen: oleic acid, a lipid 'fluidizer,' and Triton X-100, a lipid 'extractor.' The Raman measurements showed different rates of penetration of retinol from these systems. Retinol applied in myritol without penetration enhancer produced minimal penetration of retinol below the stratum corneum, while both penetration enhancers promoted substantial penetration, with the oleic acid found to be approximately twice as effective as the Triton X-100. The results have aided in greater understanding of the mechanism of delivery and penetration of therapeutic compounds through the skin. These *in* vivo Raman measurements have also recently been made fully quantitative.

7.2.5 Uptake of UV Filter Compounds from Sunscreen Formulations

In vitro Raman studies of the UV filter compounds used in commercial sun protection formulations have been reported in the literature for some time (e.g., Asher, 1984; Beyere et al., 2003). Early *in vivo* results now beginning to appear have been dramatic in their clarity, suggesting a potential for eventual widespread utility of *in vivo* Raman spectroscopy in this and similar applications. Since the chemical UV-A and UV-B absorbing compounds approved for sunscreen use are intended, ideally, to provide immediate UV protection, they are formulated to promote rapid uptake into the stratum corneum. Similar principles apply to the design of sunscreen and transdermal drug delivery (TDD) formulations, with the key difference that TDD is intended to transport APIs through the skin, whereas sunscreens are meant to remain within it. Results from one are nevertheless suggestive as models for possible studies of the other. Certainly, the *in vivo* measurement requirements are very similar.

Nature has supplied us with a favorable situation for Raman studies of UV absorbers as a class. These compounds tend to give very strong Raman signals, making the sunscreen-approved UV-A and UV-B absorbers excellent candidates for *in vivo* study by Raman spectroscopy. The strong Raman signals arise from the so-called (pre-)resonance Raman effect. (Happily, the same physical phenomenon also produces very strong Raman signals for the retinoids, as discussed in the previous section.)

Beyere et al. (2003) studied common sunscreen active ingredients with both UV absorption spectroscopy and with Raman spectroscopy using 406.7 nm laser excitation. Raman spectra of the UV filter compounds octyl *N,N*-dimethyl-*p*-aminobenzoic acid (ODPABA), octyl *p*-methoxycinnamate (OMC, also known as Parsol MCX®), octyl salicylate (OCS), dibenzoylmethane (DBM), and oxybenzone (BZ3) were presented and a range of solvents was tested, providing a useful set of reference spectra.

Van der Pol and Caspers (2006) presented the first real-time *in vivo* kinetic measurements of the process of absorption in the stratum corneum of sunscreen actives in a formulation. Approximately, 5 μl of a standard sunscreen formulation (P3: high SPF standard, Bayer standard C202/101) was gently rubbed onto a $3 \times 3\,\text{cm}^2$ area on the volar aspect of the forearm. Raman spectra were measured before and after

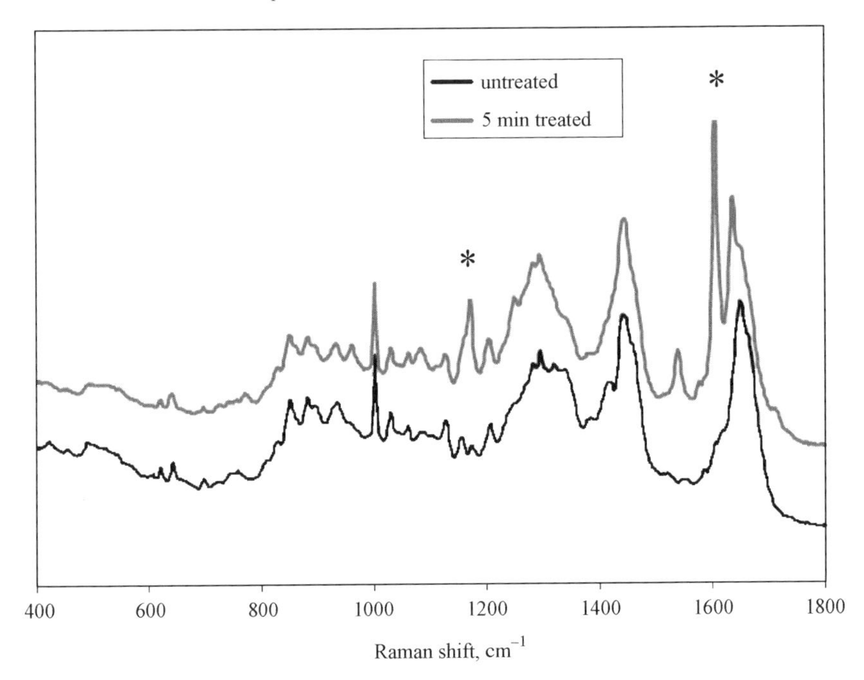

FIGURE 7.11 *In vivo* Raman spectra recorded at 5 μm below the surface of the stratum corneum of human skin (volar aspect of the forearm). Untreated (black spectrum) and treated (gray spectrum). The measurement time per spectrum was 10 s.

application. Figure 7.11 shows the resulting *in vivo* Raman spectra. Two major peaks due to the sunscreen actives are indicated. These peaks, located at about 1172 and 1605 cm^{-1}, are due to Parsol MCX (octyl *p*-methoxycinnamate), matching the peaks observed by Beyere et al. (2003).

From Raman spectra recorded at different depths and based on the peak intensity of the 1605 cm^{-1}, a concentration profile can be calculated for the relative concentration of the sunscreen active (with respect to the keratin content). A typical profile for Parsol MCX, obtained 5 min after application, is shown in Fig. 7.12. The profile shown is the average of four repeat measurements.

The absorption of the sunscreen actives into the stratum corneum was found to be very rapid. This was assessed in experiments in which Raman spectra were measured directly after application of the SPF standard. The laser focus/detection volume was held at a depth of 10 μm below the skin surface, while measuring a Raman spectrum every 20 s. From these spectra kinetic profiles of relative Parsol MCX content versus time (at a depth of 10 μm) were constructed (Fig. 7.13).

Initially, the signal due to Parsol MCX is absent at 10 μm depth. Then, after about 1.5 min, the signal becomes visible and reaches its maximum about 3 min after application. Then the signal starts to decrease, indicating product depletion

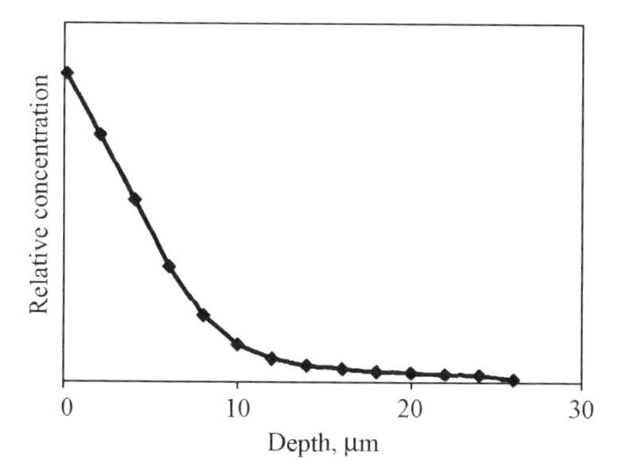

FIGURE 7.12 Relative concentration profile for Parsol MCX, 5 min after application of the SPF standard sample.

at the surface, while axial and/or lateral diffusion continues to move the sunscreen active away from the measurement location.

This work suggests strong potential for this measurement capability to enable the development of more detailed understanding of how fast and how far these compounds penetrate upon application, how long they persist in the skin, and possibly what metabolic or other processes eventually deplete them. This appears to be a fertile area for further research.

7.2.6 Raman Monitoring of Iontophoresis

Iontophoresis is a process by which drugs are driven into the skin under the influence of an electric field. Iontophoresis as a means of drug delivery was probably first described by Glass et al. (1980).

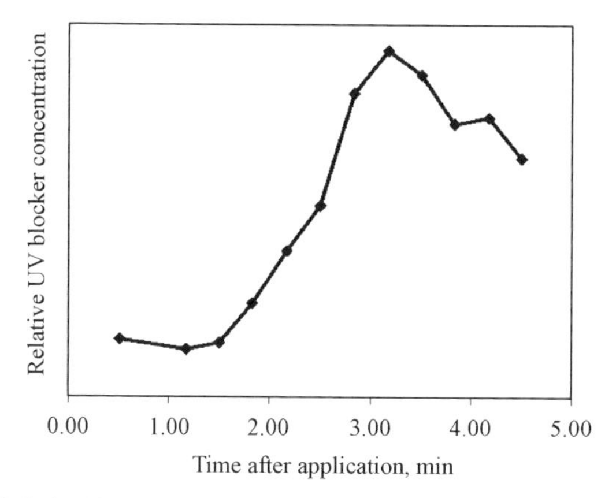

FIGURE 7.13 Relative Parsol MCX concentration at a depth of 10 μm below the stratum corneum surface, plotted versus time from application of the SPF formulation.

Iontophoresis, of course, not only moves active ingredients into the skin, but also skin constituents within the skin and even extracts skin components (reverse iontophoresis). A well-known example of the latter effect is the Glucowatch® (www.glucowatch.com), a glucose monitoring system for diabetes patients (Rao et al., 1995). The technique is based on extraction of glucose by reverse iontophoresis with automatic detection in an electrochemical sensor cell after specific enzymatic oxidation of the glucose.

The first example of the use of *in vivo* Raman microspectroscopy to study iontophoresis appeared recently (Wascotte et al., 2007). The research reported was aimed at measurement of the amount and distribution of urea in the stratum corneum before and after iontophoresis. Monitoring of urea (from serum) by noninvasive reverse iontophoresis is applied during dialysis of renal failure patients. However, previous experiments have showed high initial urea fluxes, followed by a lower steady state flux after about 30 min, which complicates interpretation of the data.

In a series of Raman measurements on four healthy volunteers and using 30 min of iontophoresis time, it was shown that the high initial urea flux is due to removal of urea that is initially present in the stratum corneum and, in the majority of cases, not related to urea from serum. This means that the noninvasive monitoring of urea can be realized using reverse iontophoresis, provided current is passed during at least a 30-min period in advance in order to empty the "urea reservoir" of the stratum corneum. In Fig. 7.14 the concentration profile of urea, as measured by Raman spectroscopy, is shown. The profile is the average of 16 repeat measurements divided over the four volunteers.

The urea concentration was shown to be significantly lower for the depth range 0–6 μm ($p = 0.05$). From the same Raman measurements concentration profiles for lactate were also calculated (not shown). Lactate content was also found to decrease upon 30 min of iontophoresis. In this case, the significant depth range was 0–20 μm ($p = 0.05$).

The measurements described above are to our knowledge the first results to be published in which Raman microspectroscopy has been used to study the processes occurring during or after iontophoresis. These results show that this technique promises to be a fruitful tool for research in the area. Many different effects of iontophoretic processes are likely to be monitorable by *in vivo* Raman microspectroscopy. These include direct measurement of the concentration profiles of the delivered drugs, reordering of lipids in the stratum corneum, alteration of the water distribution as a function of depth, transport of skin constituents towards or away from the skin surface, and changes in barrier properties. Using appropriate experimental design, both short- and long-term changes can be studied.

7.2.7 Monitoring Effects of Medicinal Skin Treatments

Since Raman microspectroscopy is noninvasive and nondestructive, the technique can be used to investigate *in vivo* the effects of pharmaceuticals applied to tissue. A pioneering example of such a study monitored the presence and amount of

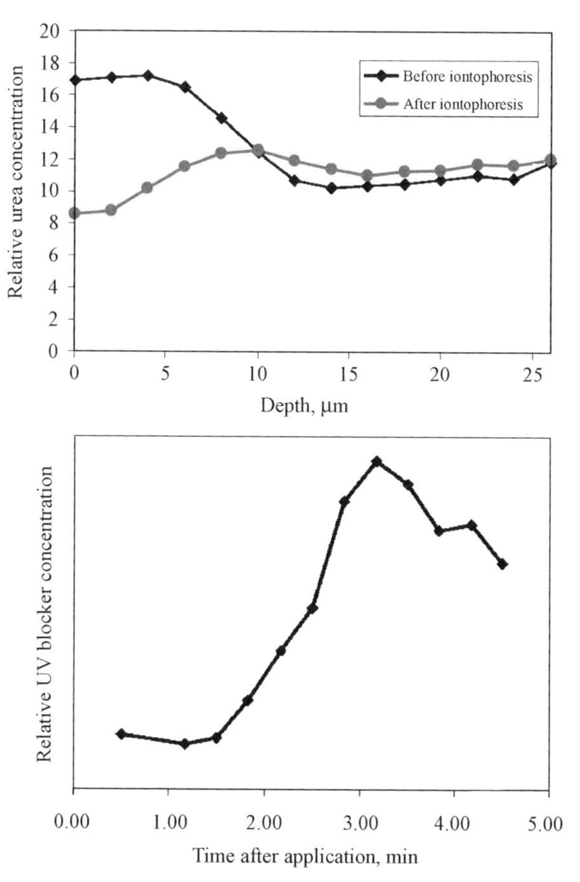

FIGURE 7.14 *In vivo* concentration profile of urea in the human stratum corneum before and after 30-min iontophoresis treatment. The profiles are the average of 16 repeat measurements over four volunteers.

hydrogen peroxide in skin as related to vitiligo, allowing deduction of the effectiveness of medical treatments for these conditions (Schallreuter et al., 2002).

Schallreuter et al. (1999) have been studying vitiligo, a common depigmentation disorder, in relation to oxidative stress. In previous work Schallreuter found extremely high concentrations of hydrogen peroxide (H_2O_2) together with low catalase levels in the epidermis of vitiligo patients (Schallreuter et al., 1999). The Raman spectrum of H_2O_2 shows an intense Raman peak due to the O–O stretch vibration at 875 cm^{-1}, which can be clearly identified in the epidermis of vitiligo patients. Using *in vitro* FT-Raman spectroscopy, the Schallreuter group was able to detect and track the decomposition of H_2O_2 over time by water from the Dead Sea, which is known for its beneficial effect on vitiligo.

They used the same technique *in vivo* for a group of 22 vitiligo patients in whom they monitored H_2O_2 levels after bathing in Dead Sea water. For comparison with a known pharmaceutical treatment, they also measured H_2O_2 removal by a topically applied pseudocatalase cream. Their results showed clearly the removal of H_2O_2 by Dead Sea water, but at slower rate than with the cream. The *in vivo* measurements showed about 50% removal of H_2O_2 by 15-min bathing in Dead Sea water versus almost 100% removal by a 15-min treatment with pseudocatalase cream.

7.2.8 Neutraceuticals

Carotenoids are thought to play an important role in the antioxidant defense system of the skin. Various studies have assessed the role of carotenoids in the development of cancer in different tissues, including the skin.

The standard analytical technique for measuring carotenoids is HPLC. However, for determination of carotenoids in the skin, HPLC would require removal of large tissue volumes. Until Raman spectroscopy came into play, this factor, plus the complexity of the available *in vitro* methods, prevented quantification of carotenoid levels in the skin.

Hata et al. (2002) used *in vitro* and *in vivo* Raman spectroscopy to attack this problem. Resonant Raman excitation at 488 nm was used to boost the carotene signals and elevate them above the relatively intense fluorescent background signal of the skin that is generated by Raman excitation wavelengths in this range. A specially developed hand-held probe enabled *in vivo* measurements of skin carotenoids on different sun-exposed and sun-protected body areas of healthy skin as well as histologically confirmed basal cell carcinoma (BCC) and clinically proven actinic keratosis (AK). Repeated measurements on the same healthy skin area showed good reproducibility of the detected carotenoid levels, whereas intra-individual differences among the six healthy volunteers were large. Large differences were also observed in the carotenoid level between anatomical regions. The comparison between healthy, cancerous (BCC), and precancerous (AK) skin showed significantly lower carotenoid levels in lesional skin as compared to healthy skin. These correlations suggest the possibility of a causal relationship between carotenoid levels in the skin and the presence or absence of skin cancer.

Another interesting finding was that carotenoids are mainly present in the stratum corneum, as was determined by the removal of skin layers by sequential tape stripping in combination with *in vivo* Raman measurements. This supports the hypothesized role of carotenoids in the initial protective barrier to UV insult and development of skin cancer. This work represents a promising early step in opening the door to the study of the depth-related distribution of clinically important molecular skin constituents within the skin by *in vivo* confocal Raman spectroscopy.

More recently, dual-wavelength resonance Raman spectroscopy has been applied by Lademann and coworkers (Darvin et al., 2005; Ermakov et al., 2004) to distinguish the two most common carotenoids, β-carotene and lycopene, in human skin. Owing to the antioxidant protection of the carotenoids, in particular protection from UV induced free radicals in the skin, there is strong interest in ways to influence skin carotenoid levels by systemic or topical administration. For example, supplemental β-carotene is commonly used to treat various photosensitivity diseases such as erythropoietic protoporphyria (Mathews-Roth, 1993). In view of this, noninvasive methods to monitor the uptake of different supplemental carotenoids in the skin are needed to add to understanding of the treatment and its effects.

β-Carotene and lycopene are thought to play different roles in the body. The Raman spectra of these molecules are quite similar. However, β-carotene has an electronic absorption band near 488 nm, whereas lycopene has a second absorption band near 514.5 nm. As a result, the Raman lines for β-carotene and lycopene have different scattering efficiencies at excitation wavelengths of 488 and 514.5 nm. Darvin et al. (2005) used this to their advantage to differentiate between the two. The use of dual-wavelength excitation from a multiline Argon ion laser enabled *in vivo* quantification of both β-carotene and lycopene in the skin. These experiments clearly illustrated large variations in the β-carotene to lycopene ratio in a small group of subjects, which may be a consequence of different individual uptake capabilities for carotene and lycopene.

The above-described research results demonstrate the scientific validity of these measurements. Subsequent informal observations in the our laboratory have shown that these compounds can be routinely detected and measured in the stratum corneum with Raman microspectroscopy. We have observed that the abundance seems to correlate with the consumption of tomatoes or carrots, for example. We conclude that routine practical application of Raman spectroscopy for this application is now possible.

7.2.9 Future Areas of Application Development and Research

A discussion of applications of Raman spectroscopy to biological tissue, especially to the skin, that have utility or potential utility in the pharmaceutical arena should not neglect mention of the combination of confocal video microscopy (CVM) with Raman spectroscopy. Using NIR laser irradiation, CVM, also known as confocal laser scanning microscopy (CLSM), is capable of imaging skin both at the surface and at defined depths below the surface at a resolution similar to that of confocal Raman spectroscopy. These techniques are therefore very compatible, and have been combined (Caspers et al., 2003). The combination enables the targeting of specific subsurface skin features for Raman analysis. In addition, molecular concentration profiles measured by Raman spectroscopy can be directly related to skin architecture. This combined capability is not yet commercially available, but its potential utility for research in transdermal drug delivery, treatment of skin disorders, and other such topics in pharmaceutical research suggests that it is likely to become available in the foreseeable future.

Another area where development can be expected is in the extraction of information from some of the more subtle but already well-known features of Raman fingerprint region measurements. For example, when using Raman spectroscopy to study drug molecules in different matrices, such as in topical formulations or in biological tissues, it can be expected that the details of some spectral features will vary due to the influence of the molecular environment. Intermolecular effects of matrices on the spectral features of analyte molecules are well known. For example, the work briefly described earlier on UV filter compounds (Beyere et al., 2003) concluded that these compounds exhibit observable solvent effects, leading to small shifts and intensity changes in the Raman spectra, correlated to the solvent properties. The same effects may also apply to the Raman spectra of active ingredients when dissolved in the skin lipids and fluids. These effects are not large enough to prevent the spectra from providing unique "fingerprint" identification of the compounds for purposes of identifying and tracking the compounds in the skin. In future, however, it may be possible to use these small effects to study more subtle phenomena of interest in the skin. Such effects can already be simulated quantum-chemically (e.g., with a density functional calculation), aiding in greater understanding of solvent–drug and drug–matrix interactions. Bondesson et al. (2007) have presented illustrative calculations on changes in spectral features (IR and Raman) of aspirin, caffeine, and ibuprofen molecules. They propose such calculations as a tool for finding structure–property relationships, connecting changes in spectra to changes in molecular structure.

7.3 SUMMARY AND DISCUSSION

In vivo confocal Raman microspectroscopy is a novel method that provides detailed information about the molecular composition of the skin. Many applications in this area to date have centered on the study of the stratum corneum. However, the method is capable of measurements to a depth of greater than 150 μm into the skin—well into the dermis. Some of the examples described here show that studies of phenomena in the dermis are indeed viable.

In the past 5 years, *in vivo* confocal Raman spectroscopy has made a major leap forward in sensitivity, speed of measurement, and ease of use. Raman technology has now reached a level of refinement where it can be applied in routine clinical studies. It has become fast enough to perform measurements on numbers of subjects ranging from several up to several dozen per day, depending on the complexity of the study. The interface between the scientist and the instrument has reached a stage of development where routine operation of the equipment by a laboratory technician is practical.

Although the Raman technique has now been shown to be routinely useful in clinical settings, it is, like all measurement techniques, subject to certain limitations. First, it should be recognized that *in vivo* Raman is not a trace analysis technique. Limits of detection range from the 100 ppm level in favorable cases such as the retinoids to the 0.1–0.2% range for a great many compounds (caffeine is an example), to 1–2% for less favorable cases. Indeed, some compounds are not Raman sensitive at all, since the Raman effect depends on the electronic structure of the scattering molecule. Also, some classes of molecules of interest in biological tissue, such as some proteins, exhibit closely similar spectra and are difficult to differentiate.

The use of microscopic resolution capability leads to many measurements being made at a single location to generate composition depth profiles, while other techniques, such as electrical conductivity, for example, normally take only a single data point at a given location. This means that even with fast instrumentation,

Raman measurements may be time consuming compared to other commonly used methods of *in vivo* skin analysis. That is, however, simply the price paid to obtain much greater information content. A related general issue is that *in vivo* tissue analysis usually requires considerable replication, by measurement of multiple locations and on multiple subjects, to achieve needed statistical accuracy given normal biological variability. This is, of course, a characteristic inherent to any human *in vivo* measurements, and not specific to Raman.

Finally, Raman instrumentation for *in vivo* skin analysis is highly specialized and therefore expensive. However, as *in vivo* Raman microspectroscopy comes into more general use, the cost of the instruments eventually can be expected to drop as volume efficiencies are realized by manufacturers.

These limitations are well compensated by the richness of information achievable and the unique ability to measure the same area of skin repeatedly and with microscopic spatial detail, allowing entirely new kinds of information to be gathered. This unique information has been unavailable to date from other noninvasive techniques. As illustrated by the examples given here, the information content of *in vivo* Raman measurements is much greater than from other noninvasive skin biophysical measurement techniques. The detailed information available on chemical composition, the ability to make spatially resolved measurements on the stratum corneum, and the ability to make these measurements *in vivo*, can provide insights into the mode of action of API that have not been previously available.

Raman spectroscopy, already an extraordinarily versatile and useful analytical technique, has now arrived at a stage of refinement that allows practical measurements directly on intact biological tissue, including *in vivo* measurements. This new capability opens up vistas for exploration that hold a great promise for pharmaceutical research.

REFERENCES

Asher SA, 1984. Ultraviolet resonance Raman spectrometry for detection and speciation of trace polycyclic aromatic hydrocarbons. *Anal. Chem.*, **56**(4), 720–724.

Attas M, Posthumus T, Schattka B, Sowa M, Mantsch H, Zhang SL, 2002. Long-wavelength near infrared spectroscopic imaging for in-vivo skin hydration measurements. *Vib. Spectrosc.*, **28**(1), 37–43.

Beyere LH, Yarasi S, Loppnow GR, 2003. Solvent effects on sunscreen active ingredients using Raman spectroscopy. *J. Raman Spectrosc.*, **34**, 743–750.

Bondesson L, Mikkelsen KV, Luo Y, Garberg P, Ågren H, 2007. Hydrogen bonding effects on infrared and Raman spectra of drug molecules. *Spectrochim. Acta part A: Mol. Bio. Spectro.* **66**, 213–224.

Caspers PJ, Williams AC, Carter EA, Edwards HGM, Barry BW, Bruining HA, Puppels GJ, 2002. Monitoring the penetration enhancer dimethyl sulfoxide in human stratum corneum by in vivo confocal Raman spectroscopy. *Pharm. Res.*, **19**, 1577–1580.

Caspers PJ, Lucassen GW, Puppels GJ, 2003. Combined in vivo confocal Raman spectroscopy and confocal microscopy of human skin, *Biophys. J.*, **85**, 572–580.

Caspers PJ, 2003. In vivo skin characterization by confocal Raman microspectroscopy. Ph.D. Thesis, Erasmus University Medical Center, Rotterdam, NL.

Darvin ME, Gersonde I, Albrecht HJ, Gonchukov SA, Sterry W, Lademann J, 2005. Determination of beta carotene and lycopene concentrations in human skin using resonance Raman spectroscopy. *Laser Physics.*, **15**(2):295–299.

Dennis AC, McGarvey JJ, Woolfson AD, McCafferty DF, Moss GP, 2004. A Raman spectroscopic investigation of bioadhesive tetracaine local anaesthetic formulations. *Int. J. Pharm.*, **279**, 43–50.

Ermakov IV, Ermakova MR, Gellermann W, Lademann J, 2004. Noninvasive selective detection of lycopene and beta-carotene in human skin using Raman spectroscopy. *J. Biomed. Opt.*, **9**(2):332–338.

Failloux N, Baron M–H, Abdul-Malak N, Perrier E, 2004. Contribution of encapsulation on the biodispensibility of retinol. *Int. J. Cosm. Sci.*, **26**(2):71–77.

Glass JM, Stephen RL, Jacobsen SC, 1980. The quality and distribution of radio labelled dexamethasone delivered to tissues by iontophoresis. *Int. J. Dermatol.*, **19**, 515–519.

Hanlon EB, Manoharan R, Koo T-W, Shafer KE, Motz JT, Fitzmaurice M, Kramer JR, Itzkan I, Dasari RR, Feld MS, 2000. Prospects for in vivo Raman spectroscopy. *Phys. Med. Biol.*, **45**, R1–R59.

Hata TR, Scholz TA, Ermakov IV, McClane RW, Khachik F, Gellermann W, Pershing LK, 2000. Non-invasive Raman spectroscopic detection of carotenoids in human skin. *J. Invest. Dermatol.*, **115**(3):441–448.

Mathews-Roth MM. Carotenoids in erythropoietic protoporphyria and other photosensitivity diseases, 1993. *Ann. N.Y. Acad. Sci.*, **691**, 127–138.

Niederreither K, Subbarayan V, Dollé P, Chambon P, 1999. Embryonic retinoic acid synthesis is essential for early mouse post-implantation development. *Nat. Genet.*, **21**, 444–448.

Notingher I, Imhof RE, 2004. Mid-Infrared in vivo depth-profiling of topical chemicals on skin. *Skin Res. Technol.*, **10**, 113–121.

Perlmann T, 2002. Retinoid metabolism: A balancing act. *Nat. Genet.*, **31**, 7–8.

Pudney PDA, Melot M, Caspers PJ, van der Pol A, Puppels GJ, 2007. An In-vivo confocal Raman study of the delivery of retinol to the skin. *Appl. Spectrosc.*, **61**(8) (in press).

Rao G, Guy RH, Glikfeld P, LaCourse WR, Leung L, Tamada J, Potts RO, Azimi N, 1995. Reverse iontophoresis: Noninvasive glucose monitoring in vivo in humans. *Pharm. Res.*, **12**, 1322–1326.

Schallreuter KU, Moore J, Behrens-Williams S, Panske A, Harari M, Rokos H, Wood JM, 2002. In vitro and in vivo identification of 'pseudocatalase' activity in Dead Sea water using Fourier transform Raman spectroscopy. *J. Raman. Spectrosc.*, **33**(7):586–592.

Schallreuter KU, Moore J, Wood JM, 1999. In vivo and in vitro evidence for hydrogen peroxide (H_2O_2) accumulation in the epidermis of patients with vitiligo and its successful removal by a UVB-activated pseudocatalase. *J. Invest. Dermatol. Symp. Proc.*, **4**, 91–96.

Sieg A, Crowther J, Blenkiron P, Marcott C, Matts PJ, 2006. Confocal Raman microspectroscopy: Measuring the effects of topical moisturizers on stratum corneum water gradient in vivo. *Proc. SPIE*, **6093**, 60930N.

Song Y, Xiao C, Mendelsohn R, Zheng T, Strekowski L, Michniak B, 2005. Investigation of iminosulfuranes as novel transdermal penetration enhancers: Enhancement activity and cytotoxicity. *Pharm. Res.*, **22**(11):1918–1925.

Tanojo H, Junginger HE, Bodde HE, 1997. In vivo human skin permeability enhancement by oleic acid: Transepidermal water loss and Fourier-transform infrared spectroscopy studies. *J. Control. Release*, **47**, 31–39.

Van der Pol A, Caspers PJ, 2005. Comparison of penetration of an API using a penetration enhancer and by derivatization. Personal Communication. Rotterdam, NL: River Diagnostics BV and PhytoGeniX BV.

Van der Pol A, Caspers PJ, 2006. Penetration of UV blockers monitored in vivo by confocal Raman microspectroscopy. Perspectives in Percutaneous Penetration Conference, la Grande Motte, France. Abstracts, **10A**, 105.

Van der Pol A, de Sterke J, Caspers P, 2006. Quantitative in vivo monitoring of topically applied active pharmaceutical ingredients and penetration enhancers. Application Note #009. River Diagnostics BV, Rotterdamm, NL.

Wascotte V, Caspers PJ, de Sterke J, Jadoul M, Guy RH, Préat V, 2007. Assessment of the "Skin Reservoir" of Urea by Confocal Raman Microspectroscopy and Reverse lontophoresis *in vivo*, *Pharmac. Res. (in press).*

Williams AC, Barry BW, 2004. Penetration Enhancers. *Adv. Drug. Deliv. Rev.*, **56**, 603–618.

Xiao C, Moore DJ, Flach CR, Mendelsohn R, 2005a. Permeation of dimyristoylphosphatidylcholine into skin: Structural and spatial information from IR and Raman microscopic imaging. *Vib. Spectrosc.*, **38**, 151–158.

Xiao C, Moore DJ, Rerek ME, Flach CR, Mendelsohn R, 2005b. Feasibility of tracking phospholipid permeation into skin: using infrared and Raman microscopic imaging. *J. Invest. Dermatol.*, **124**, 622–632.

Woo Y-A, Ahn J-W, Chun I-K, Kim H-J, 2001. Development of a method for the determination of human skin moisture using a portable near-infrared system. *Anal. Chem.*, **73**, 4964–4971.

8

RAMAN MICROSPECTROSCOPY AND IMAGING OF ACTIVE PHARMACEUTICAL INGREDIENTS IN CELLS

Jian Ling
Southwest Research Institute, San Antonio, TX, USA

8.1 INTRODUCTION

A general problem identified in the pharmaceutical industry is the need to reduce the cost and time of bringing new drugs to the clinic. At present, a large number of drug candidates are produced through rational drug design, combinatorial and computational chemistry, and high throughput screening. Many candidate drugs fail after many millions of dollars are spent for animal and clinical studies. A significant number of these failures are attributed to absorption, distribution, metabolism, elimination, and toxicity (ADMET) deficiencies. Consequently, there is a tremendous need for predictive approaches to evaluate the toxicity and efficacy of candidate drugs *in vitro* during early stages of drug discovery. The goal of *in vitro* study is to define a drug's pharmacology, for example, to understand the mechanism of action, the specific drug targets, the cellular uptake, the membrane transport, and the mechanisms of drug resistance. Now, there is a trend toward development of human-cell-based (e.g., Caco-2 cells, MDCK cells) *in vitro* assays to study the ADMET parameters at an early stage of drug development (Lesney, 2004). Although cell-based assays cannot eliminate clinical trials, they can provide an early indication of drug activity as well as insight into mechanisms of drug resistance and sensitivity before costly animal studies and clinical trials are performed.

Pharmaceutical Applications of Raman Spectroscopy, Edited by Slobodan Šašić

Early understanding of drug action at the cellular level is a cost-effective way of evaluating drug efficacy at the early stages of drug development (Dellinger et al., 1998; Kerns, et al., 1998).

This chapter briefly reviews the current approaches and their limitations of cellular drug imaging. The unique role of Raman imaging as a potential label-free imaging tool is described. The challenges in Raman imaging and the techniques to improve the sensitivity of Raman imaging are discussed. This chapter also discusses the postimage processing methods that are necessary for the extract of weak Raman signals. Finally, two examples are given to show the applications of Raman imaging to visualize subcellular drug distribution in living and fixed cells.

8.2 CURRENT APPROACHES TO DRUG IMAGING

The current technologies for drug imaging are based on (1) fluorescence technology, (2) isotope radiography, (3) HPLC–Mass spectroscopy, and (4) infrared microscopy. Fluorescence microscopy has been used for the study of drug action at the cellular level for many years and has provided key information on drug-induced intracellular distribution and content of drug targets (Oyama et al., 1992; Baker et al., 1995; Rao et al., 1998). Recently, molecular biological techniques, involving green fluorescent protein (GFP), have been developed to measure the changes of proteins that serve as drug targets or are involved in a signal pathway affected by drug treatment. Such assays, however, are not suitable for tracking drugs within cells. Fluorescent labels typically have a high molecular weight and, when attached to a low molecular weight drug, can create experimental artifacts by interfering with the drug's normal kinetics and distribution. In addition, the fluorescent markers used in the specimen may cause undesirable pharmacological or toxicological effects. Also, the continuous loss of fluorescence intensity during measurement due to photon bleaching and the potential for photo-damage to the bio-specimen due to the use of ultraviolet wavelengths are fundamental problems of fluorescence microscopy.

Drug visualization using fluorescence microscopy has been done based on a drug's autofluorescence (Coley et al., 1993; de Lange et al., 1992; Gervasoni et al., 1991; Itoh et al., 1992; Woodburn et al., 1991). However, most studies are limited to drugs such as doxorubicin (DOX) which have relatively strong autofluorescence (Bontenbal et al., 1998; Coley et al., 1993; Crivellato et al., 1999; de Lange et al., 1992; Weaver et al., 1991). Unfortunately, the autofluorescence of most drugs is weak and unspecific (broad bandwidth); therefore, molecular-specific images often cannot be acquired by autofluorescence imaging.

Isotope radiography is another method available for the visualization of subcellular drug distribution for the study of cellular pharmacokinetics (Evans et al., 1985; Puffer et al., 1979). Radioactive isotopes such as ^{3}H, ^{14}C, ^{32}P, ^{35}S, and ^{131}I are used to label the drugs to be visualized. These isotopes replace their stable counterparts in the molecules so that the labeled molecules can be traced by the radiation due to the isotope decay. The radiation is recorded on a thin film to show the distribution of the drugs. Isotope radiography is popular for two reasons: (1) the labeled drugs do not change their chemical properties because the isotopes have the same number of electrons as their stable counterparts, and (2) the isotope radiography is very sensitive, as in favorable circumstances, it can detect fewer than 1000 copies of a molecule in a sample. However, isotope radiography has several drawbacks. The samples have to be fixed before the image can be taken and the exposure time is usually long, requiring sometimes even several days. These problems make it difficult to use isotope radiography to trace chemicals in living cells. Isotope radiography also has problems due to drug metabolization. The radiation pollution and the high cost of making a special radioactive compound are also problems for the isotope radiography method.

HPLC–Mass spectroscopy has been used in label-free drug tracking in some *in vitro* assays, such as the Caco-2 cells assay. However, as it is an invasive technology, it is not suitable for visualizing drugs inside cells.

Infrared microscopy, like Raman, is a chemical imaging technique that can provide molecular-specific images. An infrared image of a sample is obtained by imaging the transmitted or scattered infrared radiation. Molecular selectivity is obtained by tuning the wavelength to a vibrational energy level of a selected molecule (or drug) in the sample. Since infrared imaging is derived from a material's intrinsic vibrational energy level, no external markers, dyes, or labels are required to contrast the infrared image. However, the spatial resolution of the image is usually several times the wavelength of infrared radiation. This is usually 10–20 μm, which is too large to resolve structures at the subcellular level. In addition, water has a high absorbance throughout the infrared region; therefore, many samples of biological interest are opaque in the infrared due to the presence of water. Consequently, it is often difficult and sometimes impossible to obtain images of many biological molecules by infrared microscopy. The main advantage of Raman images, as described in the following section, is that measurements can be carried out in aqueous solution.

8.3 RAMAN SPECTROSCOPY AND RAMAN IMAGING

The Raman scattering was discovered by Raman in 1928 (Raman, 1928). Raman scattering from a complex molecule consists of a series of Raman shifts (shown as Raman peaks) corresponding to the characteristic frequencies of different vibrational modes of the molecule. The unique combination of the shifted frequencies and their relative intensities forms the Raman spectrum of the molecule, which can be used as a fingerprint signal to identify and trace the molecule without other labeling. In general, Raman peaks have relatively narrow bands, typically 10–20 cm^{-1} in width. Thus, the Raman spectrum has higher specificity than fluorescence and infrared spectra in the identification of different molecules.

Figure 8.1 illustrates the chemical structure and Raman spectrum of pure paclitaxel, an anticancer drug often used to treat breast cancer, ovarian cancer, and non-small cell lung cancer. The most significant Raman peaks of paclitaxel are at 617,

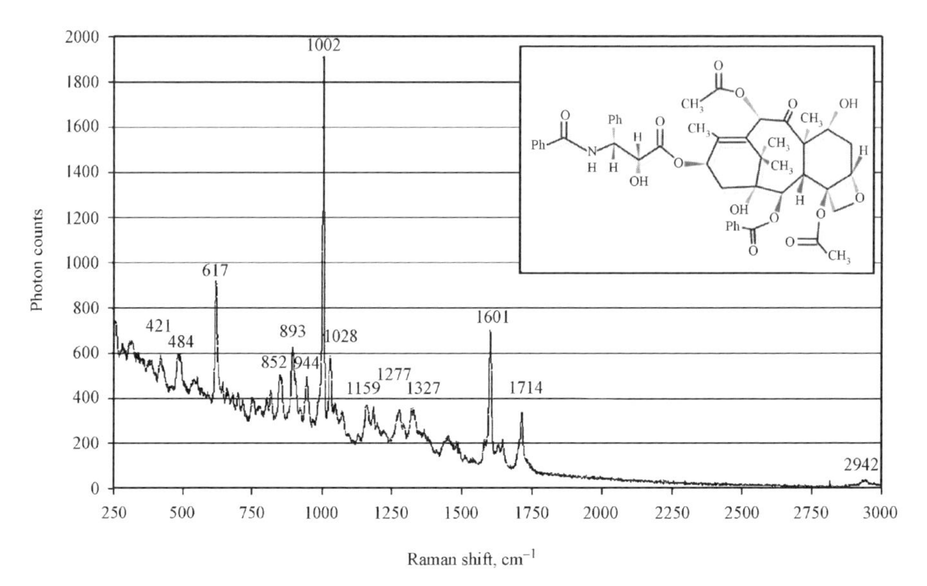

FIGURE 8.1 Raman spectrum of paclitaxel (neat powder), taken with 20× lens and exposure time of 30 s. Inset: chemical structure of paclitaxel. Reproduced from "Direct Raman imaging techniques for study of the subcellular distribution of drug" (*Applied Optics*, Vol. 41, No. 2) with permission of the Optical Society of America.

1002, and 1601 cm^{-1}. The Raman peak at 617 cm^{-1} is due to deformation of benzene rings in the structure. The Raman peak at 1002 cm^{-1} is due to the sp^3 hybridized carbon–carbon (C–C) ring vibration. The Raman peak at 1601 cm^{-1} is due to the carbon–carbon double bond (C=C) stretching vibration.

Raman images, acquired at fingerprint Raman frequencies of a molecule, can further provide an overview of the spatial arrangement of a particular type of molecule within a heterogeneous specimen. Raman imaging requires no external markers, dyes, or labels as required in fluorescence imaging (Brenan et al., 1996). This makes the sample preparation much simpler for the experiment. At the same time, the mechanism of action of the drug is minimally disturbed during imaging. Raman imaging is also suitable for imaging biomolecules in aqueous solution. In addition, the near-infrared excitation used in Raman imaging of the present work has a number of advantages for biological systems, such as producing less laser-induced fluorescence and photo-thermal degradation, and allowing better perspective depth (>1 mm) into a sample (Chinked, 1996; Manoharan et al., 1996).

There are three different methods to generate a Raman image classified by the illumination schemes. The first method is the point laser-scanning Raman imaging. This scheme acquires Raman spectra at every image pixel, and therefore it is also called spectral imaging. The advantage of this method is that all spectral features can be used to analyze the chemical distribution within the sample. The main disadvantage of the method is the very low speed of image acquisition. Because the Raman signal is weak and takes a relatively long time to obtain the spectrum for each pixel, the scanning time for an entire image is considerable. At present, laser

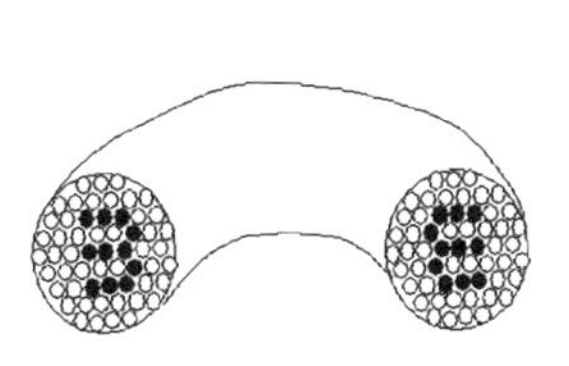

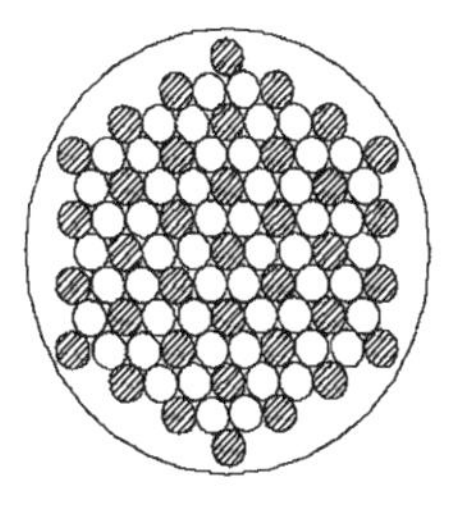

FIGURE 8.2 Spectral Raman imaging acquired using imaging fiber bundle. The excitation fiber (hatched circles) and collection fibers (unhatched circles) are evenly arranged together. Each collection fiber is a pixel of a spectral image.

scanning Raman imaging has not achieved the speed comparable to laser scanning fluorescence imaging and therefore is not suitable for live cell imaging. An alternative but fast way of acquiring pixelwise spectral imaging is to use fiber bundles as illustrated in Figure 8.2 (Gift et al., 1999; Ma and Ben-Amotz, 1997; McClain et al., 1999). Excitation fibers and collection fibers are evenly arranged together. Each collection fiber is a pixel in the image. As all the fibers can collect spectral data simultaneously instead of scanning though the imaging area, the acquisition speed for a spectral image is much faster than scanning imaging. The challenge of this method is that "in-line" filters have to be incorporated into individual fibers to block the fluorescence and Raman background from the fibers themselves.

The second imaging scheme is the line illumination method, which is the trade-off between point illumination and global illumination. This method records one line of the Raman signal. Moving the sample in a direction perpendicular to the laser line using a motorized stage records the line signal repeatedly. This scheme provides much faster recording speed than the point illumination scheme, but it still cannot compare to the global illumination scheme.

Global illumination is achieved by expanding the laser beam to cover a large sample area (Fig. 8.3). The global illuminating Raman imaging technique is also called direct Raman imaging or wide-field Raman imaging, in contrast to scanning Raman imaging. The photons scattering from the whole focused plane pass simultaneously through a holographic notch filter to remove the Rayleigh component. A follow-up bandpass filter then selects a fingerprint Raman peak that represents the molecule to be imaged. The filtered signal projects its intensity onto a charge-coupled detector (CCD), resulting in a two-dimensional image of specific molecular distribution in the sample. The intensity of the signal is related to the concentration of the molecules at a specific location. The time needed to record a direct Raman image is equivalent to the scanning time for one pixel in the point illumination method and, therefore, direct imaging is more suitable for imaging molecules in a living sample. The spatial resolution of a direct image can approach the diffraction limit imposed by the objective lens and the excitation wavelength. For example, a 782-nm illumination laser is able to record an image with spatial resolution as small as 0.53 μm (under the assumption that the numerical aperture (NA) of the objective lens is 0.9).

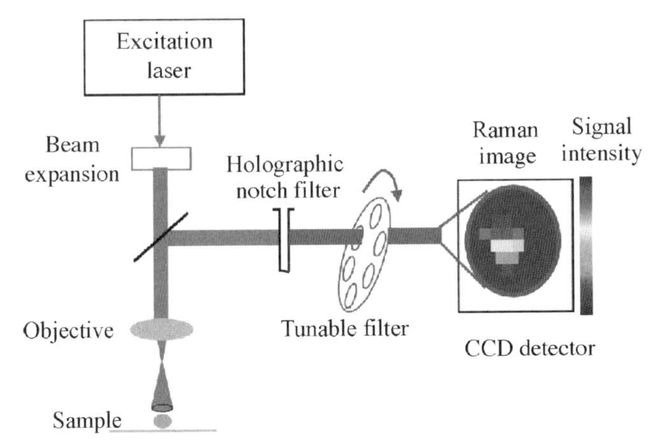

FIGURE 8.3 Schematic diagram of a CCD-based direct Raman imaging system. This system is able to record the two-dimensional distribution of a specific type of molecule in a sample. The pseudocolor image in the figure indicates the signal intensity. The red area has higher intensity, indicating higher concentration of the molecule. The dark blue area has low intensity, indicating less concentration of the molecule. Reproduced from "Direct Raman imaging techniques for study of the subcellular distribution of drug" (*Applied Optics*, Vol. 41, No. 2) with permission of the Optical Society of America. See color plates.

With the inventions of acousto-optic tunable filters (AOTF) (Schaeberle et al., 1996) and liquid crystal tunable filters (LCTF; Kline and Treado, 1997; Morris et al., 1994), direct Raman imaging is now able to acquire multiple Raman peak frequencies within seconds. Therefore, direct Raman imaging has the potential to provide an efficient way of obtaining high definition and multifrequency Raman images. This chapter will focus on the use of direct Raman imaging to visualize drug distributions in living tumor cells.

8.4 RAMAN MICROSPECTROSCOPY AND IMAGING FOR DRUG RESEARCH

Raman technology has been applied to study therapeutic drugs and their biomolecular targets. In early 1980, researchers began to use shifts in Raman bands and changes in band intensity to study the kinetics and mechanisms of drug–target interactions. Durig et al. (1980) studied the binding of methotrexate (MTX), an anticancer agent, to *L. casei* dihydrofolate reductase with Raman spectroscopy. The study supported the hypothesis of significant conformational changes in the *L. casei* dihydrofolate reductase upon the formation of MTX–dihydrofolate reductase complex. The resonance Raman (RR data obtained by Ozaki et al. (1981) revealed that MTX binds with its pteridine ring protonated. The rearrangement of the pteridine electrons accompanying protonation may account, in part, for the high affinity of MTX for the enzyme. Using RR spectroscopy and surface enhancement Raman spectroscopy, Butler et al. (1994) and Chourpa et al. (1996) studied amsacrine, a drug used to treat leukemia, to elucidate the interaction mechanism with its target DNA. Another example of drug–target interaction, reported by Feofanov et al. (1996), demonstrated that camptothecin and its metabolite interact differently with its target, DNA topoisomerase I. Raman microspectroscopy was recently used to study drug–cell interactions and multidrug resistance of cells (Krishna et al., 2006; Murali Krishna et al., 2005; Owen et al., 2006).

Raman imaging has been applied to study drug actions at the cellular level. Manfait's group has studied the antitumor drug (DOX) in living cancer cells (Nabiev et al., 1991; Sharonov et al., 1994). The results of the measurements on the population of cells containing DOX in nuclei or in the cytoplasm are well correlated with surface enhancement Raman measurements on single cells treated with DOX.

Freeman et al. (1998) mapped the location of photosensitization in single cells. This provided detailed information on how the photodynamic therapy works at the cellular level. Direct Raman imaging was recently applied to drug visualization. Sijtsema and Arzhantsev used Raman imaging to localize the photodynamic therapy agent cobalt octacarboxy phthalocyanine [CoOCP] and Cobalt phthalocyanines in fixed K562 leukemia cells. The three-dimensional drug distribution was illustrated and the intracellular drug level was estimated (Arzhantsev et al., 1999; Sijtsema et al., 1998). Feofanov et al. (2000) applied RR confocal spectral imaging techniques for the study of intracellular accumulation, localization, and retention of theraphthal. Ling et al. (2002) studied the subcellular distribution of paclitaxel in living breast tumor cells, and sulindac sulfide in fixed prostate tumor cells (Ling 2001). The use of Raman microspectroscopy to study living cells is reviewed by Notingher and Hench (2006).

These studies have indicated the potential application of Raman microspectroscopy and imaging for drug research. The major challenge in this application is how to increase the drug detection sensitivity or, in other words, how to improve the Raman signal-to-noise ratio (SNR). The Raman signal is naturally weak and the background, especially the autofluorescence background from the complex biological samples, can be quite high, even under near-infrared excitation. In many cases fluorescence is also induced by contaminants and solvents. The shot noise generated from a large fluorescence background can overwhelm the Raman signals. In addition, the imaging system will further degrade the Raman signals.

The following sections will briefly review the Raman signal intensity, fluorescence background, SNR, and the techniques that have been developed to enhance the Raman signal or reduce the fluorescence background. A Raman imaging model will be used to describe the degradation of Raman signals during the imaging processes.

8.5 RAMAN INTENSITY, FLUORESCENCE BACKGROUND, AND SNR

The Raman signal intensity (S_{Rm}) can be described by the following equation:

$$S_{\mathrm{Rm}} = I_0 \left(\frac{d\sigma}{d\Omega}\right) C, \qquad \left(\frac{d\sigma}{d\Omega}\right) = \frac{A(\lambda)J(\lambda)}{\lambda^4} \tag{8.1}$$

where I_0 is the laser excitation intensity, $(d\sigma/d\Omega)$ is the Raman scattering cross-section of the target molecules, $A(\lambda)$ is the self-absorption of the molecule, $J(\lambda)$ is a molar scattering parameter, $\lambda(=\lambda_0 \pm \lambda_{\mathrm{Rm}})$ is the excitation wavelength λ_0 plus the Raman shift λ_{Rm}, and C is the concentration of the molecules. The Raman signal is inherently very weak due to the low scattering cross-section (typically around 10^{-24} to 10^{-29} cm^2 $\mathrm{molecule}^{-1}$ sr^{-1}) for a Raman peak under green laser excitation. It is often said that for every 1 million photons scattered back from a substance, 999,000 are from Rayleigh scattering, 999 are from fluorescence, and only one photon contributes to the Raman scattering (Kincade, 1998). The Raman scattering cross-section is inversely proportional to the fourth power of the excitation wavelength: the shorter the excitation wavelength, the higher the Raman signal. The Raman signal is also proportional to the excitation light intensity as well as to the molecular concentration.

For the detection of drug molecules in biological samples, strong noise background (B), especially from the broadband autofluorescence of the biological molecules (e.g., tryptophan, flavins, porphyrins, collagen, and elastin) will further increase the difficulty of drug detection. Assume that a laser beam illuminates a biological specimen containing drug molecules. At the focal plane, the total scattering intensity (S) contains both Raman signal (S_{Rm}) from the drug and background autofluorescence (B) from the biological specimen:

$$S(x, y, z) = S_{\mathrm{Rm}}(x, y, z) + B(x, y, z), \tag{8.2}$$

where x, y, and z are the spatial coordinates. The SNR for Raman signal detection can be described as

$$\mathrm{SNR} = \frac{S_{\mathrm{Rm}}}{(S_{\mathrm{Rm}} + B)^{1/2}}, \tag{8.3}$$

It is often called the shot-noise limitation equation. It indicates that the shot noise $(S_{\mathrm{Rm}} + B)^{1/2}$ is proportional to the intensity of the Raman signal as well as the background. Under a large autofluorescence background, the SNR is reduced significantly. A low SNR decreases the sensitivity of detecting the Raman signal.

In an imaging system (e.g., a microscope), the total scattering signals get further degraded before they reach the CCD detector. In direct Raman imaging, a recorded image $g(x, y, z)$ can be described with the following equation

$$g(x, y, z) = \{h(x, y, z) * [S_{\mathrm{Rm}}(x, y, z) + B(x, y, z)]\} \times I(x, y) \times t + n(x, y, z), \tag{8.4}$$

where $*$ is the linear convolution operator, t is the exposure time, and I is the excitation intensity. Usually, the illumination intensity at the focal plane is not uniform but is dependent on the locations (x, y) due to the imperfections in the laser expanding system. This lack of homogeneity of the illumination, indicated by $I(x, y)$, causes a nonuniform illumination effect in the recorded images. The $h(x, y, z)$ is the microscope's point spread function (PSF), of which the Fourier transform

(FT) is the microscope's optical transfer function (OTF). In direct Raman imaging, a wide-field image is the sum of in-focus information from the focal plane as well as out-of-focus information from the neighborhood planes due to the finite recording aperture and the limited depth of focus of the imaging system. In other words, the scattering signal is blurred by the microscopic system. This blur is characterized in terms of the microscope's PSF $h(x, y, z)$. The blurred image is further degraded by a signal-independent additive noise $n(x, y, z)$, which includes the CCD's dark noise and reading noise. Several measurement techniques that could improve the Raman detection SNR will be discussed in Section 8.6. Section 8.7 will discuss how to use postimage processing to enhance the Raman images.

8.6 TECHNIQUES TO IMPROVE SNR IN RAMAN IMAGING

According to Equation 8.3, there are two ways to improve the SNR: one is to enhance the signal, the other is to reduce the noise background.

8.6.1 RR Scattering

The RR phenomenon was observed by Shorygin in 1953 (Brandmuller and Kiefer, 1978). This technique is often used to enhance the Raman signal. When an excitation light source (usually with UV or visible wavelength) matches the energy corresponding to the electric transition of a particular chromophoric group in a target molecule, the intensities of Raman bands originating from this chromophore are selectively enhanced by a factor of 10^3 to 10^5. In other words, when the excitation source is tuned into the absorption bands of a molecule, $A(\lambda)$ in Equation 8.1 is increased considerably, so that the Raman signal is greatly enhanced. This enhancement is due to the large change in polarizability during the electronic transition of molecules. The use of RR spectroscopy to study chromophore-containing biological molecules in living cells has been reviewed (Puppels and Greve, 1996). Application of the RR technique is limited by the availability of the chromophoric group in the target molecules.

8.6.2 FT Raman Scattering

FT Raman scattering is a technique for reducing fluorescence interference. Conventional Raman spectroscopy measures intensity versus frequency. FT-Raman, on the other hand, uses a Michelson interferometer to measure the intensity of many wavelengths simultaneously. This time-domain spectrum is then converted to a conventional frequency-domain spectrum by FT using a computer program. This technique primarily uses 1064 nm near-infrared light from a Nd:YAG laser as the excitation source. Therefore, the background fluorescence is almost completely eliminated. However, because the Raman intensity follows a $1/\lambda^4$ relationship, this technique suffers low sensitivity. Moreover, the InGaAs detector used in the FT-Raman exhibits substantial noise. Thus, FT-Raman usually requires long data collection times (typically 30 min), making it difficult to be used in tracking chemical changes in live cells.

8.6.3 Coherent Anti-Stokes Raman Scattering (CARS)

CARS scattering is another special technique used to avoid the fluorescence background (Duncan et al., 1985). When two high energy laser beams, with frequencies ν_1 and ν_2 ($\nu_1 > \nu_2$), collinearly irradiate a sample, the two beams interact with each other coherently to produce a strong scattered light of frequency $2\nu_1 - \nu_2$. If ν_2 is tuned such that $\nu_1 - \nu_2 = \nu_v$, where ν_v is a frequency of the Raman band, then a strong light of $\nu_1 + \nu_v$ is emitted. Since this emitted light is higher than ν_1 or ν_2, it is on the anti-Stokes side of the excitation source, whereas the fluorescence is on the Stokes side. Thus, the Raman spectrum separates itself from the fluorescence and is not affected by the fluorescence background. Compared to conventional Raman, CARS is also a highly sensitive probe, especially for probing lipid molecules in living cells (Cheng and Xie, 2004). However, to obtain a CARS signal from a sample requires high reproducibility of laser intensity for wavelength scanning, and high stability of pulse-to-pulse intensity; thus, the cost of a CARS instrument is high. The current CARS instrument still has lower spectral resolution than conventional Raman spectroscopy, and high nonresonant background from the solvent. These limitations so far prevent the CARS application from detecting drugs in a biological specimen.

8.6.4 Surface-Enhanced Raman Scattering (SERS)

Discoveries in the late 1970s indicated that the Raman scattering efficiency can be enhanced by factors of up to 10^6 when a molecule is adsorbed on or near a spherical metal surface. Colloidal gold or silver particles, with sizes on the order of tens of nanometers, have been demonstrated as a high sensitive substrate for SERS (Kneipp et al., 1998a,b). Although the mechanism of SERS is not completely understood, the current development of nanotechnology will make this technique extensively available. SERS is a powerful technique to be used together with direct Raman imaging for the detection of a drug in a biological specimen. SERS technology has been used in living cell studies (Breuzard et al., 2004; Eliasson et al., 2005; Vo-Dinh et al., 2005).

8.6.5 Time-Resolved Fluorescence-Rejection Raman Spectroscopy

Fluorescence is a broadband signal (like a baseline) compared to the Raman signal (see Fig. 8.4). The conventional method is to subtract the baseline during postdata processing. In Raman spectroscopy, the baseline is first estimated by low pass filters or approximating polynomials, and then is subtracted from a spectrum (Mahadevan-Jansen and Richards-Kortum, 1996). In Raman imaging, two images are often acquired. One is taken at the Raman peak f_a; and the other is taken at a neighboring frequency f_b, as the background image (see Fig. 8.4). Then the background image can be directly subtracted from the Raman image (Ling, 2001). Another

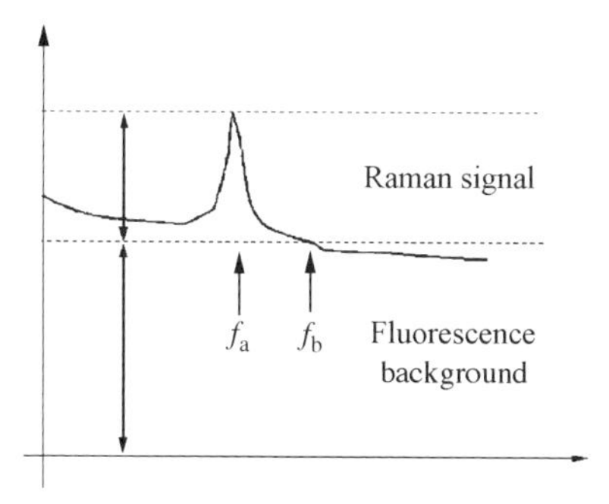

FIGURE 8.4 Raman signal and fluorescence background. Reproduced from "Direct Raman imaging techniques for study of the subcellular distribution of drug" (*Applied Optics*, Vol. 41, No. 2) with permission of the Optical Society of America.

similar fluorescence rejection method is called the shifted excitation Raman difference spectroscopy (SERDS) technique, in which two spectra or images are acquired, one with excitation frequency f, the other with shifted excitation frequency $f + \delta$. The difference between these two spectra or images eliminates the fluorescence background.

According to the shot-noise limitation Equation 8.3, if the fluorescence background is large, the shot noise generated by the background will overwhelm the Raman signal. In other words, conventional methods cannot recover the Raman signals lost in the noise. Better methods are to avoid the fluorescence background (e.g., using CARS technology) or to block the fluorescence background (e.g., using time-resolved fluorescence-rejection technology as described in the following paragraph).

Time-resolved fluorescence-rejection technology uses temporal information to discriminate between Raman and fluorescence photons. The short-lived Raman signal is generated almost instantaneously (typically within 10 ps) upon excitation by a light pulse, but autofluorescence usually appears between 1 and 100 nanoseconds (Fig. 8.5). The technique of using a pulsed laser in combination with gated detection has been applied since the 1970s to reduce the autofluorescence background (Gustafson and Lytle, 1982; Harris et al., 1976; Kamogawa, Fujii et al., 1988; Van Duyne et al., 1974). Figure 8.5 illustrates the schematic diagram of the mechanism of the time-resolved fluorescence-rejection Raman spectroscopy.

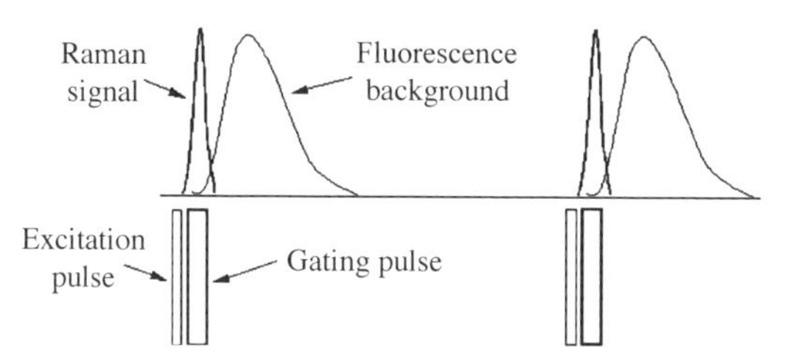

FIGURE 8.5 Schematic diagram of the time-resolved fluorescence-rejection Raman spectroscopy.

Upon excitation, a delayed gating pulse allows only the Raman signal to pass to the detector and blocks the following fluorescence background. Successful use of the time-resolved fluorescence-rejection technique requires an ultra fast, high throughput gating system as well as the laser pulse lock-in technology. The recent development (Matousek et al., 1999, 2002) of picosecond Kerr gating effectively blocks the fluorescence background (by a factor of 10^{-5}). The combination of Kerr gating with the conventional SERDS technique (Shreve et al., 1992) further increases the fluorescence rejection ratio to 10^{-6}. SERDS helps to reduce the residual fluorescence background generated by short-lived fluorescence. Rejection of the fluorescence background will greatly reduce the shot-noise generated by the background and therefore significantly increase the SNR. Equivalently, the sensitivity of Raman signal detection will be improved significantly. At present, a Raman imaging microscope with the time-resolved fluorescence-rejection technology has not yet been developed.

8.7 ENHANCED RAMAN IMAGES WITH POSTPROCESSING

In order to extract correct and useful information from Raman images that contain weak Raman signals, postimage processing is also necessary. The purpose of Raman image processing is to determine the Raman signal $S_{\mathrm{Rm}}(x, y, z)$ of the imaging area in Equation 8.4 from the recorded image $g(x, y, z)$. To determine $S_{\mathrm{Rm}}(x, y, z)$ it is necessary to reduce the noise $n(x, y, z)$ from the image $g(x, y, z)$, correct the non-uniform illumination $I(x, y)$, deconvolve with the point-spread function $h(x, y, z)$, and subtract the fluorescence background $B(x, y, z)$.

8.7.1 Noise Reduction Using Anisotropic Median-Diffusion Filter

A Raman image as a molecular image displays the distribution of molecules. Such an image can be modeled as a piecewise-smooth image that can be divided into several regions. The molecular concentrations or Raman intensities within each region are small and gradual. The intensities between the different regions are quite dissimilar because of the significant difference in molecular concentration. An anisotropic median-diffusion filter is especially useful to enhance such molecular images with low SNR. A quantitative study has shown that an anisotropic median-diffusion filter can effectively reduce the additive Gaussian noise $n(x, y, z)$ in Equation 8.4 without blurring the edge signals on the images. The enhanced images are very close to the original images in correlation, mean luminance, and contrast (Ling, 2001). Implementation of the filter is described in Ling et al. (2002).

8.7.2 Correction of Nonuniform Illumination

After reducing the noise, the nonuniform illumination $I(x, y)$ in Equation 8.4 needs to be corrected. A flat field image $r(x, y)$ is often obtained as a reference image to correct the lateral nonuniform illumination. (The axial nonuniformity of illumination

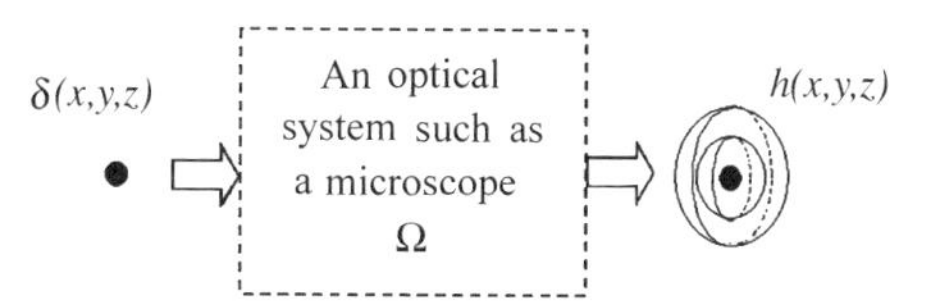

FIGURE 8.6 Point spread function $h(x,y,z)$ generated from a point source $\delta(x, y, z)$ passing through an optical system.

is actually included in the microscope's PSF and, therefore, deconvolution of the Raman image with PSF discussed in the following section will correct the illumination nonuniformity in the axial direction.) The reference image is normalized by its own median value to give the compensating image $c(x, y)$

$$c(x, y) = \frac{r(x, y)}{\text{Median}[r(x, y)]}. \tag{8.5}$$

Dividing $c(x, y)$ by the denoised Raman image g_s(x,y,z) yields the compensated image g_c(x,y,z):

$$g_c(x, y, z) = \frac{g_s(x, y, z)}{c(x, y)} = \{h(x, y, z) * [K(x, y, z) + K_0(x, y, z)]\}Ct, \tag{8.6}$$

where C is a constant that is related to the average illumination. The above compensation algorithm (Eqs. 8.5 and 8.6) keeps the median intensity of the image unchanged. It is important to perform the noise reduction described in Section 8.7.1 before this compensation; otherwise, the noise may be amplified after the compensation algorithm.

8.7.3 Three-Dimensional Image Deconvolution

The objective of three-dimensional deconvolution is to restore the blurred image by using the PSF of the imaging system. Deconvolution will reduce the in-focus-plane blurs caused by the limited aperture of the system as well as the out-of-focus-plane blurs resulting from the limited depth of field of the system.

Both PSF and OTF seek to describe how the image of an ideal point object is spread out in three dimensions by passing through a microscope or any optical system. As illustrated in Fig. 8.6, a point light source $\delta(x, y, z)$ will spread out to form a smeared point image $h(x, y, z)$, referred to as PSF, when passing through an optical system. For an object $f(x, y, z)$ that is an integration of the point sources, the smeared image can be computed as

$$g(x, y, z) = h(x, y, z) * f(x, y, z), \tag{8.7}$$

where * indicates the convolution operation. In the frequency domain, the above relationship can be written as

$$G(u, v, w) = H(u, v, w) \cdot F(u, v, w), \tag{8.8}$$

where $G(u, v, w)$, $H(u, v, w)$, and $F(u, v, w)$ are the three-dimensional FTs of $g(x, y, z)$, $h(x, y, z)$, and $f(x, y, z)$, respectively. $H(u, v, w)$ is the OTF of the optical system.

The PSF or OTF completely determines the overall performance of a linear and shift-invariant optical system. Estimation of the OTF of the microscope in use is necessary in order to predict and correct the distortion during the image formation. In practice, a small-diameter microsphere (smaller than the theoretical diffraction limit of the microscope) is often used as a point source for the estimation of OTF. Detailed procedures have been demonstrated in the literature (Hiraoka et al., 1990; Shaw and Rawlins, 1991; Ling, 2001). An example of the three-dimensional OTF of a Raman microscope is illustrated in Fig. 8.7.

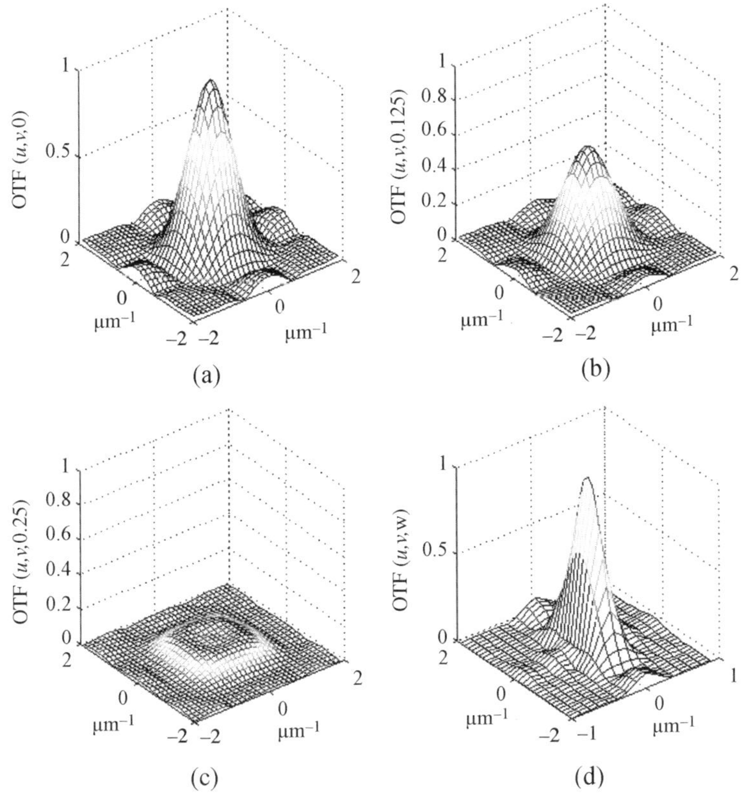

FIGURE 8.7 Three-dimensional OTF of a Raman microscope system with a water immersion lens. The OTF – *H(u,v,w)* is displayed at **(a)** $w = 0\ \mu m^{-1}$, **(b)** $w = 0.125\ \mu m^{-1}$, and **(c)** $w = 0.25\ \mu m^{-1}$. **(d)** The *H(u,v,w)* is also displayed by the cross section at *H(0,v,w)*.

With the knowledge of PSF and OTF, many three-dimensional deconvolution algorithms can be used. These include the inverse filter, the Wiener filter, the Nearest-Neighbor deconvolution, the constrained iterative method, and the expectation-maximization maximum-likelihood (EM-ML) deconvolution. When restoring an image with all the information at the neighborhood planes available and no noise present, the deconvolution results from all these algorithms are very similar. However, in practice, some amount of noise always exists in a recorded image, even after smoothing by the anisotropic median-diffusion filter. In addition, there are often only a few images that can be recorded at different defocus planes within a limited period of time. Especially in the case of Raman imaging of living cells, it may only be possible to get one image (at a specific focal plane) to represent the drug distributions. This is again due to the relatively long exposure time required for Raman imaging.

Under this no-neighborhood condition, the three-dimensional deconvolution can be performed after replicating the recorded image as the neighborhood images. This simplification is based on the assumption that there is no abrupt change among the neighborhood images. However, if the assumption is not true, the three-dimensional deconvolution will not effectively remove the blurred information from the neighborhood planes.

Different deconvolution algorithms were compared on a three-dimensional cell model under the no-neighborhood condition (Ling, 2001). The EM-ML deconvolution achieved better results compared to other algorithms. The EM-ML deconvolution was derived from the Bayesian theory with Poisson noise model. An iterative algorithm of the deconvolution, developed by Richardson (1972) and Lucy (1974), is described as follows

$$S^{(n+1)}(x,y,z) = \left[\frac{g_c(x,y,z)}{h(x,y,z) * S^{(n)}(x,y,z)} * h(x,y,z)\right] S^{(n)}(x,y,z),$$

$$\text{and,} \quad S^{(n+1)}(x,y,z) \geq 0, \tag{8.9}$$

where $g_c(x,y,z)$ is a recorded two-dimensional image after denoising and the correction of nonuniform illumination as described above, and $S^{(n)}(x,y,z)$ is the restored images at nth iteration. The nonnegative constraint is applied to the restored images after each iteration.

It is also important to subtract the constant background or the "DC" value from the images prior to performing the deconvolution. Studies (Holmes and Liu, 1989; van Kempen and van Vliet, 2000) have shown that the background intensity has critical influence on the performance of the EM-ML deconvolution. This is because the existence of the background makes the nonnegative constraint less effective. In this study, the image background could be the fluorescence from the aqueous solution, which can be assumed uniform across the image. This background was subtracted from $g_c(x,y,z)$ before deconvolution.

8.7.4 Elimination of Fluorescence Background from Biological Specimen

After deconvolution, Equation 8.4 became:

$$S(x, y, z) = S_{\mathrm{Rm}}(x, y, z) + B(x, y, z). \tag{8.10}$$

The next processing step is to subtract the fluorescence background $B(x, y, z)$ from the biological specimen. The fluorescence background is usually nonuniform due to the heterogeneity of the specimen. Estimation of this fluorescence background is often done according to the Raman spectrum properties discussed in Fig. 8.4. That is, two images are acquired; one taken at a neighboring frequency f_{b} is the fluorescence background image, which will be directly subtracted from the Raman image taken at frequency f_{a} (Ling, 2001). This neighborhood background image can go through the same processing as discussed above to get the heterogeneous fluorescence background $B(x, y, z)$.

8.8 RAMAN IMAGING OF INTRACELLULAR DISTRIBUTION OF PACLITAXEL IN LIVING CELLS

This section provides an example of applying Raman imaging techniques to visualize the subcellular distributions of an anticancer drug, paclitaxel, within a living tumor cell (Ling, 2001). Detailed steps for identifying paclitaxel's characteristic Raman band, preparing the cell specimen, setting up the Raman microscope, acquiring Raman images, and data processing are described.

8.8.1 Paclitaxel and Its Characteristic Raman Band

Paclitaxel's unique mechanism of antitumor action was identified in 1979 by Dr. Susan Horowitz at Albert Einstein College of Medicine (21CEP, 1996–2003), 16 years after the compound was discovered to have antitumor activity. Paclitaxel "attacks" cell skeleton called microtubules, which plays an important role in cell division. With exposure to paclitaxel, cancer cells cannot grow and divide, but instead go to programmed death (apoptosis). Paclitaxel was selected for this study because its interactions with cellular molecules have been well studied. This knowledge will help to examine the results and determine the capability of the Raman imaging technology.

The chemical structure and Raman spectrum of powder paclitaxel are illustrated in Fig. 8.1. The most significant Raman peaks are at 617, 1002, and 1601 cm^{-1}. Because the powder paclitaxel is not soluble in water, it cannot be used to treat cells directly. According to the clinical formula of the drug (Bristol-Myers Squibb Company, Princeton, NJ), paclitaxel was first dissolved in dehydrate ethanol alcohol and cremophor EL (polyoxyethylated castor oil) and then further diluted with phosphate buffered saline (PBS) solution. With the mix of ethanol and cremophor oil, the Raman spectrum of the paclitaxel solution was affected significantly. Figure 8.8 illustrates the Raman spectrum (without fluorescence baseline correction) of a 0.3 mg/ml (350 μM) paclitaxel solution. The strong fluorescence baseline from

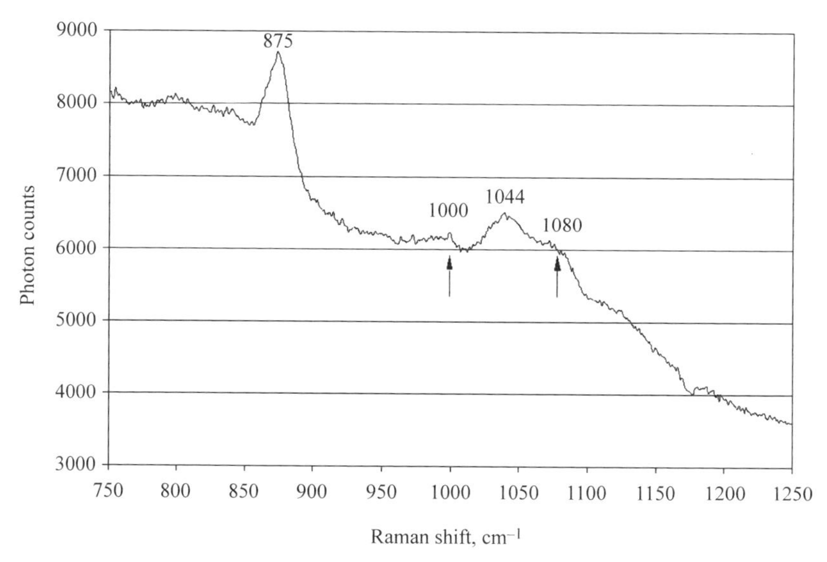

FIGURE 8.8 Raman spectrum of 0.3 mg/ml (or 350 μM) paclitaxel working solution, which shows the paclitaxel's 1000 cm^{-1} Raman peak. The spectrum was taken with 60× W/IR lens and exposure time of 300 s. Reproduced from "Direct Raman imaging techniques for study of the subcellular distribution of drug" (*Applied Optics*, Vol. 41, No. 2) with permission of the Optical Society of America.

cremophor oil and ethanol swamped most of the Raman peaks of paclitaxel, but left the peak at 1002 cm^{-1} (shift to 1000 cm^{-1}). Fortunately, neither ethanol nor cremophor oil has a Raman peak around 1000 cm^{-1} (Fig. 8.9).

8.8.2 Cell Preparation and Cell Raman Background

A human breast tumor cell line, MDA-435, was used in this study. Approximately 10^5 MDA-435 breast tumor cells were cultured on a gold-coated Petri dish and allowed to stabilize for 24 h in RPMI-1640 medium supplemented with 5% fetal bovine serum. After stabilization, the cells adhered to the bottom of the Petri dish.

A regular polystyrene Petri dish emits a strong Raman signal near 1000 cm^{-1}. To eliminate this interference from polystyrene, gold (a material with no Raman signals) was used to precoat the Petri dishes for imaging. A 200-Angstrom-thick gold coating effectively blocked the polystyrene Raman signals. The Petri dishes were resterilized before culturing the cells.

In order to detect paclitaxel in a cell, Raman and fluorescent signals from the cell itself also need to be investigated beforehand. Figure 8.10 shows the Raman spectrum of the cytoplasm at one local spot of an MDA-435 breast tumor cell. The Raman spectra at other locations of the cell cytoplasm and cell nucleus are very similar to this spectrum. The carbon–carbon stretching mode (at the Raman peak about 1003 cm^{-1}) from the molecules (proteins) inside the cell is also present on the cell spectrum. This peak is very close to the 1000 cm^{-1} Raman band used for

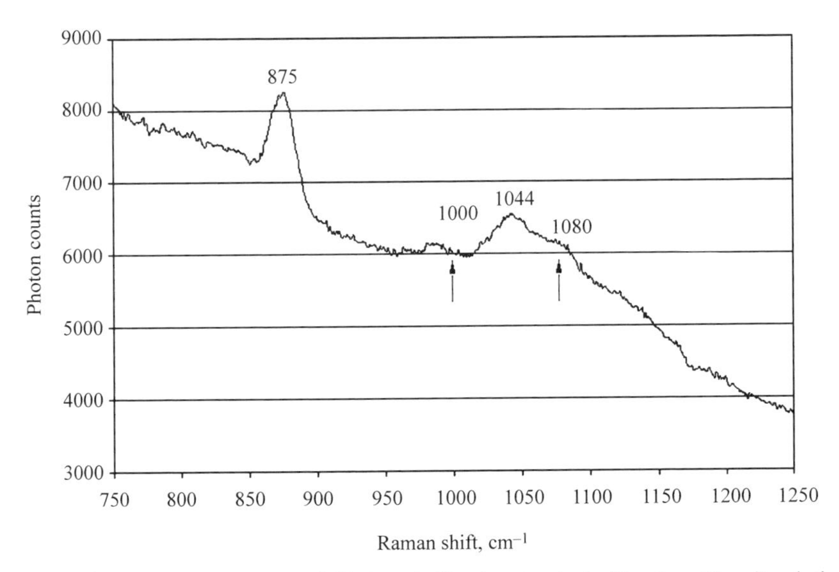

FIGURE 8.9 Raman spectrum of "0.3 mg/ml" solvent-only (without paclitaxel) solution, which does not show the paclitaxel's 1000 cm^{-1} Raman peak. The spectrum was taken with 60× W/IR lens and exposure time of 300 s.

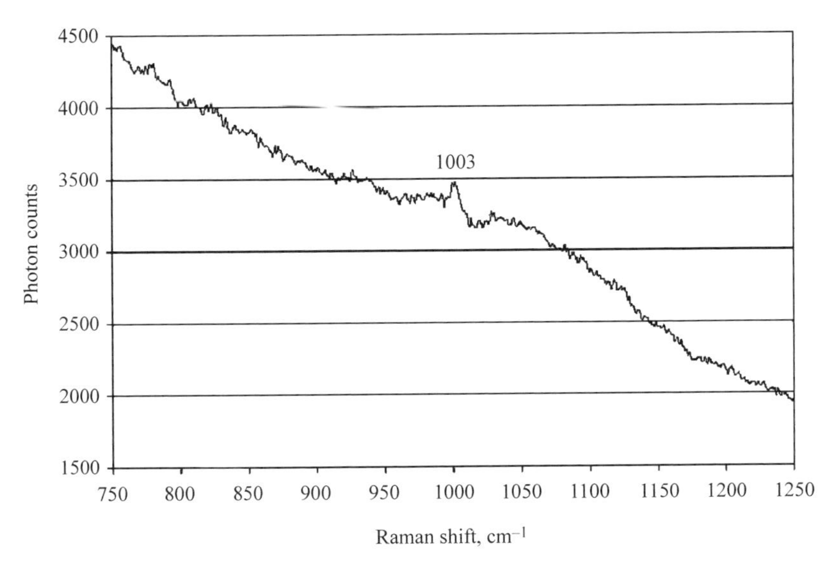

FIGURE 8.10 Raman spectrum of cytoplasm from an MDA-435 tumor cell, taken with 60× W/IR lens and exposure time of 300 s. The Raman spectrum of the cell nucleus shows a similar pattern. Reproduced from "Direct Raman imaging techniques for study of the subcellular distribution of drug" (*Applied Optics*, Vol. 41, No. 2) with permission of the Optical Society of America.

detecting the paclitaxel. Fortunately, the Raman signal at this peak from the cell is relatively weaker than that from the paclitaxel; thus, the intrinsic Raman signal from the cell has a small contribution to the Raman image at 1000 cm^{-1}. To further distinguish the changes of Raman signals before and after drug treatment, the Raman images before the cell exposure to the paclitaxel solution were used as the control case to be compared with the Raman images after the drug treatment. In summary, one Raman image will be taken at 1000 cm^{-1} to detect the paclitaxel. Another Raman image was taken at 1080 cm^{-1} to correct the contribution of the fluorescent signal on the 1000 cm^{-1} image.

8.8.3 Imaging Instrumentation and Imaging Procedures

A Renishaw Model 2000 Raman spectroscopic system (Gloucestershire, UK, 1993) was used in the study. This system is capable of acquiring Raman spectra, laser-scanning Raman spectroscopic images, and direct Raman images with an expanded laser beam. A Ti Sapphire laser system (Lexel Laser, Inc., CA) was established for the Raman system to replace the 30 mW diode laser originally supplied with the Raman system. The diode laser source is suitable for Raman spectroscopy on biological samples. However, when the laser is used for direct imaging, the beam must be expanded and spread over thousands of pixels. In such applications, it provides inadequate illumination power. In addition, the diode laser source has a line-shaped beam so that the imaging area suffers severe nonuniform illumination. The Ti Sapphire laser, pumped by a 7 W argon–ion laser (Lexel 95-7), emits near-infrared wavelength with maximum power of 1 W. The Ti Sapphire laser also has a Gaussian beam shape, which greatly improves the beam quality after beam expansion. In this study, the laser was tuned to 782 nm to match the holographic notch filter in the Raman system.

The Raman system can achieve spectral resolution of 1 cm^{-1} for spectrum measurement with a grating system. For direct imaging, the dielectric filter has a bandwidth of 10–20 cm^{-1}. The Raman system is placed in a dark room and stabilized on an antivibration table (Vibraplane Air Suspension System, Kinetic Systems, Inc., Boston, MA).

A 60× Olympus water-immersion, high infrared transmission (71%) objective lens (1-UM571 LUMPLFL 60× W/IR, Olympus, Japan) was used to obtain images of living cells incubated in aqueous solution. This lens has an NA of 0.90 and a depth of field (DOF) of 1.2 μm. The diffraction-limited optical resolution of this lens can be calculated by Abbe's equation:

$$s = \frac{0.61\lambda}{\mathrm{NA}}. \tag{8.11}$$

In this study, the excitation wavelength is $\lambda = 782$, and the maximum resolution that can be achieved is about 0.53 μm. The resolution of the system is also affected by the magnification of the microscope, the pixel size of the CCD camera, and the sampling rate. The complete optical characteristic of the Raman system is determined by its PSF (or OTF), which was estimated in this study by using small polystyrene microspheres (0.2 μm in diameter) as the point light source (Ling, 2001). This PSF was used in three-dimensional deconvolution.

In this study, the live-cell Raman imaging experiment consists of the following steps:

Step 1: Set up and calibrate the Raman instrument. The laser system was allowed to warm up and the total illumination power at the sample was stabilized around 10 mW. Experiments have shown that the cells are tolerant to this power with the expanded beam. The Raman system was calibrated with silicon wafer or polystyrene microsphere before each experiment to correct the day-to-day changes on the system. A flat-field Raman image was taken at 1000 cm^{-1} as the reference image for correction of the non-uniform illumination.

Step 2: Replace nutrition medium. The cells in the RPMI nutrition medium were washed (four times) with phosphate-buffered saline (PBS). PBS was used during imaging to reduce the fluorescence background from the nutrition medium. For long-term live cell imaging, however, the cell should be kept in nutrition medium. A live-cell imaging incubator, for example the FCS system from Bioptechs, Inc., is also necessary to maintain cells at 37°C.

Step 3: Select a cell as imaging target. A single cell, with its size well covered by the imaging area, was selected as the sample. The focal plane was set approximately at the middle of the cell.

Step 4: Take control cellular images before drug treatment. Three cell images were taken before paclitaxel treatment. A white light image of the cell illustrates the cell structure. A corresponding Raman image of the same cell is at the 1000 cm^{-1} Raman band. This image includes the cellular contributions in this Raman band and the fluorescence background across the image. A second Raman image at 1080 cm^{-1} was taken which has the contribution from fluorescent signals, only. These three images form a record representing the control situation.

Step 5: Start drug treatment. PBS was replaced by the 0.3 mg/ml paclitaxel solution to start the drug treatment. The paclitaxel solution at this concentration was found suitable for this study to detect paclitaxel in cells. The cells were exposed to paclitaxel for 1 h.

Step 6: Take cellular images during drug treatment. Several images were acquired during the 1-h drug treatment. Each record contains the same kind of images as described in step 4. However, at this time the Raman image at 1000 cm^{-1} contains Raman signals from paclitaxel as well as the background signals contributed from the solvent. The images after postprocessing will show the cellular distributions of paclitaxel during the drug treatment.

Step 7: Stop drug treatment. After 1 h of drug treatment, the paclitaxel solution was washed out using PBS. The cells were returned to the PBS medium for more cellular imaging.

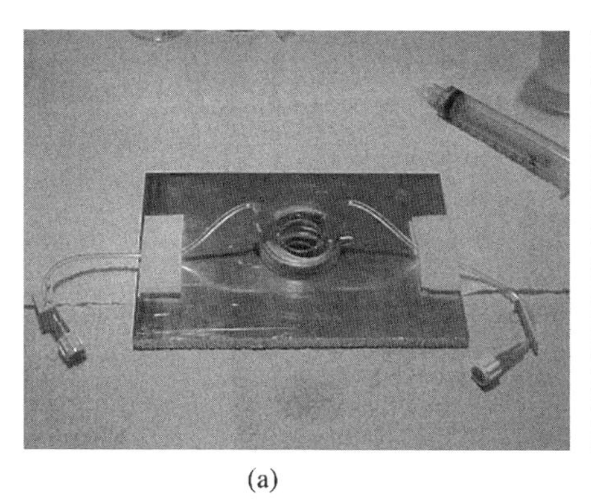

(a)

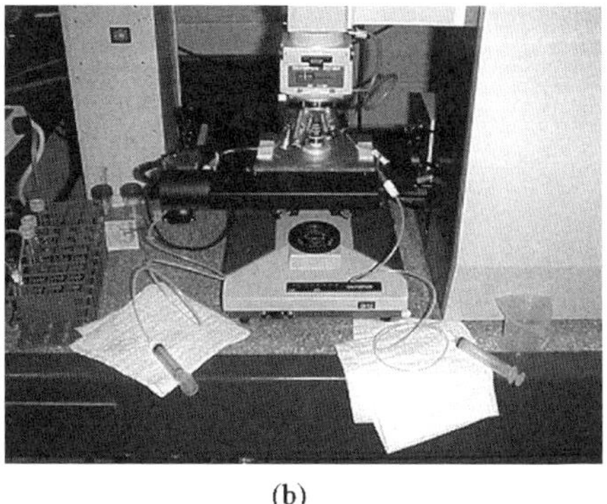

(b)

FIGURE 8.11 (**a**) A solution delivery system. (**b**) The solution delivery system used during imaging to keep the cells focused.

Step 8: Continue taking images. Serials of image records were acquired after the drug treatment. Although the drug solution was washed out, the paclitaxel molecules that entered the cell remained there and continued to interact with the cell. Thus the Raman images at 1000 cm^{-1} continued to have signals from the paclitaxel and to show the drug distributions within the cell.

During the solution exchange in Steps 5 and 7, the imaging position must remain unchanged. A solution delivery system was designed as illustrated in Fig. 8.11. This system allows the exchange of solutions through inlet/outlet syringes and at the same time keeps the cell still in focus under the microscope. A similar solution exchange system for live-cell imaging is also available commercially (e.g., from Bioptechs, Inc.).

8.8.4 Data Processing and Analysis

As described above, each image record contains a white light image and two Raman images at 1000 and 1080 cm^{-1}. An example of such a record plus the direct difference of two Raman images are shown in Fig. 8.12. It is obviously difficult to get useful information from the raw data before the postprocessing, because the raw data have many distortions (severe noise, nonuniform illumination, strong fluorescence background, and optical system blur). The image processing algorithms discussed above were applied to the images step by step as illustrated in the following paragraphs.

Figure 8.13a and b illustrates the Raman images after reducing the noise using the anisotropic median-diffusion filter. The standard deviation of the image gradient was used as the threshold in the diffusion coefficient. A 3×3 window was used for the median filter.

Figure 8.13c and d shows the Raman images with the nonuniform illumination corrected. A Raman image of a flat surface was recorded before each experiment as the reference illumination.

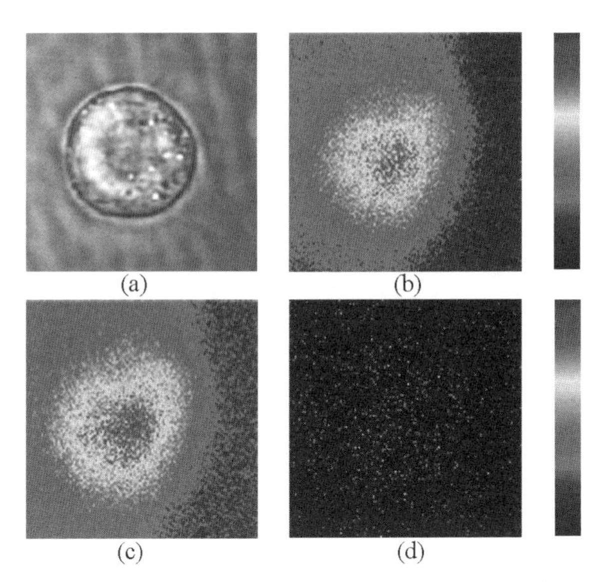

FIGURE 8.12 Example of the image record. **(a)** White light image of an MDA-435 breast cancer cell. **(b)** Raman image of the cell taken at 1000 cm^{-1} Raman band with 60× W/IR lens and exposure time of 300 seconds. **(c)** Raman image of the cell taken at 1080 cm^{-1} Raman band with 60× W/IR lens and exposure time of 300 s. **(d)** Difference of **(b)** and **(c)** before processing. The color bar indicates the relative Raman signal intensity increasing from bottom to top. Reproduced from "Direct Raman imaging techniques for study of the subcellular distribution of drug" (*Applied Optics*, Vol. 41, No. 2) with permission of the Optical Society of America. See color plates.

Figure 8.13e and f show the images after subtracting their constant background. The majority of the background is the fluorescence intensity contributed from the PBS solution; therefore, it is uniform across the image, like a "DC" component. The fluorescence background contributed from the intracellular structure, however, is not uniform due to the heterogeneity of the cell. That part of the fluorescence is referred to as the "AC" component in this chapter.

A simple way of eliminating the fluorescent "DC" component is to subtract the average value of the image and then to set all the negative values to zero. This simplified method is suitable only when the background occupies most of the area of an image. In this case, the average value of the image is close to the "DC" background. The subtraction of average value also enhanced the image, because only the intensity (fluorescent "AC" component plus Raman signal) higher than the average value was left on the image. The Raman image in Fig. 8.13e contains the Raman signal at the 1000 cm^{-1} band as well as the fluorescent "AC" component in this band. The image in Fig. 8.13f, however, contains the same fluorescent "AC" component as in the 1000 cm^{-1} band but without the Raman signal. Most important, the background subtraction makes the three-dimensional restoration more effective as discussed in the deconvolution section above.

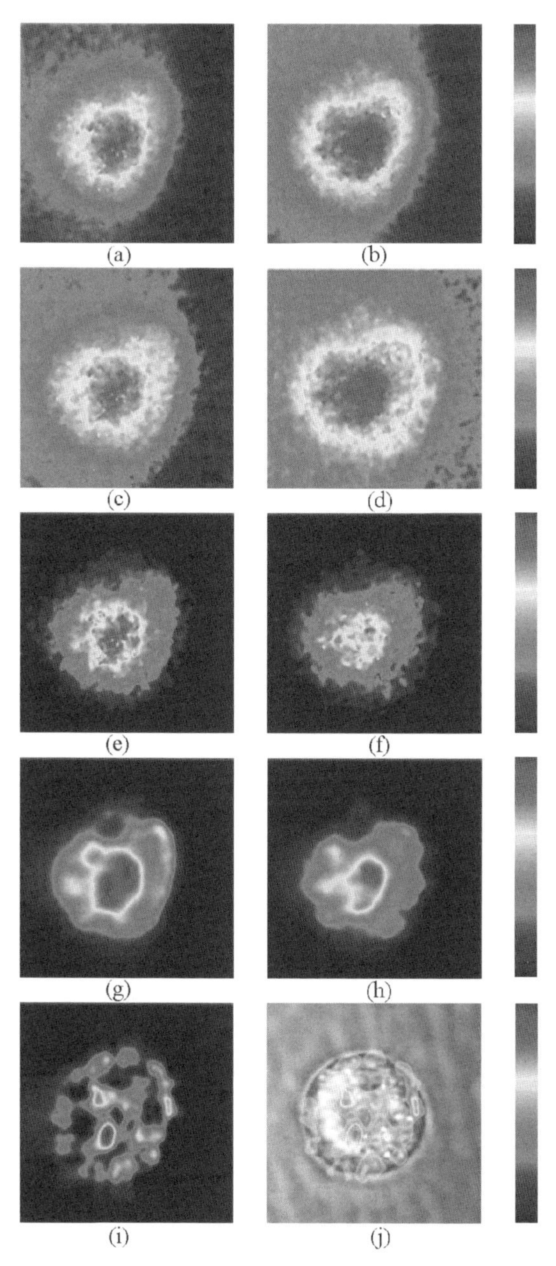

FIGURE 8.13 Postprocessing of Raman images in Fig. 8.12b and c illustrated in left and right columns, respectively. (**a** and **b**) Smoothed images. (**c** and **d**) Nonuniform illumination corrected images. (**e** and **f**) Constant background subtracted images. (**g** and **h**) Three-dimensional blur restored images. (**i**) Fluorescent signal eliminated image. (**j**) Overlay of image of (**i**) on image in Fig. 8.12a. The color bars indicate the relative Raman signal intensity increasing from bottom to top. Reproduced from "Direct Raman imaging techniques for study of the subcellular distribution of drugs" (*Applied Optics*, Vol. 41, No. 2) with permission of the Optical Society of America. See color plates.

Figure 8.13g and h illustrates the restored images using the EM-ML three-dimensional deconvolution algorithm. The OTF of the Raman system, determined through measurement, was used for the deconvolution.

Finally, the Raman image at 1080 cm^{-1} in Figure 8.13h was subtracted from the Raman image at 1000 cm^{-1} in Fig. 8.13g. The image after eliminating the fluorescent "AC" component is shown in Fig. 8.13i, which illustrates the paclitaxel distributions. The superimposition of the paclitaxel distribution image and the white light image in Fig. 8.13a is shown in Fig. 8.13j, which illustrates the paclitaxel distribution in the cell.

MD-435 tumor cells were exposed to the 0.3 mg/ml (or 350 μM) paclitaxel solution for 1 h. The white light images and Raman images were obtained before, during, and after the paclitaxel treatment (Fig. 8.14). All the Raman images were taken using a 60× water immersion lens with exposure time of 300 seconds and all were processed the same way using the method described above.

The first row in Fig. 8.14 illustrates the images before drug treatment. These images show the 1000 cm^{-1} Raman signals contributed from the molecules of the cell itself. From the overlay image, the intrinsic Raman signals appear outside the cell. This cannot be due to alignment because the white light image and the Raman image were registered with 1 μm microspheres before the experiment. Most probably, the problem is due to the imaging focal plane not being at the largest cross-area of the cell. A halo appears outside the cell, which may be the real cell boundary. The strong Raman signal in the corresponding Raman image could be the contribution from the neighborhood planes. This neighborhood information may not be removed effectively by the three-dimensional deconvolution due to the no-neighborhood condition (see Section 8.7.3). Nevertheless, this image does show that the original Raman intensities inside this cell are relatively low.

The second and third rows in Fig. 8.14 illustrate the images 10 and 45 min during exposure of the cell to paclitaxel. These images suggest that the paclitaxel were accumulated outside the cell membrane and gradually diffused into the cell. The relatively low Raman intensities in these two figures are probably because the cell is not in the PBS solution but in the drug solution, which contributes to a higher background than the PBS solution. After subtracting a higher average value from the image, its intensity became lower. In other words, the Raman intensities shown

FIGURE 8.14 Images before, during, and after an MDA-435 breast cancer cell was exposed to the paclitaxel agent. The first row illustrates the images before drug treatment. The second and third rows illustrate the images 10 and 45 min during the drug treatment. The fourth to seventh rows illustrate the images 10 min, 1.75 h, 4 h, and 4.5 h after the drug treatment. The left column shows the white light images of the cell that show the cell structure; the center column shows the Raman images of the cell that show the intensity distribution in the 1000 cm^{-1} Raman band; and the right column shows the overlay of images in the left and center columns. The red arrows point to the cell nucleus region and the blue arrows point to the cell blebbing region. The color bar indicates the relative Raman signal intensity increasing from bottom to top. Reproduced from "Direct Raman imaging techniques for study of the subcellular distribution of drugs" (*Applied Optics*, Vol. 41, No. 2) with permission of the Optical Society of America. See color plates.

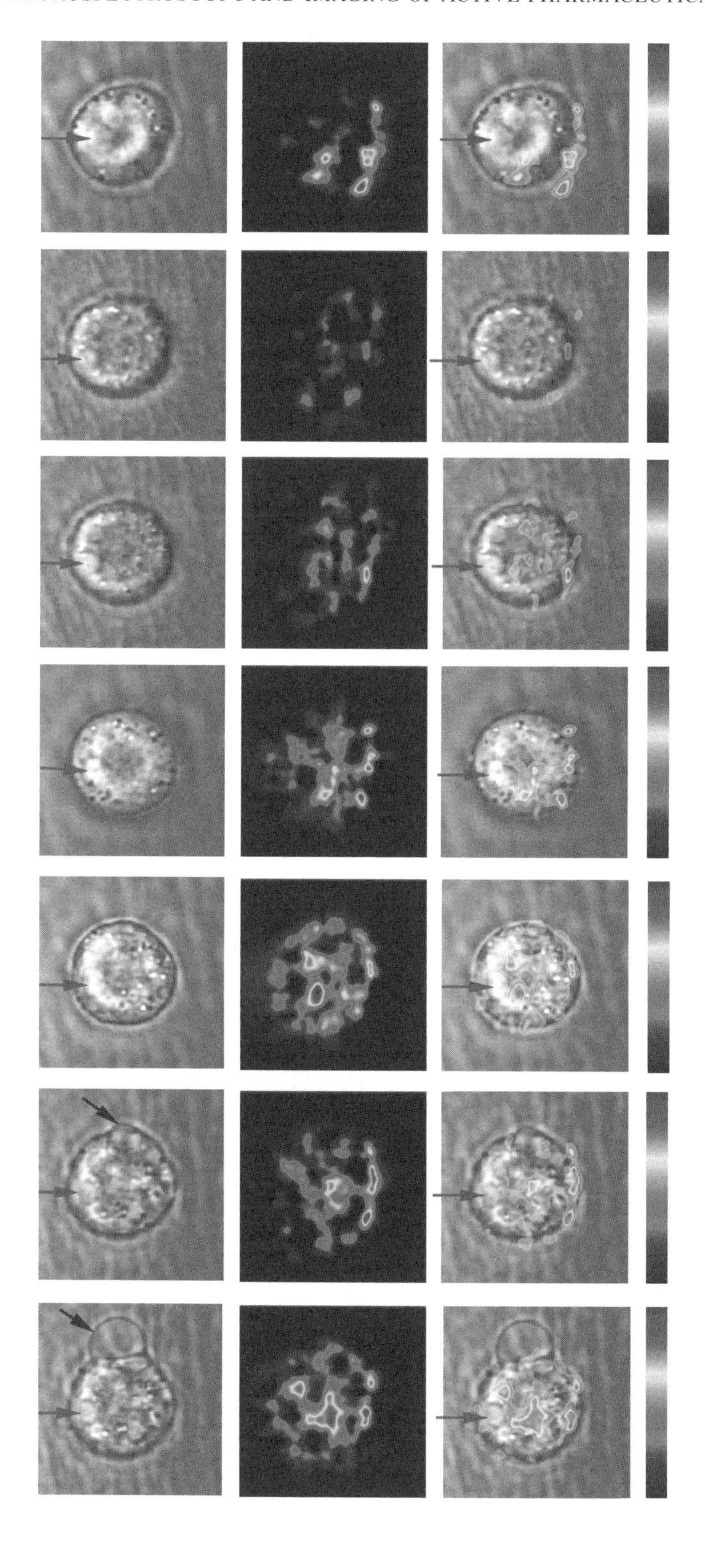

in the figures are relative after subtracting their average value. Therefore, the quantitative information was not preserved.

The fourth through seventh rows in Fig. 8.14 illustrate the images 10 min, 1.75 h, 4 h, and 4.5 h after the drug treatment (after the paclitaxel agent was washed out). These images show that the Raman intensities are relatively higher in the center area as well as near the cell membrane. However, there is no intensity in the cell nucleus area. The Raman signal is directly related to the molecular concentration: the higher the intensity, the higher the molecular concentration. Therefore, these figures suggest that paclitaxel is more concentrated near the center of the cell as well as near the cell membrane, but less concentrated in the cell nucleus. The finding of paclitaxel distributions from the Raman images is explained by the binding characteristics of the paclitaxel and its molecular target – the microtubules.

As described above, paclitaxel stabilizes the cell microtubules that play an important role in cell division. Microtubules are long and hollow tubes of protein that grow out from a small structure near the center of the cell, called centrosome, and extend out toward the cell periphery (Fig. 8.15a). Microtubules can rapidly disassemble in one location and reassemble in another. When a cell enters mitosis (division), the microtubules disassemble and then reassemble into an intricate structure called the mitotic spindle (Fig. 8.15b). The mitotic spindle provides the machinery that will segregate the chromosomes equally into the two daughter cells just before a cell divides (Alberts et al., 1997). The action of paclitaxel is to bind tightly to the growth end of the microtubules (Fig. 8.15c). In this way, paclitaxel prevents the microtubules from losing subunits (i.e., depolymerization). Since new subunits can still be added (i.e., polymerization), the microtubules can grow but cannot shrink. For the spindle to work, the microtubules must be able not only to assemble but also to disassemble. Thus, paclitaxel prevents the mitotic spindle from functioning normally and the dividing cell is arrested in mitosis (Alberts et al., 1997).

The binding mechanism of paclitaxel suggests that the high paclitaxel concentration in the center area of the cell might be the location of the centrosome (further

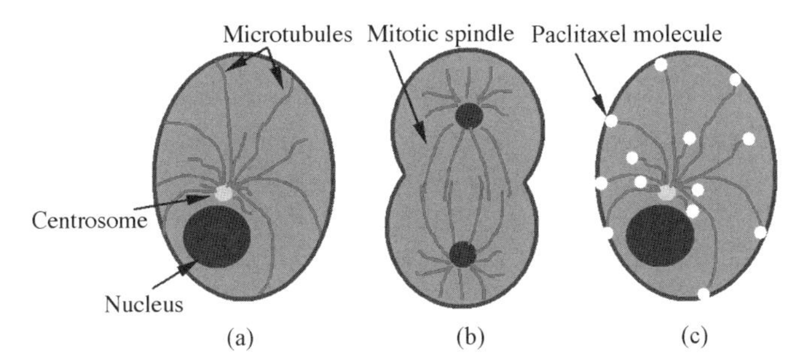

FIGURE 8.15 Distribution and functions of cell microtubules. **(a)** Microtubules grow out from the centrosome and extend to the cell membrane. **(b)** In the dividing cell, microtubules form a mitotic spindle to facilitate nucleus splitting. **(c)** Paclitaxel binds to the growth end of the microtubules. Reproduced from "Direct Raman imaging techniques for study of the subcellular distribution of drug" (*Applied Optics*, Vol. 41, No. 2) with permission of the Optical Society of America.

study is needed to prove this). The relatively high paclitaxel concentration near the cell membranes is probably because the growth-ends of the microtubules extend to the membrane. These patterns of paclitaxel distribution were also observed in the studies of fluorescence imaging (Morrison et al., 1997; Rao et al., 1998).

From Fig. 8.14, it was also found that the cell started blebbing around 4 h after exposure to the paclitaxel solution, and the blebs progressively increased in size. Previous studies (Jain and Trump, 1997; Jurkowitz-Alexander et al., 1992; Laster and Mackenzie, 1996; Lemasters et al., 1990; Malorni et al., 1990, 1991; Zahrebelski et al., 1995) have shown that cell blebbing often indicates the start of cell apoptosis (programmed death of the cell). Promotion of the assembly of microtubules after binding with paclitaxel might cause the cell blebbing.

Although the concentration of the paclitaxel solution used in this study was much higher (10–30-fold) than the regular clinical concentration, it may not indicate that the drug cannot be visualized at lower concentrations. In this study, the cells were exposed to the paclitaxel solution for only a short period. The drug was washed out after 1 h. In the clinical situation, however, the cells are exposed to a lower concentration of drug for a longer time. The intensity of Raman image is related to the local molecular concentrations. If, after a period of time, the drug can be accumulated locally (at the microtubules' growth-end), the drug can still be imaged even if the treatment drug concentration is low.

Future experiments will focus on the low drug concentration and long-term drug treatment. A temperature-control incubator with proper perfusion of nutrition medium will be used for long-term live-cell imaging. Future study will further improve the resolution of white light images so that fine subcellular structures will be visualized more clearly and will therefore improve the drug localization.

8.9 RAMAN IMAGING OF INTRACELLULAR DISTRIBUTION OF SULINDAC SULFIDE IN FIXED CELLS

8.9.1 Sulindac Sulfide and Its Characteristic Raman Band

Sulindac is a nonsteroidal antiinflammatory drug (NSAID). NSAIDs have shown potential chemoprevention, but their mechanisms of action are poorly understood. Sulindac (sulfoxide) is a prodrug that requires metabolic activation for activity. As shown in Fig. 8.16, the parent drug is reversibly reduced in the colon by bacterial microflora to a sulfide that is a potent inhibitor of both COX-1 and COX-2 and fully responsible for the antiinflammatory properties of sulindac (Duggan et al., 1977). Sulindac sulfide and sulfoxide are irreversibly oxidized in the liver to generate the sulfone, which is glucuronidated and excreted in the bile and concentrated in feces. The sulfone does not inhibit either COX-1 or COX-2 and is considered an inactive metabolite with regard to the antiinflammatory activity of sulindac. However, sulindac sulfone displays antineoplastic properties and is being developed for cancer chemoprevention because of its improved toxicity profile over sulindac. Sulindac generates the sulfide, which often causes GI toxicity (Piazza et al., 1997). Recent

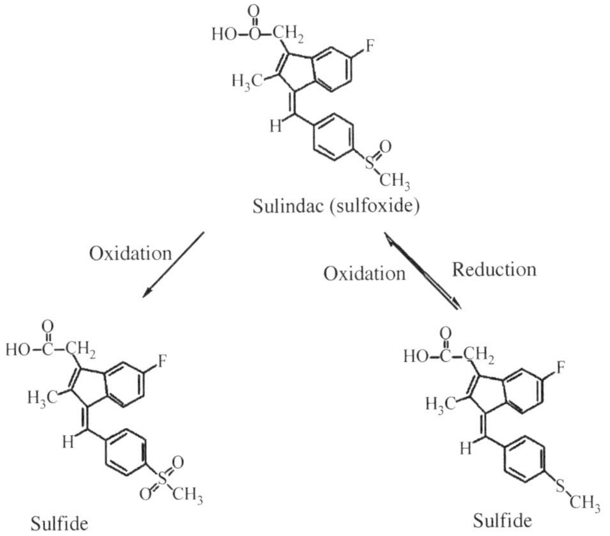

Sulindac (sulfoxide)

Oxidation

Oxidation Reduction

Sulfide

Sulfide

FIGURE 8.16 Metabolism of the anti-inflammatory drug sulindac to a sulfide (COX inhibitor) or sulfone (non-COX inhibitor). Reproduced from "Raman imaging microscopy—a potential cost-effective tool for drug development" (*Americal Pharmaceutical Review*, Vol. 8, No. 4) with permission of Russell Publishing, LLC.

studies suggest that the mechanism of apoptosis induction by sulindac sulfone involves inhibition of phosphodiesterase to result in the elevation of cGMP (Thompson et al., 2000). It is unclear, however, whether sulindac sulfide or sulfone induces apoptosis by a similar mechanism.

The Raman spectrum of sulindac sulfide is shown in Fig. 8.17 (top). The significant Raman peak at 1614 cm^{-1} and its companion at 1585 cm^{-1} are due to the C=C double bond stretching. Sulindac sulfone has a unique Raman signal compared to sulindac sulfide as shown in Fig. 8.17 (bottom). The C=C double bond vibration in sulfone shifts to 1639 cm^{-1}, slightly higher than that in sulfide. Its companion at around 1588 cm^{-1} is attenuated significantly. However, Raman peaks at 1134 and 697 cm^{-1} stand out from the Raman spectrum of sulindac sulfone, which shows that Raman spectroscopy has the capability to differentiate metabolites. This capability might allow tracking of the two metabolites sulfide and sulfone simultaneously in cells or tissue.

8.9.2 Cell Preparation and Cell Raman Background

Human LNCaP prostate tumor cells that display androgen-dependent growth properties were used in this study. Approximately 10^5 LNCaP cells were cultured on a gold-coated Petri dish and allowed to stabilize for 24 h in RPMI-1640 medium supplemented with 5% fetal bovine serum. After stabilization, the cells adhered to the bottom of the Petri dish.

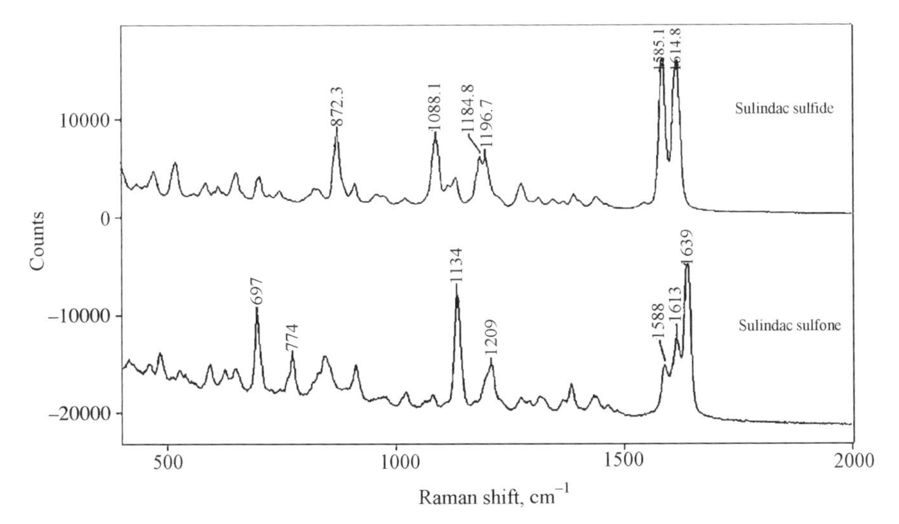

FIGURE 8.17 Raman spectra of two metabolites: sulindac sulfide and sulfone. Note that the different functional groups of the two metabolites can be differentiated by Raman spectra. This molecular specificity of the Raman signal can be used to detect metabolites in cells. Reproduced from "Raman imaging microscopy—a potential cost-effective tool for drug development" (*Americal Pharmaceutical Review*, Vol. 8, No. 4) with permission of Russell Publishing, LLC.

Sulindac sulfide was solubilized in DMSO and diluted with RPMI-1640 cell culture medium supplemented with 5% fetal bovine serum. The prepared drug concentration was 50 μM, which is a clinically relevant concentration of this drug. In fixed cells study, long exposure time is allowed and therefore the detection of low, clinically relevant concentrations of a drug was possible using the current instrumentation.

The LNCaP cells were incubated with 50 μM sulindac sulfide and treated for 24 h. Then the cells were washed twice with PBS to remove the unbound drug and followed with 1% paraformaldehyde fixation. The cells were rehydrated before imaging. LNCaP cells without drug treatment were also prepared for comparison under Raman imaging.

8.9.3 Imaging Instrumentation and Imaging Procedures

The same Raman imaging microscope was used in this study as in the live cell imaging discussed above. In addition, the same 60× water-immersion lens was used to image the rehydrated fixed cells.

Sulindac sulfide-treated and nontreated control cells were both recorded with Raman images at 1614 and 1700 cm^{-1}. The Raman image at 1614 cm^{-1} was tuned to a peak of sulindac sulfide, while the Raman image at 1700 cm^{-1} was used to correct the fluorescence background on the 1614 cm^{-1} image. Each Raman image took 10 min to acquire.

8.9.4 Data Processing and Analysis

The Raman images were processed using the same methods as described in Section 8.8.4. The Raman images of LNCaP cells are illustrated in Fig. 8.18. The Raman images of the drug-treated cell (top panel) obviously show higher Raman intensity. The weak signals presented by the control cell represent cellular background (lower panel).

From Fig. 8.18, nuclear localization of sulindac sulfide was observed. Although the available bright field image did not provide optimal contrast, it is clear that the drug was localized to the nucleus of the cell. The nuclear localization of sulindac sulfide is a novel finding and might provide new insights into the mechanism of action. Nuclear localization of sulindac sulfide suggests that NSAIDs may have a direct effect on gene transcription. This study again indicated that Raman imaging is promising for use in investigation of the mechanism of new drug candidates without external labels.

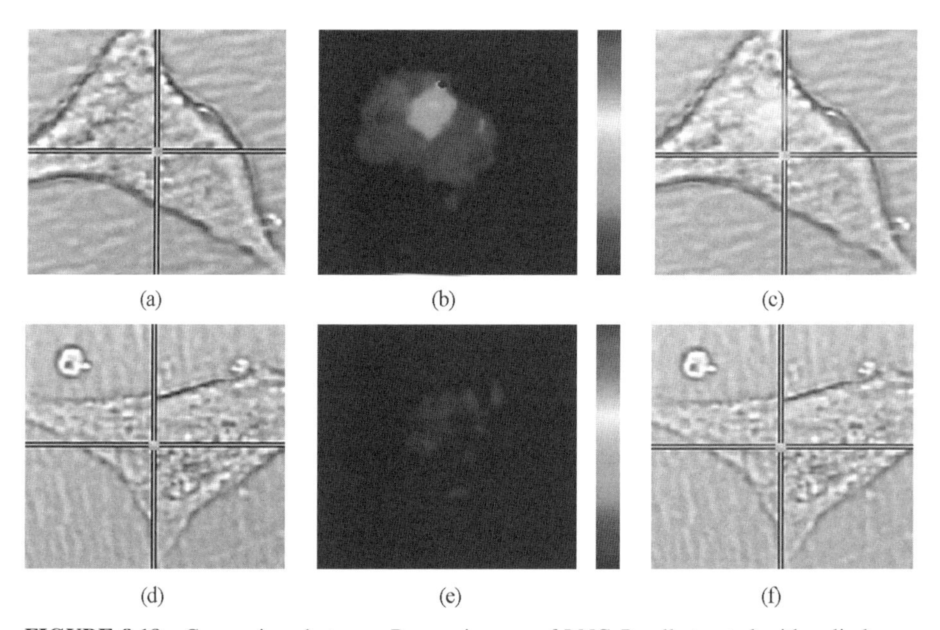

FIGURE 8.18 Comparison between Raman images of LNCaP cells treated with sulindac sulfide (top panel) and controlled cells (bottom panel). **(a)** Bright field micrograph of cell treated with sulindac sulfide. **(b)** Corresponding Raman image of **(a)**. **(c)** Overlay of images in **(a)** and **(b)**. **(d)** Bright field image of cell treated with drug solvent alone. **(e)** Corresponding Raman image of **(d)**. **(f)** Overlay of images in **(d)** and **(e)**. The color bar indicates the relative Raman signal intensity that increases from bottom to top. Reproduced from "Raman imaging microscopy—a potential cost-effective tool for drug development" (*Americal Pharmaceutical Review*, Vol. 8, No. 4) with permission of Russell Publishing, LLC. See color plates.

8.10 CONCLUSIONS AND FUTURE OUTLOOK

This chapter has explored the potential of using label-free Raman imaging to visualize the subcellular distribution of active pharmaceutical ingredients. To implement the idea, however, it needs to overcome the natural low sensitivity of the Raman technology. Raman techniques such as RR and SERS, which are especially useful in Raman imaging to improve SNR, were discussed. This chapter also pointed out that low SNR Raman images are further degraded by the optical collection system. Postimage processing methods to enhance Raman images and to extract weak Raman signals were described.

The direct Raman imaging and data analysis techniques were then applied to the visualization of drugs in both living and fixed cells. The examples indicate the potential of Raman imaging for label-free drug visualization. Similar approaches may also be suitable for the detection of drugs in biopsy tissues. However, in order to detect clinically relevant drug levels in live cells or tissue, the sensitivity of Raman imaging microscopes needs to be improved significantly. In addition, Raman imaging microscopes need to be equipped with advanced tunable filters so that Raman images can be acquired at multiple Raman frequencies of a drug. This will increase the specificity of drug detection. Raman and fluorescence imaging may also be integrated into the same microscope, so that while Raman imaging visualizes a drug, fluorescence imaging visualizes the drug target—the proteins. Raman imaging microscopes with high sensitivity and specificity are expected to be a useful tool for the design of cost-effective cell-based assays, which can be used to study drug mechanism, resistance, metabolism, and pharmacokinetics. To accelerate the conversion of Raman technology into a useful tool for drug research and development, efforts from a multidisciplinary team including pharmacists, chemists, Raman experts, and engineers are indispensable.

REFERENCES

21CEP, 1996–2003. Paclitaxel (Taxel). www.21cecpharm.com.

Alberts B, Raff M, 1997. Cytoskeleton. In *Essential Cell Biology—An Introduction to the Molecular Biology of the Cell*. Garland Publishing New York, pp. 513–546.

Arzhantsev SY, Chikishev AY, Koroteev NI, Greve J, Otto C, Sijtsema N M, 1999. Localization study of Co-phthalocyanines in cells by Raman micro(spectro)scopy. *J. Raman Spectrosc.*, **30**(3), 205–208.

Baker SD, Wadkins RM, Stewart CF, Beck WT, Danks MK 1995. Cell cycle analysis of amount and distribution of nuclear DNA topoisomerase I as determined by fluorescence digital imaging microscopy. *Cytometry*, **19**(2), 134–145.

Bontenbal M, Sieuwerts AM, Peters HA, van Putten WL, Foekens JA, Klijn JG, 1998. Uptake and distribution of doxorubicin in hormone-manipulated human breast cancer cells *in vitro*. *Breast Cancer Res. Treatment*, **51**(2), 139–148.

Brandmuller J, Kiefer W, 1978. Physicist's view, fifty years of Raman spectroscopy. *Spex Speaker* 23, 310.

Brenan CJH, Hunter IW, Korenberg MJ, 1996. Volumetric Raman spectral imaging with a confocal Raman microscope: Image modalities and applications. In *Proc. SPIE—Int. Soc. Opt. Eng.*, **2655**, 130–139.

Breuzard G, Angiboust JF, Jeannesson P, Manfait M, Millot JM, 2004. Surface-enhanced Raman scattering reveals adsorption of mitoxantrone on plasma membrane of living cells. *Biochem. Biophys. Res. Commun.*, **320**(2), 615–621.

Butler C, Cooney R, Denny WA, 1994. Resonance Raman study of the binding of the anticancer drug amsacrine to DNA. *Appl. Spectrosc.*, **48**(7), 822–826.

Cheng JX, Xie XS, 2004. Coherent anti-Stokes Raman scattering microscopy: Instrumentation, theory, and applications. *J. Phys. Chem. B*, **108**(3), 827–40.

Chinked K, 1996. Optical diagnostics image tissues and tumors. *Laser Focus World*, 71–81.

Chourpa I, Morjani H, Riou JF, Manfait M, 1996. Intracellular molecular interactions of antitumor drug amsacrine (m-AMSA) as revealed by surface-enhanced Raman spectroscopy. *FEBS Lett.*, **397**(1), 61–64.

Coley HM, Amos WB, Twentyman PR, Workman P, 1993. Examination by laser scanning confocal fluorescence imaging microscopy of the subcellular localisation of anthracyclines in parent and multidrug resistant cell lines. *Br. J. Cancer*, **67**(6), 1316–1323.

Crivellato E, Candussio L, Rosati AM, Decorti G, Klugmann FB, Mallardi F, 1999. Kinetics of doxorubicin handling in the LLC-PK1 kidney epithelial cell line is mediated by both vesicle formation and P-glycoprotein drug transport. *Histochem. J.*, **31**(10), 635–43.

de Lange, J. H., Schipper NW, Schuurhuis GJ, ten Kate TK, van Heijningen TH, Pinedo HM, Lankelma J, Baak JP, 1992. Quantification by laser scan microscopy of intracellular doxorubicin distribution. *Cytometry* **13**(6), 571–576.

Dellinger M, Geze M, Santus R, Kohen E, Kohen C, Hirschberg JG, Monti M, 1998. Imaging of cells by autofluorescence: a new tool in the probing of biopharmaceutical effects at the intracellular level. *Biotechnol. Appl. Biochem.*, **28**(Pt1), 25–32.

Duggan DE, Hooke KF, Risley EA, Shen TY, Arman CG, 1977. Identification of the biologically active form of sulindac. *J. Pharmacol. Exp. Thera.*, **201**(1), 8–13.

Duncan MD, Reintjes J, Manuccia TJ, 1985. Imaging biological compounds using the coherent anti-Stokes Raman scattering microscope. *Opt. Eng.*, **24**(2), 352–352.

Durig JR, Dunlap RB, Gerson DJ, 1980. Conformational study of methotrexate binding to *L. casei* dihydrofolate reductase by laser Raman spectroscopy. *J. Raman Spectros.*, **9**(4), 266–272.

Eliasson C, Loren A, Engelbrektsson J, Josefson M, Abrahamsson J, Abrahamsson K, 2005. Surface-enhanced Raman scattering imaging of single living lymphocytes with multivariate evaluation. *Spectrochim. Acta A - Mol. Biomol. Spectrosc.*, **61**(4), 755–760.

Evans JE, Browne TR, Kasdon DL, Szabo GK, Evans BA, Greenblatt DJ, 1985. Staggered stable isotope administration technique for study of drug distribution. *J. Clin. Pharmacol.*, **25**(4), 309–312.

Feofanov AV, Baranov AV, Fleury F, Riou JF, Nabiev IR, Manfait M, 1996. DNA topoisomerase I changes the mode of interaction between camptothecin drugs and DNA as probed by UV-resonance Raman spectroscopy. *FEBS Lett.*, **396**(2–3), 289–292.

Feofanov AV, Grichine AI, Shitova LA, Karmakova TA, Yakubovskaya RI, Egret-Charlier M Vigny, P., 2000. Confocal Raman microspectroscopy and imaging study of theraphthal in living cancer cells. *Biophys. J.*, **78**, 499–512.

Freeman TL, Cope SE, Stringer MR, Cruse-Sawyer JE, Brown SB, Batchelder DN, Birbeck K, 1998. Investigation of the subcellular localization of zinc phthalocyanines by Raman Mapping. *Appl. Spectrosc.*, **52**(10), 1257–1263.

Gervasoni JE, Jr, Fields SZ, Krishna S, Baker MA, Rosado M, Thuraisamy K, Hindenburg AA, Taub RN, 1991. Subcellular distribution of daunorubicin in *P*-glycoprotein-positive and -negative drug-resistant cell lines using laser-assisted confocal microscopy. *Cancer Res.*, **51**(18), 4955–4963.

Gift AD, Ma J, Haber KS, McClain BL, Ben-Amotz D, 1999. Near-infrared Raman imaging microscope based on fiber-bundle image compression. *J. Raman Spectrosc.*, **30**, 757–765.

Gustafson TL, Lytle FE., 1982. Time-resolved rejection of fluorescence from Raman spectra via high repetition rate gated photon counting. *Anal. Chem.*, **54**(4), 634–637.

Harris M, Chrisman RW, Lytle FE, Tobias RS, 1976. Sub-nanosecond time-resolved rejection of fluorescence from Raman spectra. *Anal. Chem.*, **48**(13), 1937–1943.

Hiraoka Y, Sedat JW, Agard DA, 1990. Determination of three-dimensional imaging properties of a light microscope. *Biophys. J.*, **57**, 325–333.

Holmes TJ, Liu YH., 1989. Richardson-Lucy/maximum likelihood image restoration algorithm for fluorescence microscopy: Further testing. *Appl. Opt.*, **28**(22), 4930–4938.

Itoh J, Osamura RY, Watanabe K, 1992. Subcellular visualization of light microscopic specimens by laser scanning microscopy and computer analysis: A new application of image analysis. *J. Histochem. Cytochem.*, **40**(7), 955–967.

Jain PT, Trump BF, 1997. Human breast cancer cell growth inhibition and deregulation of [Ca2+]i by estradiol. *Anti-Cancer Drugs*, **8**(3), 283–287.

Jurkowitz-Alexander MS, Altschuld RA, Hohl CM, Johnson JD, McDonald JS, Simmons TD, Horrocks LA, 1992. Cell swelling, blebbing, and death are dependent on ATP depletion and independent of calcium during chemical hypoxia in a glial cell line (ROC-1). *J. Neurochem.*, **59**(1), 344–352.

Kamogawa K, Fujii T, Kitagawa K, 1988. Improved fluorescence rejection in measurements of Raman spectra of fluorescent compounds. *Appl. Spectrosc.*, **42**(2), 248–254.

Kerns EH, Hill SE, Detlefsen DJ, Volk KJ, Long BH, Carboni J, Lee MS, 1998. Cellular uptake profile of paclitaxel using liquid chromatography tandem mass spectrometry. *Rapid Commun. Mass Spectrom.*, **12**(10), 620–624.

Kincade K, 1998. Raman spectroscopy enhances *in vivo* diagnosis. *Laser Focus World* (July), 83–91.

Kline NJ, Treado PJ, 1997. Raman chemical imaging of breast tissue. *J. Raman Spectrosc.*, **28**(2–3), 119–124.

Kneipp K, Kneipp H, Manoharan R, Hanlon EB, Itzkan I, Dasari RR, Feld MS, 1998a. Extremely large enhancement factors in surface-enhanced Raman scattering for molecules on colloidal gold clusters. *Appl. Spectrosc.*, **52**(12), 1493–1497.

Kneipp K, Kneipp H, Manoharan R, Itzkan I, Dasari RR, Feld MS, 1998b. Surface-enhanced Raman scattering (SERS)—A new tool for single molecule detection and identification. *Bioimaging*, **6**(2), 104–110.

Krishna CM, Kegelaer G, Adt I, Rubin S, Kartha VB, Manfait M, Sockalingum GD, 2006. Combined Fourier transform infrared and Raman spectroscopic approach for identification of multidrug resistance phenotype in cancer cell lines. *Biopolymers*, **82**(5), 462–470.

Laster SM, Mackenzie JM, Jr, 1996. Bleb formation and F-actin distribution during mitosis and tumor necrosis factor-induced apoptosis. *Microsc. Res. Tech.*, **34**(3), 272–280.

Lemasters JJ, Gores GJ, Nieminen AL, Dawson TL, Wray BE, Herman B, 1990. Multiparameter digitized video microscopy of toxic and hypoxic injury in single cells. *Environ. Health Perspect.*, **84**, 83–94.

Lesney MS, 2004. Assaying ADMET. *Mod. Drug Discov.*, 30–33.

Ling J, 2001. The development of Raman imaging microscopy to visualize drug actions in living cells. Doctoral dissertation. The University of Texas at Austin, 150.

Ling J, Weitman SD, Miller MA, Moore RV, Bovik AC, 2002. Direct Raman imaging techniques for studying the subcellular distribution of a drug. *Appl. Opt.*, **41**(28), 6006–6017.

Lucy YH, 1974. An iterative technique for the rectification of observed distributions. *Astron. J.*, **79**, 745–765.

Ma J, Ben-Amotz D, 1997. Rapid micro-Raman imaging using fiber-bundle image compression. *Appl. Spectrosc.*, **51**(12), 1845–1848.

Mahadevan-Jansen A, Richards-Kortum R, 1996. Raman spectroscopy for the detection of cancers and precancers. *J. Biomed. Opt.*, **1**(1), 31–70.

Malorni W, Fiorentini C, Paradisi S, Giuliano M, Mastrantonio P, Donelli G, 1990. Surface blebbing and cytoskeletal changes induced *in vitro* by toxin B from *Clostridium difficile*: An immunochemical and ultrastructural study. *Exp. Mol. Pathol.*, **52**(3), 340–356.

Malorni W, Iosi F, Mirabelli F, Bellomo G, 1991. Cytoskeleton as a target in menadione-induced oxidative stress in cultured mammalian cells: Alterations underlying surface bleb formation. *Chem. Biol. Interact.*, **80**(2), 217–236.

Manoharan R, Yang W, Feld, MS, 1996. Histochemical analysis of biological tissues using Raman spectroscopy. *Spectrochim. Acta A Mol. Spectrosc.*, **52A**(2), 215–249.

Matousek P, Towrie M, Parker AW, 2002. Fluorescence background suppression in Raman spectroscopy using combined Kerr gated and shifted excitation Raman difference techniques. *J. Raman Spectrosc.*, **33**, 238–242.

Matousek P, Towrie M, Stanley A, Parker AW, 1999. Efficient rejection of fluorescence from Raman spectra using picosecond Kerr gating. *Appl. Spectrosc.*, **53**(12), 1485–1489.

McClain BL, Ma J, Ben-Amotz D, 1999. Optical absorption and fluorescence spectral imaging using fiber bundle image compression. *Appl. Spectrosc.*, **53**(9), 1118–1122.

Morris HR, Hoyt CC, Treado PJ, 1994. Imaging spectrometers for fluorescence and Raman microscopy: Acousto-optic and liquid crystal tunable filters. *Appl. Spectrosc.*, **48**(7), 857–866.

Morris HR, Hoyt CC, Miller P, Treado PJ, 1996. Liquid crystal tunable filter Raman chemical imaging. *Appl. Spectrosc.*, **50**(6), 805–811.

Morrison EE, Askham JM, Clissold P, Markham AF, Meredith DM, 1997. The cellular distribution of the adenomatous polyposis coli tumour suppressor protein in neuroblastoma cells is regulated by microtubule dynamics. *Neuroscience*, **81**(2), 553–563.

Murali Krishna C, Kegelaer G, Adt I, Rubin S, Kartha VB, Manfait M, Sockalingum GD, 2005. Characterization of uterine sarcoma cell lines exhibiting MDR phenotype by vibrational spectroscopy. *Biochim. Biophys. Acta*, **1726**(2), 160–167.

Nabiev IR, Morjani H, Manfait M, 1991. Selective analysis of antitumor drug interaction with living cancer cells as probed by surface-enhanced Raman spectroscopy. *Eur. Biophys. J.*, **19**(6), 311–316.

Notingher I, Hench LL, 2006. Raman microspectroscopy: A noninvasive tool for studies of individual living cells *in vitro*. *Future Drugs Expert Rev. Med. Devices*, **3**(2), 215–234.

Owen CA, Selvakumaran J, Notingher I, Jell G, Hench LL, Stevens MM, 2006. *In-vitro* toxicology evaluation of pharmaceuticals using Raman micro-spectroscopy. *J. Cell Biochem.*, **99**(1), 178–186.

Oyama H, Nagane M, Shibui S, Nomura K, Mukai K, 1992. Intracellular distribution of CPT-11 in CPT-11-resistant cells with confocal laser scanning microscopy. *Jpn. J. Clin. Oncol.*, **22**(5), 331–334.

Ozaki Y, King RW, Carey PR, 1981. Methotrexate and folate binding to dihydrofolate reductase. Separate characterization of the pteridine and p-aminobenzoyl binding sites by resonance Raman spectroscopy. *Biochemistry*, **20**(11), 3219–3225.

Piazza GA, Alberts DS, Hixson LJ, Paranka NS, Li H, Finn T, Bogert C, Guillen JM, Brendel K, Gross PH, Sperl G, Ritchie J, Burt RW, Ellsworth L, Ahnen DJ, Pamukcu R, 1997. Sulindac sulfone inhibits azoxymethane-induced colon carcinogenesis in rats without reducing prostaglandin levels. *Cancer Res.*, **57**(14), 2909–2915.

Puffer HW, Frasher WG, Netto DJ, Marcus CS, Goldwater J, 1979. Drug distribution studies: a possible new approach using intravital microscopy and a spatially discriminate isotope detector. *Proc. Western Pharmacol. Soc.*, **22**, 53–56.

Puppels GJ, Greve J, 1996. Biomedical applications of spectroscopy. In *Advances in Spectroscopy*, edited by Clark RJH, and Hester RE, **25**, 1–47. Wiley, Chester, England.

Raman CV, 1928. The colour of the sea. *Nature*, **121**, 150.

Rao CS, Chu JJ, Liu RS, Lai YK, 1998. Synthesis and evaluation of [14C]-labelled and fluorescent-tagged paclitaxel derivatives as new biological probes. *Bioorg. Med. Chem.*, **6**(11), 2193–2204.

Richardson WH, 1972. Bayesian-based iterative method of image restoration. *J. Opt. Soc. Am.*, **62**, 55–59.

Schaeberle MD, Kalasinsky VF, Luke JL, Lewis EN, Levin IW, Treado PJ, 1996. Raman chemical imaging: Histopathology of inclusions in human breast tissue. *Anal. Chem.*, **68**(11), 1829–1833.

Sharonov S, Nabiev I, Chourpa I, Feofanov A, Valisa P, Manfait M, 1994. Confocal three-dimensional scanning laser Raman-SERS-fluorescence microprobe. Spectral imaging and high-resolution applications. *J. Raman Spectrosc.*, **25**, 699–707.

Shaw PJ, Rawlins DJ, 1991. The point-spread function of a confocal microscope - Its measurement and use in deconvolution of 3-D data. *J. Microsc. (Oxford)*, **163**, 151–165.

Shreve AP, Cherepy NJ, Mathies RA, 1992. Effective rejection of fluorescence interference in Raman spectroscopy using a shifted excitation difference technique. *Appl. Spectrosc.*, **46**(4), 707–711.

Sijtsema NM, Wouters SD, De Grauw CJ, Otto C, Greve J, 1998. Confocal direct imaging Raman microscope: Design and applications in biology. *Appl. Spectrosc.*, **52**(3), 348–355.

Thompson WJ, Piazza GA, Li H, Liu L, Fetter J, Zhu B, Sperl G, Ahnen D, Pamukcu R, 2000. Exisulind induction of apoptosis involves guanosine 3′,5′-cyclic monophosphate phosphodiesterase inhibition, protein kinase G activation, and attenuated beta-catenin. *Cancer Res.*, **60**(13), 3338–3342.

Van Duyne RP, Jeanmaire DL, Shriver DF, 1974. Mode-locked laser Raman spectroscopy—A new technique for the rejection of interfering background luminescence signals. *Anal. Chem.*, **46**(2), 213–222.

van Kempen GMP, van Vliet LJ, 2000. Background estimation in nonlinear image restoration. *J. Opt. Soc. Am. A Opt. Image Sci. Vision*, **17**(3), 425–433.

Vo-Dinh T, Yan F, Wabuyele MB, 2005. Surface-enhanced Raman scattering for medical diagnostics and biological imaging. *J. Raman Spectrosc.*, **36**, 640–647.

Weaver JL, Pine PS, Aszalos A, Schoenlein PV, Currier SJ, Padmanabhan R, Gottesman MM, 1991. Laser scanning and confocal microscopy of daunorubicin, doxorubicin, and rhodamine 123 in multidrug-resistant cells. *Exp. Cell Res.*, **196**(2), 323–329.

Woodburn KW, Vardaxis NJ, Hill JS, Kaye AH, Phillips DR, 1991. Subcellular localization of porphyrins using confocal laser scanning microscopy. *Photochem. Photobiol.*, **54**(5), 725–732.

Zahrebelski G, Nieminen AL, al-Ghoul K, Qian T, Herman B, Lemasters JJ, 1995. Progression of subcellular changes during chemical hypoxia to cultured rat hepatocytes: A laser scanning confocal microscopic study. *Hepatology*, **21**(5), 1361–1372.

INDEX

Pharmaceutical Applications of Raman Spectroscopy, Edited by Slobodan Šašić